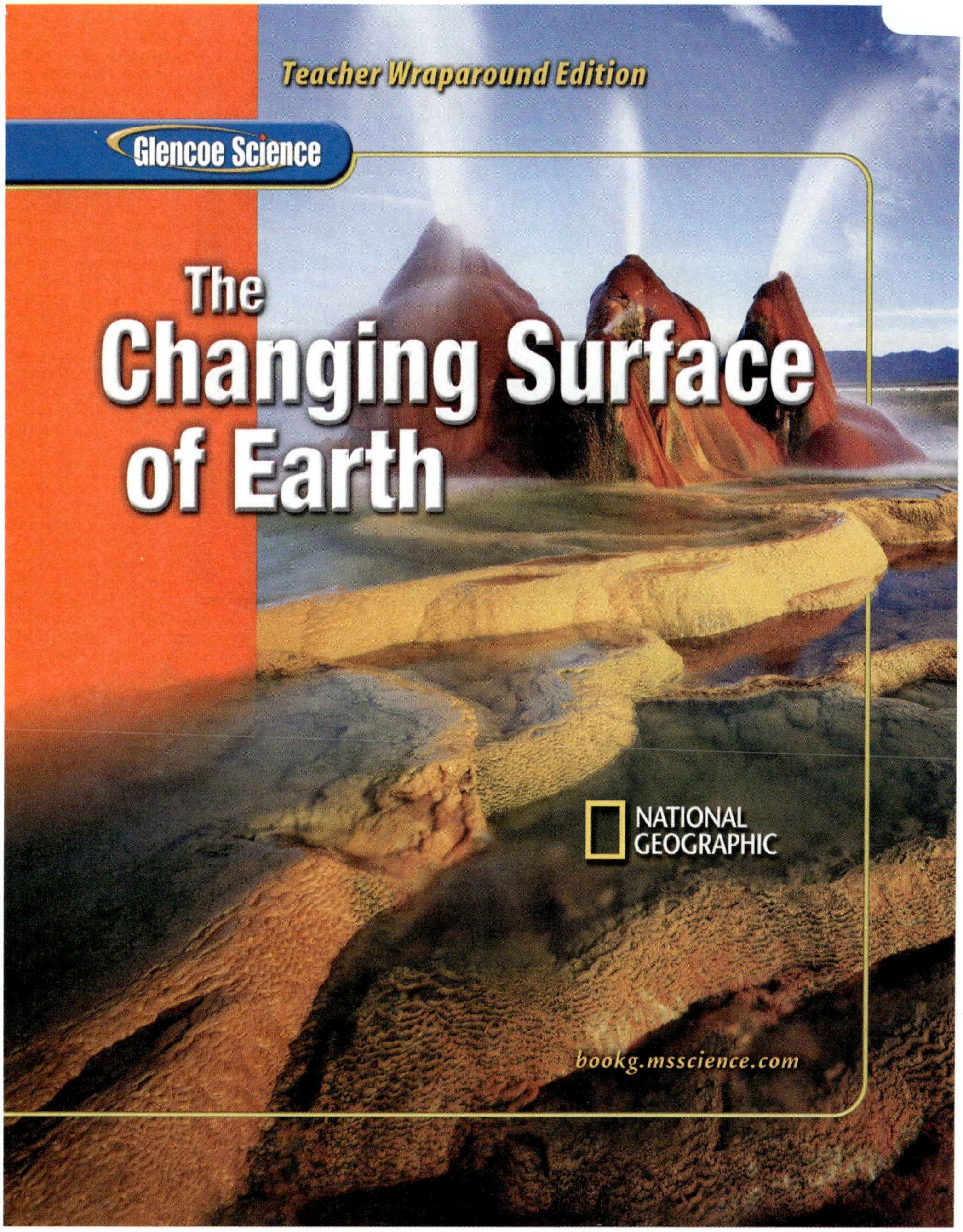

McGraw-Hill Glencoe

New York, New York Columbus, Ohio Chicago, Illinois Peoria, Illinois Woodland Hills, California

Glencoe Science

The Changing Surface of Earth

Fly Geyser is a manmade drill well, which is now a constantly spouting hotspring. It is located in the Black Rock Desert, near Gerlach, Nevada. The tufa terraces, or "natural steps," are formed by mineral deposits from the springs.

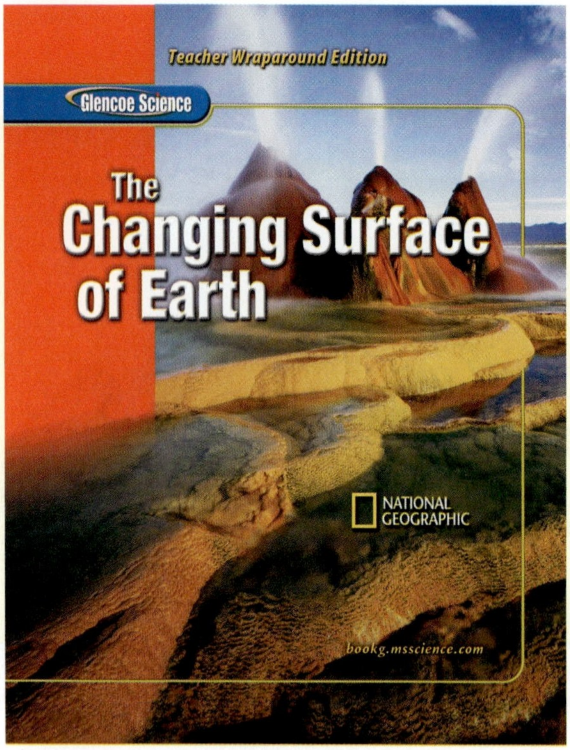

McGraw Hill Glencoe

The McGraw·Hill Companies

Send all inquiries to:
Glencoe/McGraw-Hill
8787 Orion Place
Columbus, OH 43240-4027

ISBN 0-07-861752-9 (Student Edition)
ISBN 0-07-861753-7 (Teacher Wraparound Edition)

Printed in the United States of America.

1 2 3 4 5 6 7 8 9 10 027/043 09 08 07 06 05 04

Authors

NATIONAL GEOGRAPHIC
Education Division
Washington, D.C.

Susan Leach Snyder
Retired Earth Science Teacher
Jones Middle School
Upper Arlington, OH

Ralph M. Feather Jr., PhD
Assistant Professor
Geoscience Department
Indiana University of Pennsylvania
Indiana, PA

Dinah Zike
Educational Consultant
Dinah-Might Activities, Inc.
San Antonio, TX

Series Consultants

CONTENT

William C. Keel, PhD
Department of Physics and Astronomy
University of Alabama
Tuscaloosa, AL

Robert Nierste
Science Department Head
Hendrick Middle School, Plano ISD
Plano, TX

MATH

Michael Hopper, DEng
Manager of Aircraft Certification
L-3 Communications
Greenville, TX

Teri Willard, EdD
Mathematics Curriculum Writer
Belgrade, MT

READING

Carol A. Senf, PhD
School of Literature, Communication, and Culture
Georgia Institute of Technology
Atlanta, GA

SAFETY

Aileen Duc, PhD
Science 8 Teacher
Hendrick Middle School, Plano ISD
Plano, TX

Sandra West, PhD
Department of Biology
Texas State University-San Marcos
San Marcos, TX

ACTIVITY TESTERS

Nerma Coats Henderson
Pickerington Lakeview Jr. High School
Pickerington, OH

Mary Helen Mariscal-Cholka
William D. Slider Middle School
El Paso, TX

Science Kit and Boreal Laboratories
Tonawanda, NY

Series Reviewers

Lois Burdette
Green Bank Elementary-Middle School
Green Bank, WV

Marcia Chackan
Pine Crest School
Boca Raton, FL

Karen Curry
East Wake Middle School
Raleigh, NC

Annette D'Urso Garcia
Kearney Middle School
Commerce City, CO

Nerma Coats Henderson
Pickerington Lakeview Jr. High School
Pickerington, OH

Michael Mansour
Board Member
National Middle Level Science Teacher's Association
John Page Middle School
Madison Heights, MI

Sharon Mitchell
William D. Slider Middle School
El Paso, TX

Glencoe Science 15-Book Series

Teach science your way!

With the 15 Life, Earth, and Physical Science titles in our modular series, you can select the science topics you want to cover and customize your science curriculum in any way you choose.

F Earth Materials and Processes
1 Minerals
2 Rocks
3 Earth's Energy and Mineral Resources
4 Plate Tectonics
5 Earthquakes
6 Volcanoes

G The Changing Surface of Earth
1 Views of Earth
2 Weathering and Soil
3 Erosional Forces
4 Water Erosion and Deposition
5 Clues to Earth's Past
6 Geologic Time

H The Water Planet
1 Water
2 Freshwater at Earth's Surface
3 Groundwater Resources
4 Ocean Motion
5 Oceanography

I The Air Around You
1 Atmosphere
2 Weather
3 Climate
4 Air Pollution

J Astronomy
1 Exploring Space
2 The Sun-Earth-Moon System
3 The Solar System
4 Stars and Galaxies

Multi-Level Review and Assessment

Each chapter provides five pages of review and testing to help you evaluate students' knowledge and ability to apply science concepts.

Section Review

- Summary pinpoints important concepts in the section.
- Skill-based questions promote critical thinking skills.

Study Guide

- Main idea summary of each section
- Concept mapping activity to help students visualize the main ideas

Chapter Review

- Using Vocabulary
- Checking Concepts
- Thinking Critically
- Performance Activities
- Applying Math

Standardized Test Practice

- Multiple Choice
- Short Response/ Grid In
- Open-Ended Questions

Dynamic Instruction

The consistent instructional strategies in each chapter strengthen students' learning—from the beginning of each chapter where students see "Chapter Preview," to the end where they have a chance to test the knowledge they have acquired and prepare for the next lesson.

Chapter Opener

- *Chapter Preview* introduces the main concepts.
- *Science Journal* promotes critical-thinking skills
- *Dinah Zike's Foldables*™ let students create interactive study guides.
- *Launch Labs* give students an opportunity to explore new ideas.

Section Opener

- *What You'll Learn* introduces main concepts.
- *Why It's Important* provides an answer to "Why do we have to learn this?"
- *Review Vocabulary* reviews a term that helps students better understand section content.
- *New Vocabulary* highlights new terms students will learn in the section.

Labs

- *Labs* allow students to design their own experiments or follow well-tested procedures, enabling them to learn and practice science processes.
- *MiniLABS* offer students quick and easy-to-do ways to clarify concepts and reinforce skills.

Assessment

Glencoe Science offers the Glencoe Assessment Advantage, a system designed to give you all the tools you need to prepare your students for success in any testing situation.

In the *Student Edition*

Section Review and **Applying Math** questions appear in every chapter.

Chapter Review questions help you evaluate students' knowledge and ability to apply science concepts.

Standardized Test Practice questions at the end of each chapter provide students with additional opportunities to practice their test-taking skills.

In the *Teacher Wraparound Edition*

Assessments located throughout the *Teacher Wraparound Edition* provide methods for assessing students' comprehension with Performance, Process, and Content exercises.

Teacher Classroom Resources

Performance Assessment in the Science Classroom

- Guidelines for assessing the performance of a task
- Reproducible activities for evaluating students
- Sample rubrics and checklists

***Fast File* Chapter Resources** provides six pages of assessement for every chapter including *Testing Concepts, Applying Concepts,* and *Writing Skills.*

Technology Support

MindJogger Videoquizzes are interactive video quizzes set in game show format. Each is designed for the full range of student learning styles.

Exam*View*® Pro Testmaker CD-ROM for Windows® and Macintosh® provides an easy way to create, edit, and customize your tests. Select your own test items by objective from two different levels of difficulty, or write and edit your own. Translate tests from English to Spanish and vice versa.

Rubrics

The following rubrics are sample scoring devices for short response and open-ended questions.

Short Response

Points	Description
2	The student demonstrates a thorough understanding of the science of the task. The response may contain minor flaws that do not detract from the demonstration of a thorough understanding.
1	The student has provided a response that is only partially correct.
0	The student has provided a completely incorrect solution or no response at all.

Open Ended

Points	Description
4	The student demonstrates a thorough understanding of the science of the task. The response may contain minor flaws that do not detract from the demonstration of a thorough understanding.
3	The student demonstrates an understanding of the science of the task. The response is essentially correct and demonstrates an essential but less than thorough understanding of the science.
2	The student demonstrates only a partial understanding of the science of the task. Although the student may have used the correct approach to a solution or may have provided a correct solution, the work lacks an essential understanding of the underlying science concepts.
1	The student demonstrates a very limited understanding of the science of the task. The response is incomplete and exhibits many flaws.
0	The student provides a completely incorrect solution or no response at all.

Time-Saving Teacher Resources

Glencoe Science provides an extensive array of support materials and resources designed to help you create and customize your science course quickly and easily.

FAST FILE Chapter Resources

For each chapter, Chapter Resources contain key reproducible masters along with additional teaching strategies, teacher support, and answer keys.

Teacher Wraparound Edition

The *Teacher Wraparound Edition* is your key to the teaching resources available. In addition to teaching strategies and suggestions, the *Teacher Wraparound Edition* provides a guide for all print and software materials available for each lesson.

Transparencies

Color Transparencies provides three types of transparencies for use while teaching each chapter. The *Section Focus Transparencies* are designed to generate interest and focus students' attention on the topic being presented in the section. The *Teaching Transparency* for each chapter addresses a major concept that will benefit from an extra visual learning aid. The *Assessment Transparency* for each chapter is set up to resemble standardized tests.

Exam*View*® Pro Testmaker CD-ROM

This CD-ROM will help you create, edit, and customize tests. In addition, it will help you create multiple versions of tests, translate tests from English to Spanish and vice versa, and build tests aligned with state standards.

Video Labs

These VHS cassettes contain step-by-step procedures for selected *Student Edition* labs. They also contain lab safety skills, teacher support, and troubleshooting advice.

TeacherWorks™

This CD-ROM is your all-in-one teacher resource center. Personalize a lesson plan, access resources from the *Teacher Wraparound Edition*, connect to the Internet, or make a to-do list. These are only a few of the many features that can assist you in the planning and organizing of your lessons.

StudentWorks™

This CD-ROM is a valuable resource for students to access content online and use online resources to continue learning chapter concepts.

Interactive Chalkboard with Image Bank — PowerPoint® Presentations

This CD-ROM brings Microsoft® PowerPoint® presentations right to your door. With the large number of graphics provided, students can use a visual approach to learning chapter content.

Virtual Labs CD-ROM Program

The Virtual Labs CD-ROM contains a collection of labs that allow students to complete labs that are too expensive, take too long to complete, or might be too dangerous in a classroom laboratory.

Science Online

This website is a portal to hundreds of prescreened Internet sites (Web links) that correlate to content in the text. Visit this link to find interactive activities that review chapter concepts and to access the *Student Edition* online.

McGraw Hill Learning Network mhln.com

mhln.com is an online teaching and learning space for teachers, students, and parents.

Differentiated Instruction

Teaching Strategies

Following each suggested assessment and activity, ability levels are supplied to accommodate all students. For a key to the Teaching Strategies designations, see the C page before each chapter.

Identifying Misconceptions

These short, diagnostic, and perscriptive lessons target common science misconceptions.

Multiple Learning Styles

Look for these italicized designations under various activities to help you target your lessons to each student's preferred learning style.

- ***Kinesthetic*** learners learn through touch, movement, and manipulating objects.
- ***Visual-Spatial*** learners think in terms of images, illustrations, and models.
- ***Interpersonal*** learners understand and work well with other people.
- ***Intrapersonal*** learners can analyze their own strengths and weaknesses and may prefer to work on their own.
- ***Linguistic*** learners write clearly and easily understand the written word.
- ***Logical-Mathematical*** learners understand numbers easily and have highly-developed reasoning skills.

Daily Interventions

Found at the end of each chapter section, this feature is designed to intercept students who are struggling and prescribe a system to help them get back on track. *Reteach* provides reinforcement of the section's concepts through visual activities.

Differentiated Instruction

These activities present various teaching strategies designed to help you meet the special needs of students with learning disabilities, physical challenges, visual impairment, and hearing impairment. *Challenge* activities provide opportunities for students who excel to engage in activities and research projects that extend the chapter's concepts. English-language learners in the classroom will also find exercises that bridge the gap between language barriers and the chapter content.

Cultural Diversity

These readings provide insights into the unique ways in which people of different ethnicities and cultural heritage have approached science. The intent of these features is to build awareness and appreciation for the global community in which we live.

Inquiry-Based Science

The call for more inquiry-based science by the *National Science Education Standards* has been met by Glencoe Science.

Glencoe Science recognizes the importance of conducting inquiry-based science activities in the classroom. The process of doing inquiry models actual science practice, encouraging problem-solving strategies and developing critical thinking skills. Inquiry gets students actively involved in the learning process by allowing them to determine materials, procedures, or the topics and questions they want to investigate.

Inquiry can range from a very structured activity for those students who need more guidance to a more open-ended approach where students lead the investigations. Glencoe Science recognizes that the inquiry activities suggested will not look the same in every classroom. We encourage teachers to modify the suggested activities in a manner that best supports your students.

Glencoe also provides teachers with *Alternative Inquiry Labs,* teaching strategies or suggestions for making existing labs more inquiry-based.

Research-Based Learning Strategies

Glencoe Science incorporates the most current and applicable educational research on science learning and follows recommendations from the American Association for the Advancement of Science and the National Science Teachers Association. The following research-based strategies can be found throughout the text.

Learning Strategies

The following research-based strategies can be found throughout the text:

- **Using Prior Knowledge** Glencoe Science encourages students to use their prior knowledge to learn information because this adds relevance to the material. Students are referred back to other parts of the text or to their own real-life experiences.
- **Practicing Important Tasks** By offering students an opportunity to practice important tasks using a variety of labs and activities in the *Student Edition, Teacher Wraparound Edition,* ancillaries and technology, Glencoe Science makes learning fun and relevant for students.
- **Using Visuals to Communicate, Organize, and Reinforce Learning** High-quality art and photos throughout the text communicate concepts more efficiently and reinforce learning, while allowing students to organize information.
- **Motivating Students to Achieve** Active strategies and real-world experiences motivate students to achieve. Throughout Glencoe's programs, students are encouraged to apply their knowledge in ways that will motivate them to learn.
- **Developing Decoding and Reading Comprehension Strategies** Throughout the text, students are supplied with caption questions, reading checks, and other strategies to aid in comprehension.
- **Using Study Strategies** Through the use of highlighting, outlining, note-taking, summarizing, and other such strategies, students can monitor their own progress and organize information more effectively, thereby increasing their scientific literacy. These strategies are found throughout the text and ancillaries.

The use of these strategies within Glencoe Science will help teachers to achieve the goals set forth by the *National Science Education Standards.*

White Paper

The Glencoe Science White Paper outlines the educational strategies on which this program was based. This document provides specific examples from the *Student Edition, Teacher Wraparound Edition,* ancillary program, and technology resources, highlighting extensive use of educationally sound strategies that help students learn science.

Field Research and Testing

Feedback from students, teachers, curriculum supervisors, department chairpersons, parents, learning specialists, and science content experts was invaluable in the development of this program. The following pre-publication and post-publication research was conducted.

Prior to Publication

- Detailed classroom teacher and curriculum supervisor **surveys** were conducted by independently contracted researchers.
- A **nationwide panel** of science teachers, curriculum supervisors, and department chairpersons provided countless hours of feedback and assistance throughout program development.
- A wide range of **educator and content reviewers** provided in-depth reviews of and suggestions for manuscripts and pre-publication versions of the program.
- **Face-to-face interviews** with science teachers provided insight into teachers' day-to-day challenges.

Field-Test Results

- Field-test research indicates that test scores increased among students using Glencoe Science programs.
- Nine out of ten students earned higher scores after using Glencoe programs.
- Scores improved among both male and female students.
- Scores improved among both minority and non-minority students.
- Overall, the gap between the average pre-test score and a perfect score closed by 33 percent. Stated differently, on average, **scores increased 77 percent after students used the Glencoe program.**

After Publication

- Field tests were conducted in which students and teachers used a pre-publication manuscript in the classroom.
- Follow-up interviews, observations, and surveys of Glencoe Science users provide ongoing opportunities for program development and verification of program success.

KS Studios

National Education Standards

Correlation of *Glencoe Science* to the National Science Education Standards.

Content Standard	Chapter and Section
(UCP) Unifying Concepts and Processes	
1. Systems, order, and organization	F1-1, F1-2, F1-3, F2-1, F2-2, F2-3, F2-4, F3-1, F3-2, F3-3, F4-1, F4-2, F4-3, F5-1, F5-2, F5-3, F6-1, F6-2, F6-3, G1-1, G1-2, G1-3, G2-1, G2-2, G2-3, G3-1, G3-2, G3-3, G4-1, G4-2, G4-3, G5-1, G5-2, G5-3, G6-1, G6-2, G6-3, H2-1, H2-2, H2-3, H2-4, H3-1, H3-2, H3-3, H4-1, H4-2, H4-3, H5-1, H5-2, H5-3, I1-1, I1-2, I1-3, I2-1, I2-2, I2-3, I3-1, I3-2, I3-3, I4-1, I4-2, I4-3, J1-1, J1-2, J1-3, J2-1, J2-2, J2-3, J3-1, J3-2, J3-3, J3-4, J4-1, J4-2, J4-3
2. Evidence, models, and explanation	F1-1, F1-2, F1-3, F2-1, F2-2, F2-3, F2-4, F3-1, F3-2, F3-3, F4-1, F4-2, F4-3, F5-1, F5-2, F5-3, F6-1, F6-2, F6-3, G1-1, G1-2, G1-3, G2-1, G2-2, G2-3, G3-1, G3-2, G3-3, G4-1, G4-2, G4-3, G5-1, G5-2, G5-3, G6-1, G6-2, G6-3, H1-1, H1-2, H1-3, H2-1, H2-2, H2-3, H2-4, H3-1, H3-2, H3-3, H4-1, H4-2, H4-3, H5-1, H5-2, H5-3, I1-1, I1-2, I1-3, I2-1, I2-2, I2-3, I3-1, I3-2, I3-3, I4-1, I4-2, I4-3, J1-1, J1-2, J1-3, J2-1, J2-2, J2-3, J3-1, J3-2, J3-3, J3-4, J4-1, J4-2, J4-3
3. Change, constancy, and measurement	F1-1, F1-2, F1-3, F2-1, F2-2, F2-3, F2-4, F3-1, F3-2, F3-3, F4-1, F4-2, F4-3, F5-1, F5-2, F5-3, F6-1, F6-2, F6-3, G1-1, G1-2, G1-3, G2-1, G2-2, G2-3, G3-1, G3-2, G3-3, G4-1, G4-2, G4-3, G5-1, G5-2, G5-3, G6-1, G6-2, G6-3, H1-1, H1-2, H1-3, H2-1, H2-2, H2-3, H2-4, H3-1, H3-2, H3-3, H4-1, H4-2, H4-3, H5-1, H5-2, H5-3, I1-1, I1-2, I1-3, I2-1, I2-2, I2-3, I3-1, I3-2, I3-3, I4-1, I4-2, I4-3, J1-1, J1-2, J1-3, J2-1, J2-2, J2-3, J3-1, J3-2, J3-3, J3-4, J4-1, J4-2, J4-3
4. Evolution and equilibrium	G6-1, G6-2, J4-3
5. Form and function	F1-1, F1-2, F1-3, F2-1, F2-2, F2-3, F2-4, F3-1, F3-2, F3-3, F4-1, F4-2, F4-3, F5-1, F5-2, F5-3, F6-1, F6-2, F6-3, G1-1, G1-2, G1-3, G2-1, G2-2, G2-3, G3-1, G3-2, G3-3, G4-1, G4-2, G4-3, G5-1, G5-2, G5-3, G6-1, G6-2, G6-3, H1-1, H1-2, H1-3, H2-1, H2-2, H2-3, H2-4, H3-1, H3-2, H3-3, H4-1, H4-2, H4-3, H5-1, H5-2, H5-3, I1-1, I1-2, I1-3, I2-1, I2-2, I2-3, I3-1, I3-2, I3-3, I4-1, I4-2, I4-3, J1-1, J1-2, J1-3, J2-1, J2-2, J2-3, J3-1, J3-2, J3-3, J3-4, J4-1, J4-2, J4-3
(A) Science as Inquiry	
1. Abilities necessary to do scientific inquiry	F1-1, F1-2, F1-3, F2-1, F2-2, F2-3, F2-4, F3-1, F3-2, F3-3, F4-1, F4-2, F4-3, F5-1, F5-2, F5-3, F6-1, F6-2, F6-3, G1-1, G1-2, G1-3, G2-1, G2-2, G2-3, G3-1, G3-2, G3-3, G4-1, G4-2, G4-3, G5-1, G5-2, G5-3, G6-1, G6-2, G6-3, H1-1, H1-2, H1-3, H2-1, H2-2, H2-3, H2-4, H3-1, H3-2, H3-3, H4-1, H4-2, H4-3, H5-1, H5-2, H5-3, I1-1, I1-2, I1-3, I2-1, I2-2, I2-3, I3-1, I3-2, I3-3, I4-1, I4-2, I4-3, J1-1, J1-2, J1-3, J2-1, J2-2, J2-3, J3-1, J3-2, J3-3, J3-4, J4-1, J4-2, J4-3
2. Understandings about scientific theory	F1-1, F1-2, F1-3, F2-1, F2-2, F2-3, F2-4, F3-1, F3-2, F3-3, F4-1, F4-2, F4-3, F5-1, F5-2, F5-3, F6-1, F6-2, F6-3, G1-1, G1-2, G1-3, G2-1, G2-2, G2-3, G3-1, G3-2, G3-3, G4-1, G4-2, G4-3, G5-1, G5-2, G5-3, G6-1, G6-2, G6-3, H1-1, H1-2, H1-3, H2-1, H2-2, H2-3, H2-4, H3-1, H3-2, H3-3, H4-1, H4-2, H4-3, H5-1, H5-2, H5-3, I1-1, I1-2, I1-3, I2-1, I2-2, I2-3, I3-1, I3-2, I3-3, I4-1, I4-2, I4-3, J1-1, J1-2, J1-3, J2-1, J2-2, J2-3, J3-1, J3-2, J3-3, J3-4, J4-1, J4-2, J4-3
(B) Physical Science	
1. Properties and changes of properties in matter	G2-1, G5-3, H1-1, H4-1, H5-2, I4-1, J4-3
2. Motions and forces	F1-2, F4-1, F4-2, F4-3, F5-1, F5-2, F6-1, F6-2, G2-2, G2-3, G3-3, I1-2, I1-3, I2-2, I3-3, J2-1, J2-2, J3-1, J4-3
3. Transfer of energy	H5-2, I1-2, I3-3, J4-3
(C) Life Science	
1. Structure and function in living systems	H2-1, H2-2, H2-3, H5-2
2. Reproduction and heredity	H5-2
3. Regulation and behavior	H5-1, H5-2
4. Populations and ecosystems	H2-1, H2-2, H2-3, H5-1, H5-2, I4-2
5. Diversity and adaptations of organisms	H2-1, H2-2, H2-3, H5-2
(D) Earth and Space Science	
1. Structure of the Earth system	F1-1, F2-1, F2-2, F2-3, F2-4, F3-1, F3-2, F3-3, F4-1, F4-2, F4-3, F5-1, F5-2, F5-3, F6-1, F6-2, G1-1, G1-2, G1-3, G2-1, G2-2, G4-1, G4-2, G4-3, G6-1, G6-2, G6-3, H2-1, H2-2, H3-1, H3-2, H3-3, H4-1, H4-2, H4-3, H5-1, H5-2, I1-1, I1-2, I1-3, I2-1, I2-2, I2-3, I3-1, I3-2, I3-3, J2-1, J2-3
2. Earth's history	F1-1, F1-2, F2-1, F2-2, F2-3, F2-4, F3-1, F4-1, F5-1, F6-1, G1-1, G1-3, G2-1, G2-3, H3-3, I2-3, I3-3
3. Earth and the solar system	F1-3, F2-1, F2-2, F2-3, F2-4, F3-1, F3-3, H4-3, J1-1, J1-2, J1-3, J2-3, J3-2, J4-2

Content Standard	Chapter and Section
(E) Science and Technology	
1. Abilities of technological design	H4-3
2. Understandings about science and technology	H4-3
(F) Science in Personal and Social Perspectives	
1. Personal health	F3-2, G2-2, I4-2
2. Populations, resources, and environments	F3-1, F3-2, F3-3, H1-3, H2-4, H3-2, H5-3, I4-1, I4-2, I4-3
3. Natural hazards	F4-1, F4-2, F4-3, F5-1, F5-2, F5-3, F6-1, F6-2, G3-3, G4-1, G4-3, I2-3, I4-1, I3-3
4. Risks and benefits	G2-1, G3-3, H3-2
5. Science and technology in society	F3-1, G5-1, G5-3, I4-3, J1-3
(G) History and Nature of Science	
1. Science as a human endeavor	F1-3
2. Nature of science	F1-2, F1-3, F4-3, F6-3, I4-3
3. History of science	F1-2, F1-3, F5-2, G5-3, H5-3, I4-3, J2-3, J3-1, J3-4, J4-3

How Glencoe Science Aligns with the National Science Education Standards

The correlations at the left and above show the close alignment between Glencoe Science and the grade-appropriate standards. Glencoe Science allows students to discover concepts within each of the content standards and gives students opportunities to make connections among the science disciplines. Hands-on activities and inquiry-based lessons reinforce the science processes emphasized in the standards.

How Glencoe Science Aligns with the NCTM Standards for Grades 6–8

Throughout Glencoe Science, each Applying Math activity provides students with the opportunity to practice and apply some of the mathematical concepts and applications described in the NCTM Standards. These activities serve to reinforce mathematical skills in real-life situations, thus preparing students to meet their needs in an ever-changing world.

Correlation of *Glencoe Science* to NCTM Standards

Math Standard	Page
1. Number and Operations	F-31, F-54, F-61, F-121, F-143, F-151, F-172, F-181, G-31, G-46, G-59, G-87, G-119, G-149, G-173, H-47, H-63, H-95, H-123, H-130, I-31, I-39, I-61, I-114, I-121, J-35, J-65, J-80, J-99, J-130
2. Algebra	F-54, F-61, F-143, F-151, F-172, F-181, G-31, G-119, G-149, H-47, H-95, H-123, H-130, I-39, I-61, I-114, I-121, J-65
3. Geometry	G-31, H-95, J-16, J-65
4. Measurement	G-31, H-95, I-91, I-121, J-35, J-65
5. Data Analysis and Probability	F-31, F-93, F-151, F-181, G-46, G-149, G-181, H-31, H-63, H-123, H-153, I-31, I-39, I-61, I-121, J-35, J-99, J-130
6. Problem Solving	F-31, G-46, G-87, G-149
7. Reasoning and Proof	F-54, F-143, F-172, G-46, G-105, G-173, H-108, H-130, I-39, J-16, J-80
8. Communication	F-117, F-147, G-83, I-87, J-31
9. Connections	F-54, F-61, F-121, F-143, F-151, F-172, F-181, G-31, G-59, G-119, H-47, H-63, H-95, H-123, H-130, H-153, I-31, I-39, I-61, I-91, I-114, I-121, J-35, J-65, J-80, J-99, J-130
10. Representation	G-31, G-149

Foldables™

Foldables™ are easy-to-make, three-dimensional, interactive graphic organizers that students create out of simple sheets of paper. These unique hands-on tools for studying and reviewing were created exclusively for Glencoe by education specialist Dinah Zike.

Research Behind Foldables™

According to research (Bransford, 1979; Corno, 1994), study strategies help students understand, organize, remember and apply new information presented in science textbooks. Some study strategies include concept mapping, highlighting, outlining, note taking, summarizing, and underlining (Peverly, Brobst, Graham & Shaw, 2003). Glencoe Science offers Dinah Zike's Foldables™ Study Organizer as an organizational tool and study guide for students.

Foldables™

- Build prereading skills
- Encourage active reading and writing
- Summarize content for review

FOLDABLES™ Study Organizer

Primary and Secondary Succession Make the following Foldable to help you illustrate the main ideas about succession.

STEP 1 **Fold** a vertical sheet of paper in half from top to bottom.

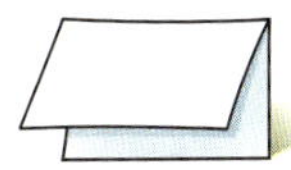

For more ideas on how to incorporate Foldables™ into your lessons consult **Dinah Zike's *Teaching Science with Foldables*™**

Educational Partnerships

NATIONAL GEOGRAPHIC

Some topics in the chapter either require or benefit from a larger, more detailed visual explanation. The National Geographic Society has created *Visualizing* features that call out an important concept from the chapter and illustrate it in a way that will inform, excite, and motivate your students.

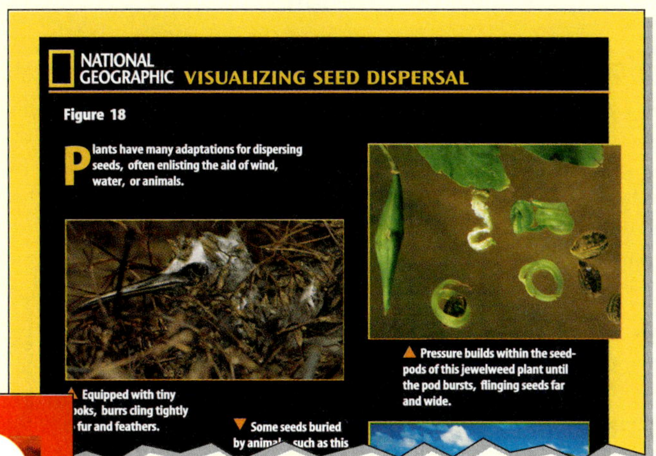

TIME

TIME magazine brings science topics and history together to further explain the chapter's main ideas and show how science relates to real life.

Safety in the Laboratory

All activities are designed to minimize dangers in the laboratory. Careful laboratory planning and management by both the instructor and the student are essential to a safe laboratory experience. **Local, state, and federal laboratory safety laws and regulations must be strictly followed.** The information provided here is one of the many resources to which you can refer for information about laboratory safety.

Classroom and Laboratory Preparation

1. Store equipment properly and securely and other thing.
   a. Clean and dry all equipment before storing.
   b. Protect electronic equipment and microscopes from dust, humidity, and extreme temperatures.
   c. Number, catalog, and organize equipment.
2. Ensure adequate work space for each student.
3. Ensure adequate classroom and storeroom ventilation.
4. Explain and post safety and evacuation guidelines along with expectations of conduct.
5. Ensure that all safety equipment is functioning properly and is clearly visible.
6. Provide hot plates as a heat source whenever possible. If gas burners are used, know where the central gas supply shutoff valve is located.
7. Ensure that each workstation has a GFCI-protected electrical source.
8. Provide safety goggles consistent with ANSI Standard Z87.1 for each student, including students who wear corrective lenses.

Before Each MiniLAB or Lab

1. Arrange the lab in such a way that equipment and supplies are clearly labeled and easily accessible.
2. Have available only equipment and supplies needed to complete the assigned investigation.
3. Review the procedure with students, emphasizing any caution statements or safety symbols that appear.
4. Be sure all students know the proper procedures to follow if an accident should occur.

After the MiniLAB or Lab

1. Be certain that students have returned all equipment and disposed of broken glassware and chemicals properly.
2. Be sure that all hot plates and electrical connections are off.

Storage of Chemicals

Be sure to store all chemicals properly. The following are guidelines commonly used. Your school, city, county, or state may have additional requirements for handling chemicals. It is the responsibility of each teacher to become informed of the rules or guidelines in effect in his or her area.

1. Separate chemicals by reaction type. Strong acids should be stored together. Likewise, strong bases should be stored together and should be separated from acids. Oxidants should be stored away from easily oxidized materials, and so on.
2. Be sure all chemicals are stored in labeled containers indicating contents, concentration, source, date purchased (or prepared), any precautions for handling and storage, and expiration date.
3. Hazardous chemicals require special storage containers and conditions. Be sure to know which chemicals those are and the accepted practices for your area. Some substances must be stored outside the building.

Disposal of Chemicals

Local, state, and federal laws regulate the proper disposal of chemicals. These laws should be consulted before chemical disposal is attempted. Although many substances encountered in the science classroom can be flushed down the drain with plenty of water, it is not safe to assume that this is always true.

DISCLAIMER

Glencoe Publishing Company makes no claims to the completeness of this discussion of laboratory safety and chemical storage. The material presented is not all-inclusive, nor does it address all of the hazards associated with handling, storing, and disposing of chemicals, or with laboratory management.

Preparation of Solutions

It is important to use safe laboratory techniques when handling all chemicals. Always check the MSDS (Material Safety Data Sheet) for each chemical before using it in the classroom. Many substances might appear harmless, but might be toxic, corrosive, or very reactive. Chemicals should never be ingested. Use proper techniques to smell any chemical, wear safety goggles and an apron in the laboratory, and observe the following precautions.

1. **Dilution of Acids and Bases** When diluting acids with water, always add the acids to the water. Never add water to acids. When sulfuric acid and sodium hydroxide are added to water, a large amount of thermal energy is released. Use extra care when handling these substances.

2. **Poisonous and Corrosive Liquids or Vapors** Use a fume hood if possible. Examples include hydrochloric acid, acetic acid, nitric acid, and ammonium hydroxide.

3. **Poisonous and Corrosive to Eyes, Lungs, and Skin** Examples include acids, bases, silver nitrate, iodine, and potassium permanganate.

Bromthymol blue: Add 0.5 g bromthymol blue powder to 500 mL distilled water to make a BTB stock solution. Dilute 40 mL BTB stock solution to 2 L with distilled water. Solution should be bright blue. If not, add one drop of NaOH at a time, swirling to mix. Check color.

Hydrochloric acid (HCL) solution: To make a 5% solution, add 13.6 mL concentrated HCl to 73 mL water while stirring. To make a 0.1*M* solution, add 1 mL concentrated hydrochloric acid to 100 mL water while stirring.

Iodine solution/Iodine stain: Dilute 1 part Lugol's solution with 15 parts water.

Lugol's solution: Dissolve 10 g potassium iodide in 100 mL distilled water. Then add and dissolve 5 g iodine. Store in dark bottle. Keeps indefinitely.

Phenolphthalein indicator: From a drug store, buy a package of any laxative that contains phenolphthalein. To make 1% solution, mash 4 tablets and pour the powder into 10 mL of rubbing alcohol. Let mixture soak for 15 minutes. Pour liquid into and store in a dropper bottle.

Potassium permanganate: For a 0.01*M* solution of potassium permanganate, dissolve 0.15 g $KMnO_4$ in 100 mL water.

Red cabbage concentrate: Put 5 leaves of red cabbage in a pot. Add 1 L of water, bring to a boil, and simmer until water turns a deep purple. Pour liquid through a strainer or piece of cheesecloth into a storage bottle. Keep refrigerated.

Salt solution: For a 3.5% salt (NaCl) solution that simulates the concentration of ocean water, dissolve 35 g of salt (NaCl) in 965 mL of water. For a 1% solution (weak), dissolve 1 g of salt (NaCl) in 99 mL of water. For a 6% solution, dissolve 6 g of salt (NaCl) in 94 mL of water.

Silver nitrate solution: To make a 10% solution, put 5 g of silver nitrate in 50 mL of distilled water.

Sugar solution: Add 1 tablespoon of sugar to 1 cup of warm water in a deep jar or flask. Stir to dissolve.

Sodium hydroxide (dilute): To make a 1% solution, dissolve 1 g NaOH in 99 mL of water.

Equipment and Materials List

Refer to the Chapter Organizer in front of each chapter for a list of equipment and materials used for each laboratory activity in the chapter.

Consumables			
Material	**Launch Lab (Chapter)**	**MiniLAB (Chapter-Section)**	**Lab (Chapter-Section)**
bowl, paper	5		
cardboard	2		3-3, 5-3
clay			5-3
Epsom salt	2		
food coloring, yellow	2		
gelatin powder		3-1	
gravel		4-2	3-3
ice block			3-2
jelly beans, red and green		5-3	
pipe cleaners			5-3
plaster of paris	5		3
rubber tubing			4-3
sand		4-2	3-2, 3-3, 4-3
sand-and-gravel mixture	3		
shells, small	5		
sod		3-3	
soil	4	2-2	2-2
steel wool		2-1	
string, cotton	2		
toothpicks			5-3
vinegar, white			2-3
wire			5-3
yarn, colored	6		
Nonconsumables			
aluminum pie pan	4		
aquarium gravel		3-1	
baking pan		3-1	
bowl, mixing			3-3
bricks		3-1	
computer probe			2-3
container, clear plastic		4-2	
deck of playing cards			6-2
glasses, drinking		4-1	
hair dryer			3-3
hot plate			3
jar	5	2-1	
magnet			4-3
mesh screen, fine			4-3
model of a landform, plastic			1-3
overhead light with reflector			3-2
paint roller pan			4-3
pans, flat			3-3
plastic, rigid			5-3
rain gauge		4-1	
rock, flat		3-1	
shoe-box lid, large	3		
sprinkling can		4-1	3-3
storage box with lid, clear-plastic			1-3
tray, metal or plastic, large			3-2
water bottle			2-2
watering can		3-1	
wind speed indicator			3-3
wood block			4-3

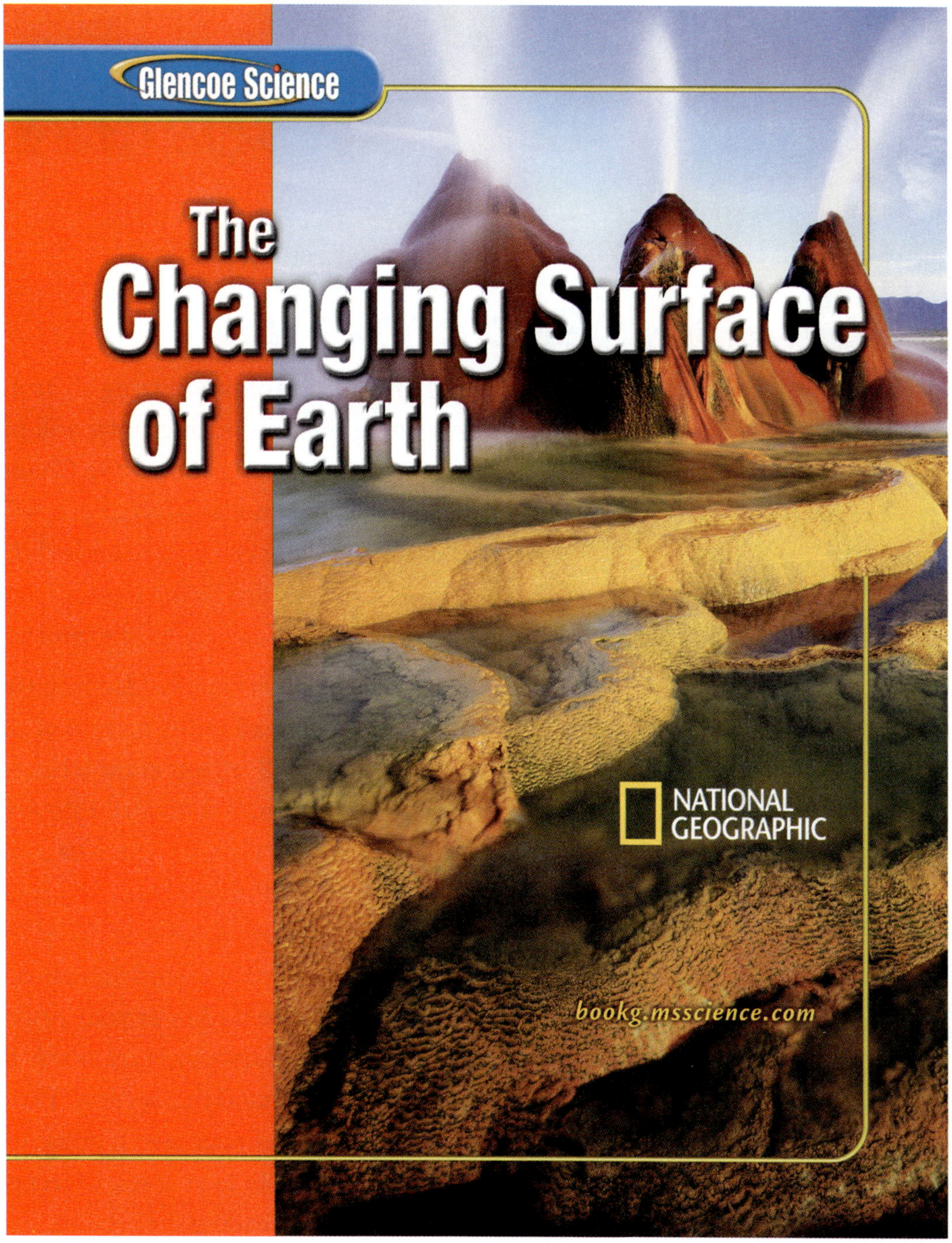

Glencoe

New York, New York   Columbus, Ohio   Chicago, Illinois   Peoria, Illinois   Woodland Hills, California

HOW TO...

Use Your Science Book

Why do I need my science book?

Have you ever been in class and not understood all of what was presented? Or, you understood everything in class, but at home, got stuck on how to answer a question? Maybe you just wondered when you were ever going to use this stuff?

These next few pages are designed to help you understand everything your science book can be used for . . . besides a paperweight!

Before You Read

- **Chapter Opener** Science is occurring all around you, and the opening photo of each chapter will preview the science you will be learning about. The **Chapter Preview** will give you an idea of what you will be learning about, and you can try the **Launch Lab** to help get your brain headed in the right direction. The **Foldables** exercise is a fun way to keep you organized.
- **Section Opener** Chapters are divided into two to four sections. The **As You Read** in the margin of the first page of each section will let you know what is most important in the section. It is divided into four parts. **What You'll Learn** will tell you the major topics you will be covering. **Why It's Important** will remind you why you are studying this in the first place! The **Review Vocabulary** word is a word you already know, either from your science studies or your prior knowledge. The **New Vocabulary** words are words that you need to learn to understand this section. These words will be in **boldfaced** print and highlighted in the section. Make a note to yourself to recognize these words as you are reading the section.

As You Read

- **Headings** Each section has a title in large red letters, and is further divided into blue titles and small red titles at the beginnings of some paragraphs. To help you study, make an outline of the headings and subheadings.
- **Margins** In the margins of your text, you will find many helpful resources. The **Science Online** exercises and **Integrate** activities help you explore the topics you are studying. **MiniLabs** reinforce the science concepts you have learned.
- **Building Skills** You also will find an **Applying Math** or **Applying Science** activity in each chapter. This gives you extra practice using your new knowledge, and helps prepare you for standardized tests.
- **Student Resources** At the end of the book you will find **Student Resources** to help you throughout your studies. These include **Science, Technology,** and **Math Skill Handbooks**, an **English/Spanish Glossary,** and an **Index.** Also, use your **Foldables** as a resource. It will help you organize information, and review before a test.
- **In Class** Remember, you can always ask your teacher to explain anything you don't understand.

FOLDABLES™ Study Organizer

Science Vocabulary Make the following Foldable to help you understand the vocabulary terms in this chapter.

STEP 1 **Fold** a vertical sheet of notebook paper from side to side.

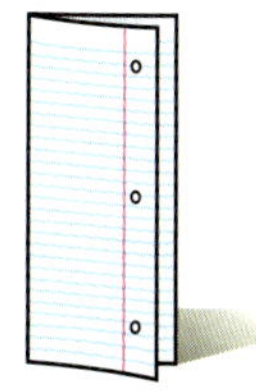

STEP 2 **Cut** along every third line of only the top layer to form tabs.

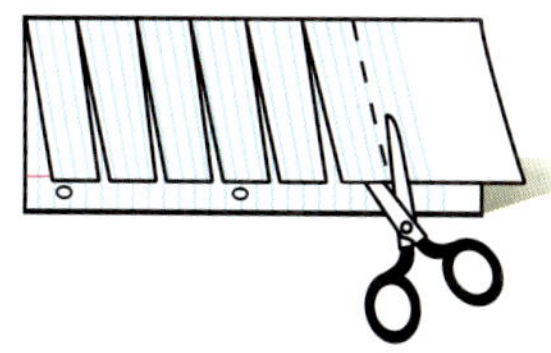

STEP 3 **Label** each tab with a vocabulary word from the chapter.

Build Vocabulary As you read the chapter, list the vocabulary words on the tabs. As you learn the definitions, write them under the tab for each vocabulary word.

In Lab

Working in the laboratory is one of the best ways to understand the concepts you are studying. Your book will be your guide through your laboratory experiences, and help you begin to think like a scientist. In it, you not only will find the steps necessary to follow the investigations, but you also will find helpful tips to make the most of your time.

- Each lab provides you with a **Real-World Question** to remind you that science is something you use every day, not just in class. This may lead to many more questions about how things happen in your world.
- Remember, experiments do not always produce the result you expect. Scientists have made many discoveries based on investigations with unexpected results. You can try the experiment again to make sure your results were accurate, or perhaps form a new hypothesis to test.
- Keeping a **Science Journal** is how scientists keep accurate records of observations and data. In your journal, you also can write any questions that may arise during your investigation. This is a great method of reminding yourself to find the answers later.

Before a Test

Admit it! You don't like to take tests! However, there *are* ways to review that make them less painful. Your book will help you be more successful taking tests if you use the resources provided to you.

- Review all of the **New Vocabulary** words and be sure you understand their definitions.
- Review the notes you've taken on your **Foldables,** in class, and in lab. Write down any question that you still need answered.
- Review the **Summaries** and **Self Check questions** at the end of each section.
- Study the concepts presented in the chapter by reading the **Study Guide** and answering the questions in the **Chapter Review.**

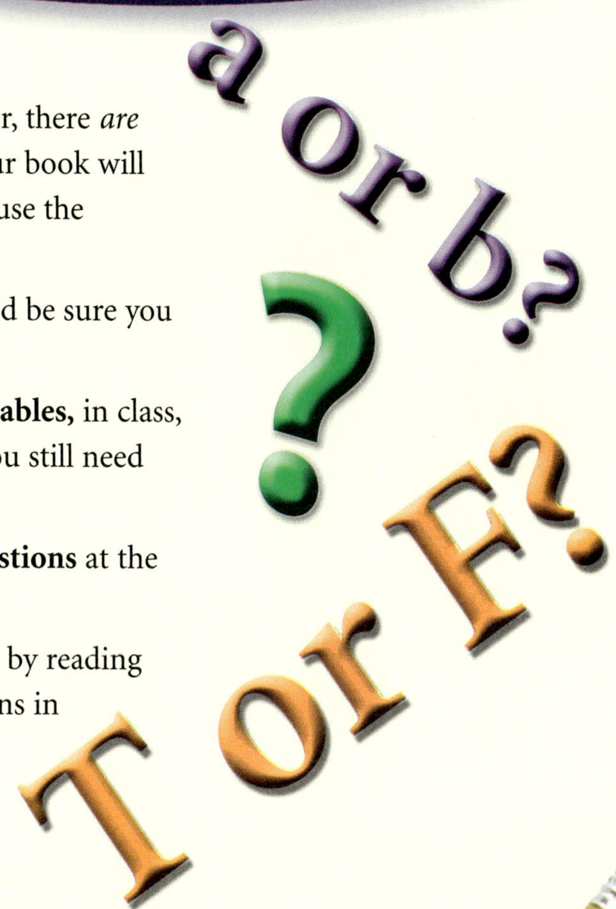

Look For...

- **Reading Checks** and **caption questions** throughout the text.
- the **Summaries** and **Self Check questions** at the end of each section.
- the **Study Guide** and **Review** at the end of each chapter.
- the **Standardized Test Practice** after each chapter.

Scavenger HUNT

Let's Get Started

To help you find the information you need quickly, use the Scavenger Hunt below to learn where things are located in Chapter 1.

1. What is the title of this chapter?
2. What will you learn in Section 1?
3. Sometimes you may ask, "Why am I learning this?" State a reason why the concepts from Section 2 are important.
4. What is the main topic presented in Section 2?
5. How many reading checks are in Section 1?
6. What is the Web address where you can find extra information?
7. What is the main heading above the sixth paragraph in Section 2?
8. There is an integration with another subject mentioned in one of the margins of the chapter. What subject is it?
9. List the new vocabulary words presented in Section 2.
10. List the safety symbols presented in the first Lab.
11. Where would you find a Self Check to be sure you understand the section?
12. Suppose you're doing the Self Check and you have a question about concept mapping. Where could you find help?
13. On what pages are the Chapter Study Guide and Chapter Review?
14. Look in the Table of Contents to find out on which page Section 2 of the chapter begins.
15. You complete the Chapter Review to study for your chapter test. Where could you find another quiz for more practice?

Teacher Advisory Board

The Teacher Advisory Board gave the authors, editorial staff, and design team feedback on the content and design of the Student Edition. They provided valuable input in the development of the 2005 edition of ***Glencoe Science.***

John Gonzales
Challenger Middle School
Tucson, AZ

Rachel Shively
Aptakisic Jr. High School
Buffalo Grove, IL

Roger Pratt
Manistique High School
Manistique, MI

Kirtina Hile
Northmor Jr. High/High School
Galion, OH

Marie Renner
Diley Middle School
Pickerington, OH

Nelson Farrier
Hamlin Middle School
Springfield, OR

Jeff Remington
Palmyra Middle School
Palmyra, PA

Erin Peters
Williamsburg Middle School
Arlington, VA

Rubidel Peoples
Meacham Middle School
Fort Worth, TX

Kristi Ramsey
Navasota Jr. High School
Navasota, TX

Student Advisory Board

The Student Advisory Board gave the authors, editorial staff, and design team feedback on the design of the Student Edition. We thank these students for their hard work and creative suggestions in making the 2005 edition of ***Glencoe Science*** student friendly.

Jack Andrews
Reynoldsburg Jr. High School
Reynoldsburg, OH

Peter Arnold
Hastings Middle School
Upper Arlington, OH

Emily Barbe
Perry Middle School
Worthington, OH

Kirsty Bateman
Hilliard Heritage Middle School
Hilliard, OH

Andre Brown
Spanish Emersion Academy
Columbus, OH

Chris Dundon
Heritage Middle School
Westerville, OH

Ryan Manafee
Monroe Middle School
Columbus, OH

Addison Owen
Davis Middle School
Dublin, OH

Teriana Patrick
Eastmoor Middle School
Columbus, OH

Ashley Ruz
Karrer Middle School
Dublin, OH

The Glencoe middle school science Student Advisory Board taking a timeout at COSI, a science museum in Columbus, Ohio.

Contents

In each chapter, look for these opportunities for review and assessment:
- Reading Checks
- Caption Questions
- Section Review
- Chapter Study Guide
- Chapter Review
- Standardized Test Practice
- Online practice at bookg.msscience.com

Contents

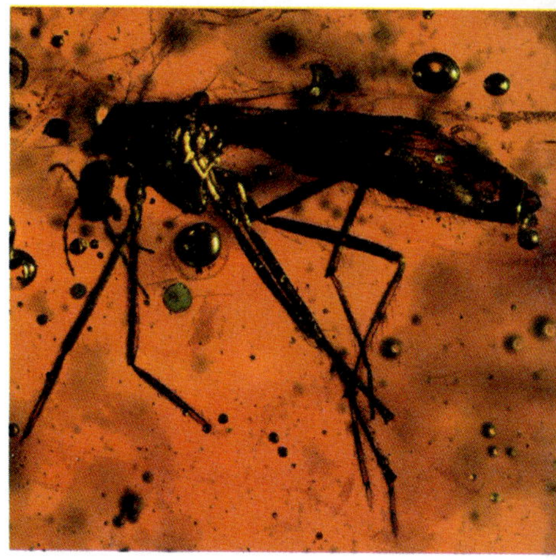

Student Resources

Cross-Curricular Readings/Labs

available as a video lab

NATIONAL GEOGRAPHIC VISUALIZING

TIME SCIENCE AND Society

TIME SCIENCE AND HISTORY

Oops! Accidents in SCIENCE

Science and Language Arts

SCIENCE Stats

Launch LAB

Mini LAB

Mini LAB Try at Home

Labs/Activities

One-Page Labs

Two-Page Labs

Design Your Own Labs

Model and Invent Labs

Use the Internet Labs

Applying Math

Applying Science

INTEGRATE

Science Online

Standardized Test Practice

Content Details

The NATURE OF SCIENCE

Land Use in Floodplains

Introduction

This feature depicts a common situation that highlights the limitations of science. Scientists can perform laboratory experiments or observe real-life situations to reach conclusions. Science can provide information that is important for making political decisions. However, science alone cannot provide answers to all societal problems.

1 Motivate

Write responses in two columns on the board as students think of the pros and cons of living on floodplains. Ask how they might react if their homes were flooded. See if the class can reach a consensus on whether development on floodplains should be outlawed.

The NATURE OF SCIENCE

Limits of Science

Land Use in Floodplains

Figure 1 Major floods devastated entire communities, such as Jefferson City, Missouri, during the Great Flood of 1993.

Imagine what it would be like to watch water rise higher and higher in your house, threatening all your possessions. That's what happened to some people in the midwest in June and July of 1993. The upper Mississippi River basin, with the land already soaked from a wet winter and spring, received almost 14 inches of rain. Even though people struggled desperately to protect their homes and businesses, whole towns were flooded and entire farms were lost, resulting in billions of dollars of damage and loss of life. And yet, after the water receded, the people moved back, homes and businesses were repaired or rebuilt, and new crops were planted.

Living on a Floodplain

Floods are the most common natural disaster in the world and occur when there is persistent heavy rainfall and the soil is water-logged. The excess water swells a river, which eventually spills out onto the surrounding flatlands. Science is limited to offering options instead of solutions to those who choose to live on floodplains—areas prone to floods. It is the task of individuals and government to develop solutions to the problem of flooding.

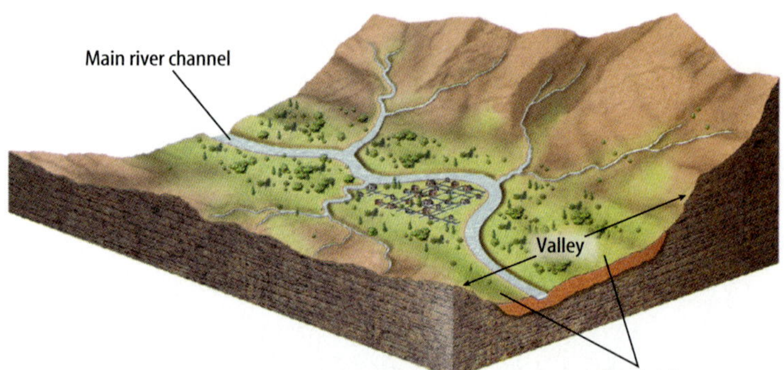

Figure 2 Although rivers usually stay within their channels, rivers cover floodplains when heavy rainfall increases their flow.

Science Journal

Living Dangerously Tell students that many cities have been built on floodplains and include neighborhoods and business districts that are prone to periodic flooding. Have students write in their science journals their thoughts on how people might justify their decisions to stay in such areas. LS **Interpersonal**

Almost 10 million homes in the United States are situated on floodplains. When floodplains are not underwater, their flat, fertile soil and proximity to water make them popular sites for towns, farms, and industrial transportation. However, when rivers flow onto floodplains, they also wash through the homes and farms that are located there.

How have scientists and engineers tried to deal with the problems of flooding on developed floodplains? One method is to use human-made structures to block or divert the flow of rivers. Many dams have been built to block water from flowing downstream during heavy rains, forcing it instead into human-made lakes. Thousands of miles of levees, or artificial embankments, have been built along rivers to keep water contained in channels.

Figure 3 A wetland can absorb floodwaters more effectively than developed land can.

Limitations

These efforts have proven to be only partly successful. Human-made systems only can contain and direct water to a certain extent. Dams have overflowed and artificial conduits, or pathways for water, have caused flooding downstream. Levees have caused drying of sponge-like, absorbent wetlands, creating land even more prone to flooding.

Because of these problems, many environmental scientists are in favor of leaving floodplains undeveloped and moving floodplain communities to higher ground with the help of the Federal Emergency Management Agency (FEMA) which provides financial assistance to owners of homes and businesses that have been flooded and wish to relocate. The United States Department of Agriculture's Wetlands Reserve Program also is working to keep floodplain wetlands from being developed and to return developed floodplains to wetlands. The program has successfully restored wetlands in Missouri, Michigan, and several other states.

Figure 4 The floodgates on this dam had to be opened after weeks of heavy rain.

2 Teach

Content Background

An extensive floodplain is the product of thousands of years of river evolution. The most mature river valleys, such as the lower Mississippi River, feature floodplains that are several kilometers wide on each side. Such rivers are close to their base level, which is the lowest point to which a stream can erode its channel. At the base level, erosion or river current really doesn't exist. Rivers might not be cutting down above base level if the stream exists in a graded condition or if it is aggrading. As a river meanders back and forth, the floodplain is formed and the valley becomes wider. Repeated flooding results in the river dropping fine-grained sediment across the valley floor. These thick deposits form exceptionally fertile soils for agriculture. Features such as natural levees, meanders, and oxbow lakes typically mark floodplains.

Discussion

What are the limitations of dams in terms of blocking floodwaters? Possible answers: The reservoir behind a dam is limited in its capacity. When that capacity is exceeded, the water is uncontrollable. A dam also could break, causing disaster below it. **How might a levee system be considered a way of shifting the flooding problem?** Answer: Levees can succeed in moving the brunt of a flood downstream. Water builds up and gains speed in a confined channel, causing worse problems farther downstream than would be experienced otherwise.

Visual Learning

Figure 4 Ask students to study this picture and consider implications downstream. Possible answers: fast currents, high water levels, flooding, erosion, property damage **Is heavy rain the only circumstance that could cause such high water?** Spring snowmelt might also cause high water if heavy snows were followed by suddenly warm weather.

Extension

Challenge students to find the locations of some of the world's highest dams. Ask them to model one of these dams in a plastic basin. Have them research the characteristics of specific river systems that have such dams and to discover their effects on the floodplains below. Ask them to explain the benefits and drawbacks of large dams. Supply poster board on which they can make informational posters to present their findings.

The NATURE OF SCIENCE

The Limits of Science

Society often asks scientists to solve problems. But there are times when the role of a scientist is limited to offering options. Science deals with facts, but there are questions it cannot answer. For example, people want answers to questions about how to prevent flooding or how to protect the homes, businesses, and farms in floodplains. When answering these questions, however, personal, political and economic factors all must be considered.

Figure 5 A system of levees is designed to keep the Mississippi River from pouring into developed areas.

What Does Science Do?

Science helps people solve problems and answer questions. By experimenting, analyzing data, and forming conclusions, people use science to gain an increased understanding of nature. The procedures used by scientists must be carefully planned to produce reliable results. Scientific methods must be observable, testable, and repeatable.

Scientists can apply scientific methods to the problem of living on floodplains. Identifying the problem of flooding is straightforward. Floodwaters are a danger to people and structures on floodplains. Scientists propose and test various solutions to the problem. Controlling the flow of river water through dams and levees can be beneficial, but it will not completely avoid flooding.

Testing Ideas

Scientists often can perform experiments in a laboratory to test ideas about a problem. For example, they might test how water flows through different types of soil or test the reliability of various levee designs. Other times scientists must rely on observations of real-life situations, and observations of past flooding, in order to reach conclusions. When studying floodplain development, scientists cannot accurately predict how water will flow on developed floodplains.

Sometimes mistakes are made in the process of applying scientific methods to solve a problem. When original hypotheses are proven inaccurate, the experience gained through experimentation can help refine or restructure ideas.

Curriculum Connection

Social Studies Have students report on the Nile River and the role of its floodplain in the development of civilization. Have students include information on the length of the Nile and the width of its floodplain. Ask them to find details about the major dam that now controls the Nile River, and the great cost in cultural relics caused by its construction. Suggest students make posters to present their findings.

Differentiated Instruction

Learning Disabled Show students how a seemingly large task—like the Curriculum Connection—can seem less daunting when it is broken into steps. Have students make a "shopping list" of the specific items of information needed to make their posters. Offer examples of list entries such as: Name of large Nile dam; or Lost Nile River historical relics.

For example, engineers who constructed early dam and levee systems did not do so carelessly. While designing systems to control the flow of water, workers cannot predict all the shortcomings and adverse effects that dams and levees might produce. What is now known about rivers and their floodplains largely comes from earlier attempts at flood control.

Figure 6 This home in Rhineland, Missouri, was moved to higher ground after the Great Flood of 1993.

What Doesn't Science Do?

Science deals with facts, but it can't tell people how to think or feel. Science exposes the dangers of flooding, but the decision of whether to accept the risks of living on a floodplain remains with those who wish to stay or move there. One solution to the problem of flooding is to leave floodplains undeveloped and to move communities out of floodplains that already have been developed. Science is not qualified to make this decision. Solutions to the problem of flooding must be balanced among different viewpoints.

Moving communities to higher ground removes them from a river—a fertile agricultural site, a scenic area, and perhaps even an industrial work site. By moving entire communities, neighborhoods are lost and historic sites could be destroyed. In addition, people might not wish to lose government funding that maintains development of floodplains.

People ultimately must make the difficult decisions concerning floodplain development. Although science can be applied to help offer possible solutions, it cannot provide the answers to philosophical and political questions that arise from the problem of flooding. Science can make recommendations but it cannot and should not dictate behavior.

Figure 7 These students are using debate as a way to learn about both sides of an issue.

You Do It

Think of an important land use decision that your community faces now, or faced recently. Research both sides of the issue. What part did science play in the debate? What about politics and economics? Take a position, and then participate in a class panel discussion of the issue.

Fun Fact

The tallest dam in the world will be Russia's Rogun Dam at 335 meters. Slated for completion in 2003, this earthen/rockfill dam will supply electricity and irrigation water. By comparison, the famous Hoover Dam on the Colorado River is only 221 meters high.

3 Assess

Use the question in You Do It to assess understanding. Answers should relate to land use and should be as specific as possible.

You Do It

Ask students to consult the microfiche files in the school or community library to find back issues of regional newspapers containing articles about a land use decision in the community. Have them ask the librarian to show them indexes to the local papers that will quickly lead them to the relevant editions and pages.

chapter 1 Organizer

Section/Objectives	Standards: National	Standards: State/Local	Labs/Features
Chapter Opener	See pp. 9T–10T for a Key to Standards.		**Launch Lab:** Describe Landforms, p. 7 **Foldables,** p. 7
Section 1 Landforms 2 sessions 1 block 1. **Discuss** differences between plains and plateaus. 2. **Describe** folded, upwarped, fault-block, and volcanic mountains.	National Content Standards: UCP.1–UCP.3, UCP.5, A.1, A.2, D.1, D.2		**MiniLAB:** Profiling the United States, p. 10 **Integrate Physics,** p. 11 **Science Online,** p. 11
Section 2 Viewpoints 2 sessions 1 block 3. **Define** latitude and longitude. 4. **Explain** how latitude and longitude are used to identify locations on Earth. 5. **Determine** the time and date in different time zones.	National Content Standards: UCP.1–UCP.3, UCP.5, A.1, A.2, D.1		**MiniLAB:** Interpreting Latitude and Longitude, p. 15 **Integrate Life Science,** p. 16
Section 3 Maps 3 sessions 1.5 blocks 6. **Compare and contrast** map projections and their uses. 7. **Analyze** information from topographic, geologic, and satellite maps.	National Content Standards: UCP.1–UCP.3, UCP.5, A.1, A.2, D.1, D.2		**Integrate Physics,** p. 20 **Visualizing Topographic Maps,** p. 21 **Science Online,** p. 22 **Applying Science:** How can you create a cross section from a geologic map?, p. 23 **Lab:** Making a Topographic Map, p. 25 **Lab:** Constructing Landforms, p. 26 **Science and History:** Location, Location, p. 28

TeacherWorks™ — Glencoe Exclusive! All-In-One Planner and Resource Center

Lab Materials	Reproducible Resources	Section Assessment	Technology
Launch Lab: globe, atlas, or world map	**Chapter *FAST FILE* Resources** Foldables Worksheet, p. 15 Directed Reading Overview, p. 17 Note-taking Worksheets, pp. 31–32	Glencoe's Assessment Advantage	TeacherWorks includes: • Interactive Teacher Edition • Lesson Planner with calendar • Access to all program blacklines • Correlations to standards • Web links
MiniLAB: piece of paper, pencil, map of the United States	**Chapter *FAST FILE* Resources** Transparency Activity, p. 42 MiniLAB, p. 3 Enrichment, p. 28 Reinforcement, p. 25 Directed Reading, p. 18 Transparency Activity, pp. 45–46 **Cultural Diversity,** p. 33	**Portfolio** Cultural Diversity, p. 10 **Performance** MiniLAB, p. 10 Applying Skills, p. 13 **Content** Section Review, p. 13	Section Focus Transparency Teaching Transparency Virtual Labs CD-ROM Guided Reading Audio Program Interactive Chalkboard CD-ROM
MiniLAB: world map *Need materials?* Contact Science Kit at 1-800-828-7777 or www.sciencekit.com on the Internet.	**Chapter *FAST FILE* Resources** Transparency Activity, p. 43 MiniLAB, p. 4 Lab Activities, pp. 9–11, 13–14 Enrichment, p. 29 Reinforcement, p. 26 Directed Reading, p. 19	**Portfolio** Challenge, p. 16 **Performance** MiniLAB, p. 15 Applying Math, p. 17 **Content** Section Review, p. 17	Section Focus Transparency Virtual Labs CD-ROM Guided Reading Audio Program Interactive Chalkboard CD-ROM
Lab: plastic model of a landform, water tinted with food coloring, transparency, transparency marker, clear plastic storage box with lid, beaker, metric ruler, tape **Lab:** U.S. Geological Survey 7.5-minute quadrangle maps, sandbox sand, rolls of brown paper towels, spray bottle filled with water, ruler	**Chapter *FAST FILE* Resources** Transparency Activity, p. 44 Enrichment, p. 30 Reinforcement, p. 27 Directed Reading, pp. 19, 20 Lab Worksheets, pp. 5–6, 7–8 **Earth Science Critical Thinking/Problem Solving,** p. 5 **Cultural Diversity,** pp. 27, 49 **Reading and Writing Skill Activities,** p. 21	**Portfolio** Challenge, p. 20 **Performance** Applying Science, p. 23 Applying Skills, p. 24 **Content** Section Review, p. 24	Section Focus Transparency Virtual Labs CD-ROM Guided Reading Audio Program Interactive Chalkboard CD-ROM Video Lab

End of Chapter Assessment

Glencoe's Assessment Advantage

Blackline Masters	Technology	Professional Series
Chapter *FAST FILE* Resources Chapter Review, pp. 35–36 Chapter Tests, pp. 37–40 **Standardized Test Practice,** pp. 7–10	MindJogger Videoquiz Virtual Labs CD-ROM Exam*View*® Pro Testmaker TeacherWorks CD-ROM Interactive Chalkboard CD-ROM	**Performance Assessment in the Science Classroom (PASC)**

Transparencies

Section Focus

Section Focus Transparency 1 — Sacred Heights

Mount Fuji, or Fujiyama, is Japan's highest mountain at 3,776 m (12,388 feet) high. Below is a photo of the mountain along with a topographical image made by a satellite. Mount Fuji is considered a sacred mountain, and its name means "everlasting life."

1. What might Mount Fuji's shape and surrounding terrain tell you about the way in which it was formed?
2. How might wind and water change the shape of a mountain?

L2

Section Focus Transparency 2 — A Lovely Planet

This is an image of Earth made from space. Even on this flat photo, you can tell Earth is a sphere. Notice the way light from the Sun changes over the surface of Earth.

1. Which parts of Earth does this image show?
2. If you were standing on one of the landmasses shown, how would you describe your location to a friend calling from North America?
3. How might a grid overlaying Earth help to indicate exact locations?

L2

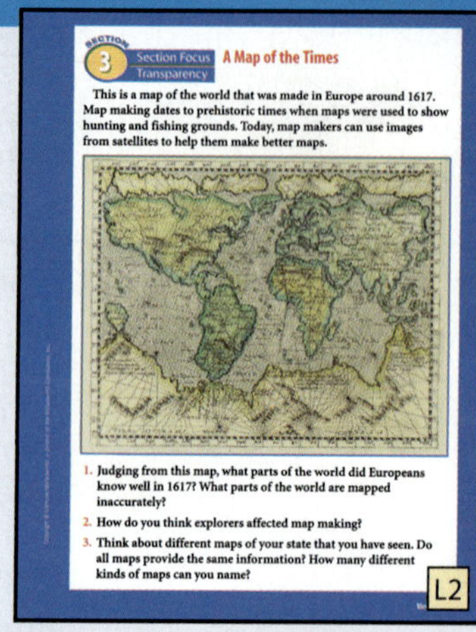

Section Focus Transparency 3 — A Map of the Times

This is a map of the world that was made in Europe around 1617. Map making dates to prehistoric times when maps were used to show hunting and fishing grounds. Today, map makers can use images from satellites to help them make better maps.

1. Judging from this map, what parts of the world did Europeans know well in 1617? What parts of the world are mapped inaccurately?
2. How do you think explorers affected map making?
3. Think about different maps of your state that you have seen. Do all maps provide the same information? How many different kinds of maps can you name?

L2

This is a representation of key blackline masters available in the Teacher Classroom Resources. See Resource Manager boxes within the chapter for additional information.

Assessment

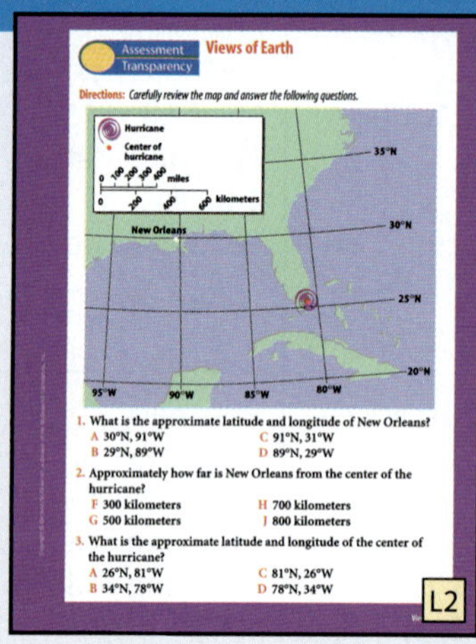

Assessment Transparency — Views of Earth

Directions: *Carefully review the map and answer the following questions.*

1. What is the approximate latitude and longitude of New Orleans?
   A 30°N, 91°W C 91°N, 31°W
   B 29°N, 89°W D 89°N, 29°W
2. Approximately how far is New Orleans from the center of the hurricane?
   F 300 kilometers H 700 kilometers
   G 500 kilometers J 800 kilometers
3. What is the approximate latitude and longitude of the center of the hurricane?
   A 26°N, 81°W C 81°N, 26°W
   B 34°N, 78°W D 78°N, 34°W

L2

Teaching

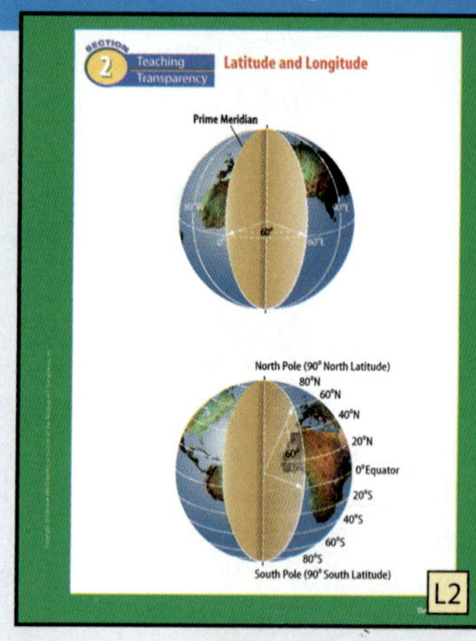

Teaching Transparency 2 — Latitude and Longitude

L2

Key to Teaching Strategies

The following designations will help you decide which activities are appropriate for your students.

- **L1** Level 1 activities should be appropriate for students with learning difficulties.
- **L2** Level 2 activities should be within the ability range of all students.
- **L3** Level 3 activities are designed for above-average students.
- **ELL** ELL activities should be within the ability range of English Language Learners.
- **COOP LEARN** Cooperative Learning activities are designed for small group work.
- **LS** Multiple Learning Styles logos, as described on page 6T, are used throughout to indicate strategies that address different learning styles.
- **P** These strategies represent student products that can be placed into a best-work portfolio.
- **PBL** Problem-Based Learning activities apply real-world situations to learning.

Hands-on Activities

Student Text Lab Worksheet

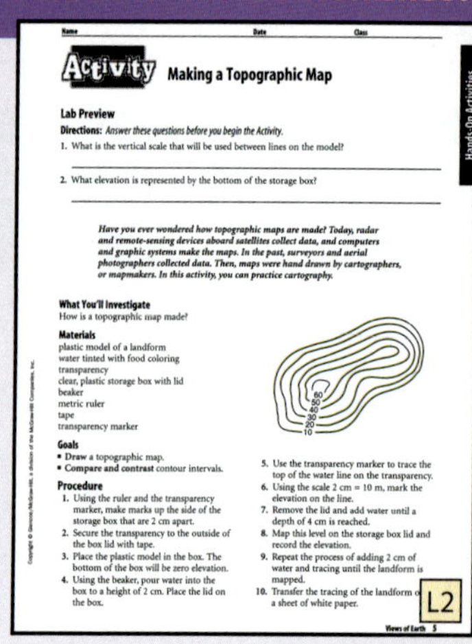

Activity — Making a Topographic Map

L2

Laboratory Activities

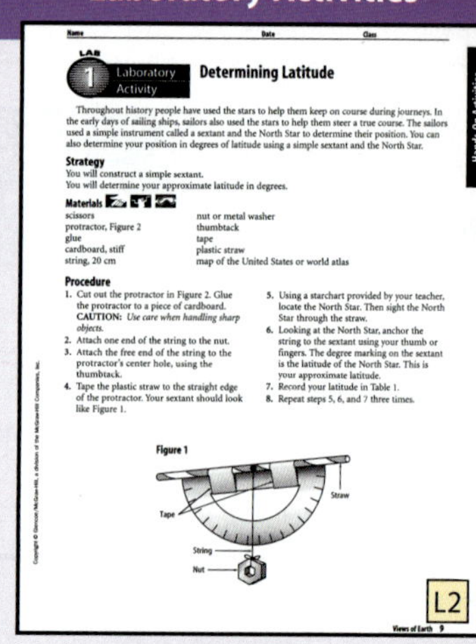

Laboratory Activity 1 — Determining Latitude

L2

Meeting Different Ability Levels

Content Outline

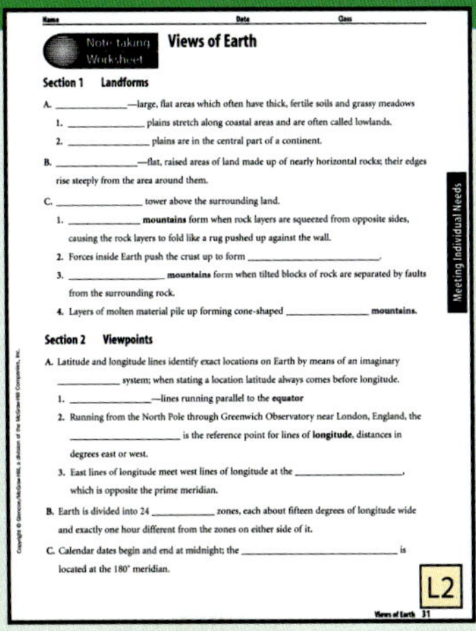

Reinforcement

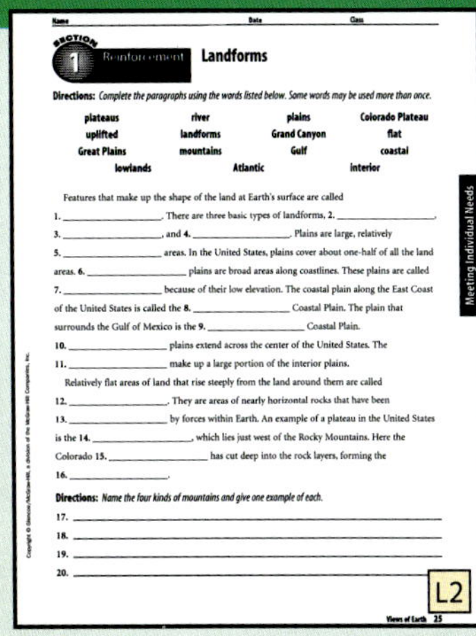

Enrichment

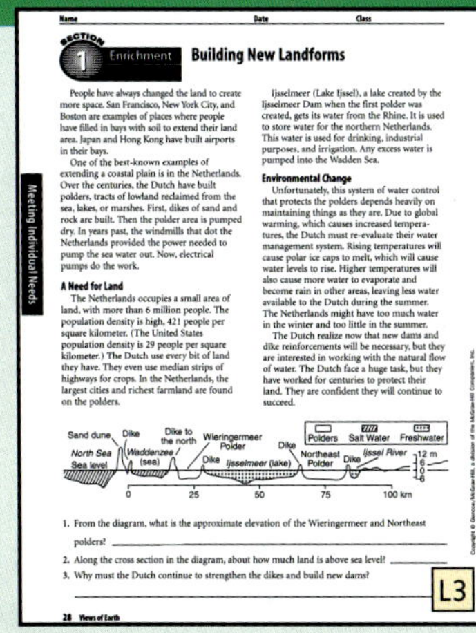

Directed Reading (English/Spanish)

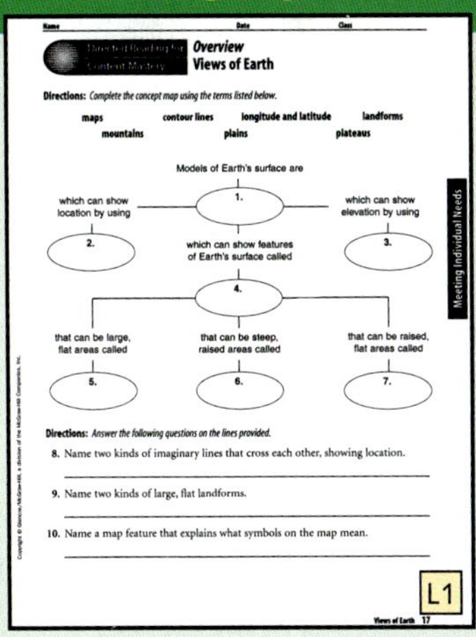

Study Guide

Reading Essentials

Assessment

Test Practice Workbook

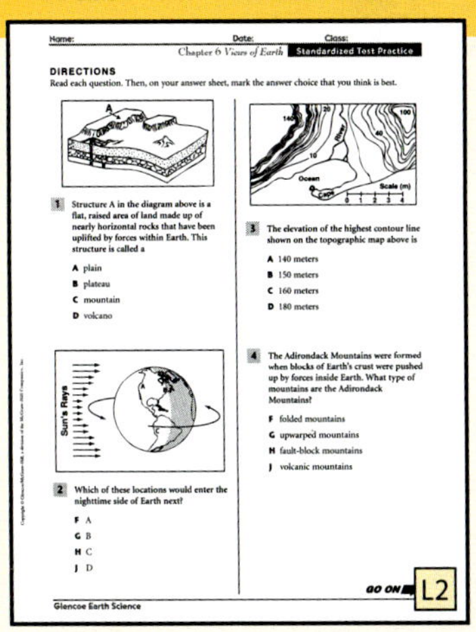

Chapter Review

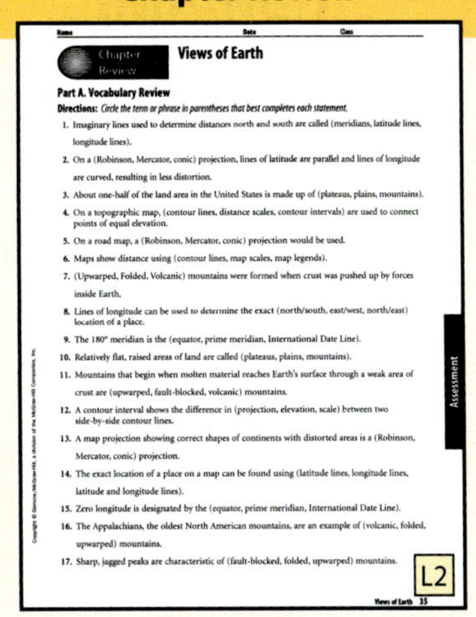

Chapter Tests

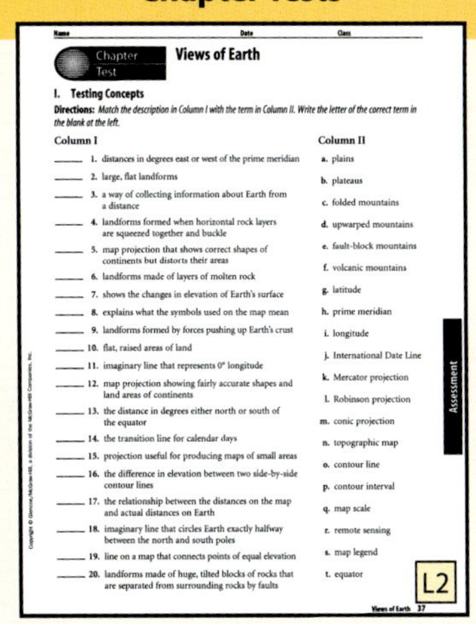

Science Content Background

section 1 Landforms

Plains and Mountains

Plains form in a variety of ways. Some are formed by the erosion of plateaus and mountains. Others are formed by the deposition of sediments into depressions. Mountains are classified according to their most dominant characteristics. These characteristics are caused by the processes involved in mountain building, such as folding, faulting, upwarping, and igneous activities.

Earth Satellite Corporation/Science Photo Library/Photo Researchers

section 2 Viewpoints

Latitude and Longitude

Latitude is defined as the angle formed between the point in the sky directly above an observer (zenith) and the plane of Earth's equator. Each degree of latitude represents a distance of about 111 km over Earth's surface. Lines of latitude form imaginary circles around Earth that are parallel to the equator. The diameter of these circles decrease with distance from the equator. At the poles, the circles are mere points.

Lines of longitude, or meridians, are not parallel. They converge toward Earth's poles. The distance between lines of longitude varies with distance from the equator. At the poles, the distance shrinks to zero.

section 3 Maps

Map Projections

A map projection is made by transferring data about Earth onto paper. Because Earth is round and paper is flat, some distortion is always present in map projections. The larger the area covered by the map, the greater the distortion. Gerhardus Mercator developed the projection named for him in 1569. In a Mercator projection, the size of Greenland is exaggerated 900 percent. When Arthur Robinson created his projection in 1963, he visualized how each country should look on a map, then used mathematics and computers to draw their shapes. The Robinson projection has a 60-percent exaggeration of the size of Greenland.

chapter content resources

Internet Resources

For additional content background, visit **bookg.msscience.com** to:

- access your book online
- find references to related articles in popular science magazines
- access Web links with related content background
- access current events with science journal topics

Print Resources

Sciencewise Book 3, by Dennis Holley, Critical Thinking Books and Software, 1999

Compass and Map Navigation by Michael Hodgson, Globe Pequot, 2000

The Island of Lost Maps; A True Story of Cartographic Crime by Miles Harvey, Broadway Books, 2001

The Mapmakers by John Noble Wilford, Vintage Books, 2001

IDENTIFYING Misconceptions

Find Out What Students Think

Students may think that . . .

Earth is not a sphere, but is flat.

This misconception is a result of students' common experience—from the ground, Earth's surface looks flat. In addition, many students have yet to grasp the idea of gravity pulling all parts of Earth, and the people and other objects on its surface, toward its center. Thus students think that if Earth were a sphere, people on the bottom would "fall off."

Demonstration

Evidence, both direct and indirect, needs to be provided to students to help them understand that Earth is spherical and that it rotates. Evidence could include

- satellite photos of Earth.

S. Nielsen/DRK Photo

- photographs showing Earth's curved shadow on the Moon during a lunar eclipse.

Promote Understanding

Activity

Provide each pair of students with construction paper, scissors, tape, and a basketball or other sports ball.

- Have students make a small representation of a sail boat using the construction paper and tape. Instruct them to tape the boat onto the basketball.
- Have students stand about 3 m from their partners. Direct the partner to hold the basketball in front of his or her body, with the boat toward the partner's body. Then have him or her rotate the ball so the boat appears to move up and over the top of the ball.
- As the ship appears, have the watching partner closely observe which parts of the boat appear first. Then have students switch roles.

Discussion

After all students have observed the approach of the boat, bring the class together and discuss their observations. Bring out the idea that as a ship disappears over the horizon the last part that one observes is the mast. As it approaches, the first thing one sees is the mast. If Earth were flat, one would see the entire boat either appear or disappear.

Assess

After completing the chapter, see *Identifying Misconceptions* in the Study Guide at the end of the chapter.

chapter 1

Chapter Vocabulary

plain, p. 8
plateau, p. 10
folded mountain, p. 11
upwarped mountain, p. 12
fault-block mountain, p. 12
volcanic mountain, p. 13
equator, p. 14
latitude, p. 14
prime meridian, p. 15
longitude, p. 15
conic projection, p. 19
topographic map, p. 20
contour line, p. 20
map scale, p. 22
map legend, p. 22

Science Journal Student answers may vary. Green or red areas may be best for building a new home. These areas tend to be less populated and less developed. Students should choose their site carefully; green areas within the city are not usually possibilities for building.

INTERACTIVE CHALKBOARD with Image Bank

PowerPoint® Presentations

This CD-ROM is an editable Microsoft® PowerPoint® presentation that includes:

- a pre-made presentation for every chapter
- interactive graphics
- animations
- audio clips
- image bank
- all new section and chapter questions
- Standardized Test Practice
- transparencies
- pre-lab questions for all labs
- Foldables directions
- links to bookg.msscience.com

chapter 1

Views of Earth

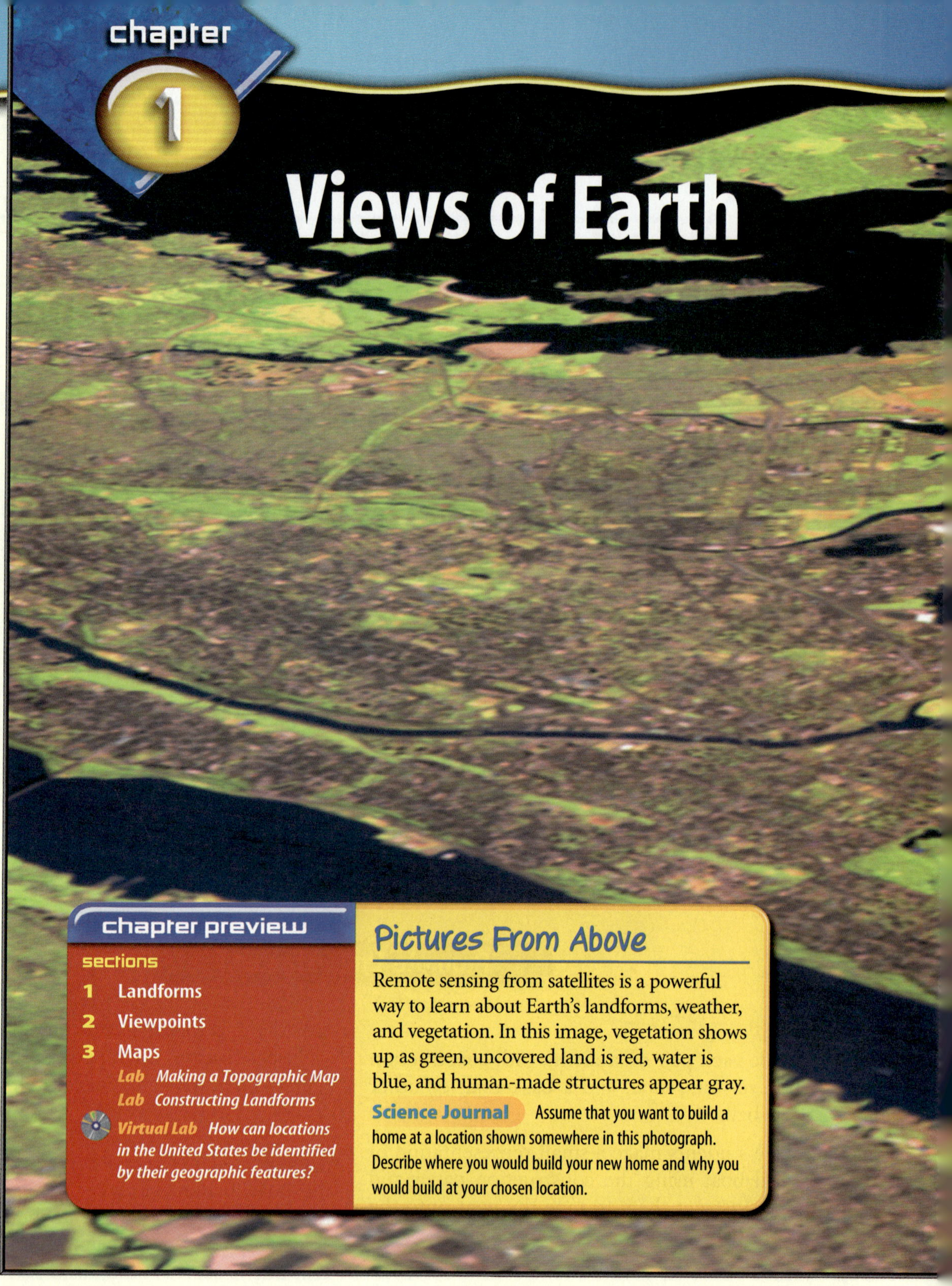

chapter preview

sections

1 Landforms
2 Viewpoints
3 Maps
Lab *Making a Topographic Map*
Lab *Constructing Landforms*
Virtual Lab *How can locations in the United States be identified by their geographic features?*

Pictures From Above

Remote sensing from satellites is a powerful way to learn about Earth's landforms, weather, and vegetation. In this image, vegetation shows up as green, uncovered land is red, water is blue, and human-made structures appear gray.

Science Journal Assume that you want to build a home at a location shown somewhere in this photograph. Describe where you would build your new home and why you would build at your chosen location.

Theme Connection

Scale and Structure The structural differences among major landforms are examined. Students also focus on understanding the scale and structure of maps.

About the Photo

Energy Images Landsat images show how much energy from the Sun (electromagnetic radiation) was reflecting off Earth's surface when the image was taken. Clear water reflects little radiation, so it looks black. Pavement and bare ground reflect a lot of radiation, so they look bright. Urban areas usually look light blue-gray. Vegetation absorbs visible light but reflects infrared, so it looks red.

Start-Up Activities

Launch LAB

Describe Landforms

Pictures of Earth from space are acquired by instruments attached to satellites. Scientists use these images to make maps because they show features of Earth's surface, such as mountains and rivers.

1. Using a globe, atlas, or a world map, locate the following features and describe their positions on Earth relative to other major features.
   a. Andes mountains
   b. Amazon, Ganges, and Mississippi Rivers
   c. Indian Ocean, the Sea of Japan, and the Baltic Sea
   d. Australia, South America, and North America
2. Provide any other details that would help someone else find them.
3. **Think Critically** Choose one country on the globe or map and describe its major physical features in your Science Journal.

Science Online — Preview this chapter's content and activities at bookg.msscience.com

FOLDABLES™ Study Organizer

Views of Earth Make the following Foldable to help identify what you already know, what you want to know, and what you learned about the views of Earth.

STEP 1 **Fold** a vertical sheet of paper from side to side. Make the front edge about 1.25 cm shorter than the back edge.

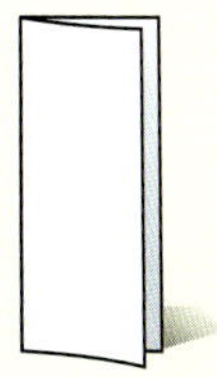

STEP 2 **Turn** lengthwise and **fold** into thirds.

STEP 3 **Unfold and cut** only the top layer along both folds to make three tabs.

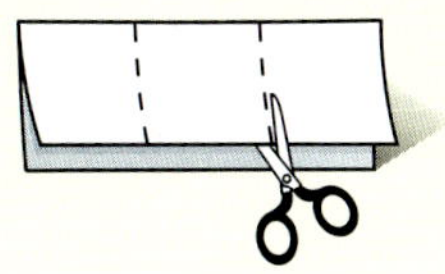

STEP 4 **Label** each tab.

Identify Questions Before you read the chapter, write what you already know about the views of Earth under the left tab of your Foldable, and write questions about what you want to know under the center tab. After you read the chapter, list what you learned under the right tab.

G ◆ 7

Launch LAB

Purpose Students reinforce map skills by using a world map or globe to observe and record the relative positions of Earth's landforms. L2 ELL LS **Visual-Spatial**

Preparation Be sure the maps and globes you use have the listed features clearly labeled.

Materials globe or world map for every three students

Teaching Strategy Have students practice describing landform locations before they begin the activity. For example, have them describe the location of the Rocky Mountains. north-south mountain range in western North America

Think Critically

Answers will vary depending on the country chosen.

Assessment

Performance Ask students to describe in writing how mountains and rivers are indicated on the globe or world map they used. Use **Performance Assessment in the Science Classroom,** p. 159.

FOLDABLES™ Study Organizer — Dinah Zike Study Fold

Student preparation materials for this Foldable are available in the Chapter *Fast File* Resources.

section 1 Landforms

1 Motivate

Bellringer

INTERACTIVE CHALKBOARD PowerPoint® Presentations

Section Focus Transparencies also are available on the Interactive Chalkboard CD-ROM. L2 ELL

Section 1 Section Focus Transparency **Sacred Heights**

Mount Fuji, or Fujiyama, is Japan's highest mountain at 3,776 m (12,388 feet) high. Below is a photo of the mountain along with a topographical image made by a satellite. Mount Fuji is considered a sacred mountain, and its name means "everlasting life."

1. What might Mount Fuji's shape and surrounding terrain tell you about the way in which it was formed?
2. How might wind and water change the shape of a mountain?

L2

Tie to Prior Knowledge

Community Landforms Have students brainstorm the names of prominent landforms in or near your community. List them on the board. Tell students they will learn about several types of landforms in this section.

as you read

What You'll Learn

- **Discuss** differences between plains and plateaus.
- **Describe** folded, upwarped, fault-block, and volcanic mountains.

Why It's Important

Landforms influence how people can use land.

Review Vocabulary
landform: a natural feature of a land surface

New Vocabulary
- plain
- plateau
- folded mountain
- upwarped mountain
- fault-block mountain
- volcanic mountain

Plains

Earth offers abundant variety—from tropics to tundras, deserts to rain forests, and freshwater mountain streams to saltwater tidal marshes. Some of Earth's most stunning features are its landforms, which can provide beautiful vistas, such as vast, flat, fertile plains; deep gorges that cut through steep walls of rock; and towering, snowcapped peaks. **Figure 1** shows the three basic types of landforms—plains, plateaus, and mountains.

Even if you haven't ever visited mountains, you might have seen hundreds of pictures of them in your lifetime. Plains are more common than mountains, but they are more difficult to visualize. **Plains** are large, flat areas, often found in the interior regions of continents. The flat land of plains is ideal for agriculture. Plains often have thick, fertile soils and abundant, grassy meadows suitable for grazing animals. Plains also are home to a variety of wildlife, including foxes, ground squirrels, and snakes. When plains are found near the ocean, they're called coastal plains. Together, interior plains and coastal plains make up half of all the land in the United States.

Figure 1 Three basic types of landforms are plains, plateaus, and mountains.

Section 1 Resource Manager

Chapter FAST FILE Resources
- Transparency Activity, pp. 42, 45–46
- Directed Reading for Content Mastery, pp. 17, 18
- Note-taking Worksheets, pp. 31–32
- MiniLAB, p. 3
- Enrichment, p. 28
- Reinforcement, p. 25

Reading and Writing Skill Activities, p. 3
Cultural Diversity, p. 33

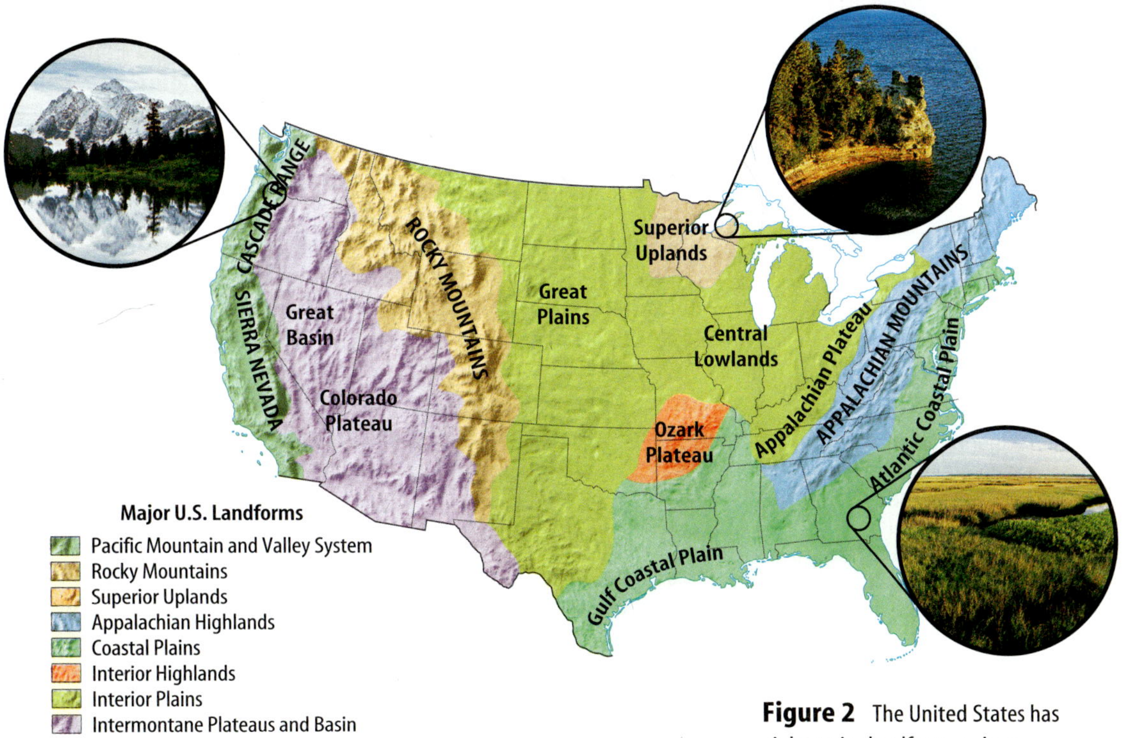

Figure 2 The United States has eight major landform regions, which include plains, mountains, and plateaus.
Describe *the region that you live in.*

Coastal Plains A coastal plain often is called a lowland because it is lower in elevation, or distance above sea level, than the land around it. You can think of the coastal plains as being the exposed portion of a continental shelf. The continental shelf is the part of a continent that extends into the ocean. The Atlantic Coastal Plain is a good example of this type of landform. It stretches along the east coast of the United States from New Jersey to Florida. This area has low rolling hills, swamps, and marshes. A marsh is a grassy wetland that usually is flooded with water.

The Atlantic Coastal Plain, shown in **Figure 2,** began forming about 70 million years ago as sediment began accumulating on the ocean floor. Sea level eventually dropped, and the seafloor was exposed. As a result, the coastal plain was born. The size of the coastal plain varies over time. That's because sea level rises and falls. During the last ice age, the coastal plain was larger than it is now because so much of Earth's water was contained in glaciers.

The Gulf Coastal Plain includes the lowlands in the southern United States that surround the Gulf of Mexico. Much of this plain was formed from sediment deposited in deltas by the many rivers that enter the Gulf of Mexico.

Reading Check *How are coastal plains formed?*

2 Teach

Activity

Relief Map Display a relief map of North America. Have students compare and contrast the different landforms. L2 LS **Visual-Spatial**

IDENTIFYING Misconceptions

Plains Locations Students may think that because plains are flat, they are always found at much lower elevations than mountains. Explain that plains can be found at any elevation. For example, Denver, Colorado, is at the western edge of the Great Plains at an elevation of 1,655 m. In comparison, the average elevation of the Appalachian Mountains is less than 2,000 m above sea level.

Reading Check

Answer Coastal plains form when sea level drops and the seafloor is exposed or from sediment deposited in deltas by rivers.

Caption Answer

Figure 2 Answers will vary depending on your location.

Curriculum Connection

Social Studies More than half the world's population lives on plains. Have students research possible explanations for this fact. Students should discover that plains are often fertile farmlands. L2 LS **Logical-Mathematical**

Differentiated Instruction

Visually Impaired While discussing the landforms in the United States, provide students who are visually impaired a three-dimensional relief map of the U.S. They will then be able to follow along with the class discussion and be able to distinguish between the different landform regions. L2

SECTION 1 Landforms G ◆ 9

Mini LAB

Purpose Students make a profile that shows the major landforms of the United States from the west coast to the east coast. L2 LS **Visual-Spatial**

Materials Figure 2, metric ruler, pencil, physical map of the United States, paper

Teaching Strategy If possible, project a transparency of **Figure 2**. With a clear transparency as an overlay, use a marking pen to show students how to make the profile.

Analysis

1. Profiles progress from mountains near the west coast, to a high elevation basin, and then to a wide set of steep mountains representing the Rockies. A wide plain should stretch from these mountains through much of the central area of the country. A range of lower mountains lies between the central plain and the coastal plain.
2. Many landforms run north-south in the United States, thus these profiles would not show as much change in elevation.

Assessment

Performance Have students write a travelogue describing the landforms they would see along the route shown in their profiles. Use **Performance Assessment in the Science Classroom,** p. 157.

Try at Home

Mini LAB

Profiling the United States

Procedure

1. Place the bottom edge of a piece of **paper** across the middle of **Figure 2,** extending from the west coast to the east coast.
2. Mark where different landforms are located along this edge.
3. Use a **map of the United States** and the **descriptions of the landforms in Section 1** to help you draw a profile, or side view, of the United States. Use steep, jagged lines to represent mountains. Low, flat lines can represent plains.

Analysis

1. Describe how your profile changed shape as you moved from west to east.
2. Describe how the shape of your profile would be different if you oriented your paper north to south.

Try at Home

Interior Plains The central portion of the United States is comprised largely of interior plains. Shown in **Figure 3,** you'll find them between the Rocky Mountains, the Appalachian Mountains, and the Gulf Coastal Plain. They include the Central Lowlands around the Missouri and Mississippi Rivers and the rolling hills of the Great Lakes area.

A large part of the interior plains is known as the Great Plains. This area lies between the Mississippi River and the Rocky Mountains. It is a flat, grassy, dry area with few trees. The Great Plains also are referred to as the high plains because of their elevation, which ranges from 350 m above sea level at the eastern border to 1,500 m in the west. The Great Plains consist of nearly horizontal layers of sedimentary rocks.

Plateaus

At somewhat higher elevations, you will find plateaus (pla TOHZ). **Plateaus** are flat, raised areas of land made up of nearly horizontal rocks that have been uplifted by forces within Earth. They are different from plains in that their edges rise steeply from the land around them. Because of this uplifting, it is common for plateaus, such as the Colorado Plateau, to be cut through by deep river valleys and canyons. The Colorado River, as shown in **Figure 3,** has cut deeply into the rock layers of the plateau, forming the Grand Canyon. Because the Colorado Plateau is located mostly in what is now a dry region, only a few rivers have developed on its surface. If you hiked around on this plateau, you would encounter a high, rugged environment.

Figure 3 Plains and plateaus are fairly flat, but plateaus have higher elevation.

Cultural Diversity

Landforms and Ancient Builders Among the mountains and canyons of northwestern New Mexico is Pueblo Bonito, a three-acre stone and mortar structure built by many generations of Anasazi Indian masons. The builders used the sandstone to fashion rectangular blocks for the walls of the structure. Mud and water mortar knits the blocks together. Pueblo Bonito contains more than 600 rooms. Like many Anasazi structures, it was abandoned about 900 years ago when a prolonged drought made living in this dry area difficult. Many of the structures are still partially standing. Have students use the Internet to research the ruins of Chaco Canyon, including Pueblo Bonito. Have them write short reports on what they find. L2 LS **Linguistic** P

Mountains

Mountains with snowcapped peaks often are shrouded in clouds and tower high above the surrounding land. If you climb them, the views are spectacular. The world's highest mountain peak is Mount Everest in the Himalaya—more than 8,800 m above sea level. By contrast, the highest mountain peaks in the United States reach just over 6,000 m. Mountains also vary in how they are formed. The four main types of mountains are folded, upwarped, fault-block, and volcanic.

Reading Check ***What is the highest mountain peak on Earth?***

Folded Mountains The Appalachian Mountains and the Rocky Mountains in Canada, shown in **Figure 4,** are comprised of folded rock layers. In **folded mountains,** the rock layers are folded like a rug that has been pushed up against a wall.

INTEGRATE Physics To form folded mountains, tremendous forces inside Earth squeeze horizontal rock layers, causing them to fold. The Appalachian Mountains formed between 480 million and 250 million years ago and are among the oldest and longest mountain ranges in North America. The Appalachians once were higher than the Rocky Mountains, but weathering and erosion have worn them down. They now are less than 2,000 m above sea level. The Ouachita (WAH shuh tah) Mountains of Arkansas are extensions of the same mountain range.

Science Online

Topic: Landforms

Visit bookg.msscience.com for Web links to information about some ways landforms affect economic development.

Activity Create four colorful postcards with captions explaining how landforms have affected economic development in your area.

Figure 4 Folded mountains form when rock layers are squeezed from opposite sides. These mountains in Banff National Park, Canada, consist of folded rock layers.

Reading Check

Answer Mount Everest

Use Science Words

Word Usage Obtain photographs of the three basic kinds of landforms: plains, plateaus, and mountains. Have students write a sentence using each term. L2 LS **Linguistic**

Fun Fact

The Colorado Plateau has been rising approximately 0.3 cm every year for the last 10 million years. It is presently 2 km higher than the area that surrounds it.

Quick Demo

Forming Folded Mountains

Materials sheet of paper

Estimated Time five minutes

Procedure Place a sheet of paper on a flat surface. Have a student place a hand on one end. Have another student place a hand on the other end. Tell the students to slide the ends of the paper toward the center. Have the class note the high folded loop that forms. L1 ELL COOP LEARN LS **Visual-Spatial**

Virtual Labs

Identifying Features *How can locations in the United States be identified by their geographic features?*

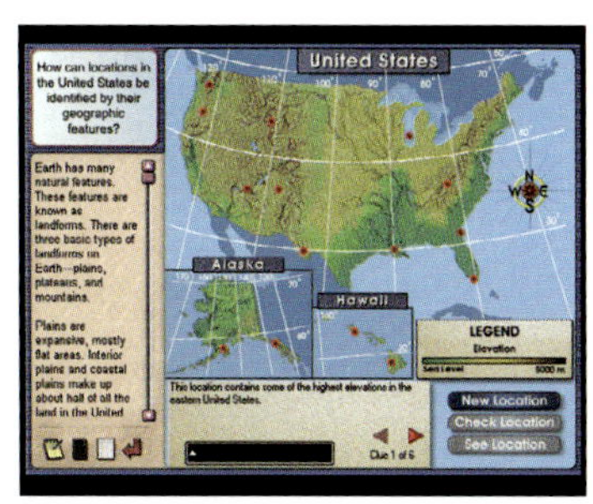

Teacher FYI

Mt. McKinley Named for President William McKinley, the highest mountain peak in the U.S. rises to 6,194 m. The Alaskan Athabaskan Indians called the peak *Denali*, or "The Great One." The preserve surrounding the mountain now bears the Indian name, Denali National Park.

SECTION 1 Landforms G ◆ 11

Make a Model

Clay Mountains Have students use clay to make models of the different kinds of mountains discussed in this section. L1 ELL LS **Kinesthetic**

Discussion

Pennsylvania Topography Tell students that the Ridge and Valley area of Pennsylvania is composed of parallel lines of rounded ridges with lowlands between them. Is this area more likely to be composed of folded or fault-block mountains? Folded mountains; resistant rock layers in the folds stand up as ridges. Some students may recognize that the mountains of Pennsylvania are part of the Appalachians, which are identified as folded mountains in the text.

Caption Answer

Figure 6 Fault-block mountains form as a result of movement along a fault. Upwarped mountains form because of a broad uplift of rock.

Activity

Volcanic Placement Have students use a world map to locate major volcanic mountain chains. Have them write a statement in their Science Journals about where most of these mountains occur. L2 LS **Visual-Spatial and Linguistic**

Figure 5 The southern Rocky Mountains are upwarped mountains that formed when crust was pushed up by forces inside Earth.

Upwarped Mountains The Adirondack Mountains in New York, the southern Rocky Mountains in Colorado and New Mexico, and the Black Hills in South Dakota are upwarped mountains. **Figure 5** shows a mountain range in Colorado. Notice the high peaks and sharp ridges that are common to this type of mountain. **Upwarped mountains** form when blocks of Earth's crust are pushed up by forces inside Earth. Over time, the soil and sedimentary rocks at the top of Earth's crust erode, exposing the hard, crystalline rock underneath. As these rocks erode, they form the peaks and ridges.

Fault-Block Mountains **Fault-block mountains** are made of huge, tilted blocks of rock that are separated from surrounding rock by faults. These faults are large fractures in rock along which mostly vertical movement has occurred. The Grand Tetons of Wyoming, shown in **Figure 6,** and the Sierra Nevada in California, are examples of fault-block mountains. As **Figure 6** shows, when these mountains formed, one block was pushed up, while the adjacent block dropped down. This mountain-building process produces majestic peaks and steep slopes.

Figure 6 Fault-block mountains such as the Grand Tetons are formed when faults occur. Some rock blocks move up, and others move down. **Describe** *the difference between fault-block mountains and upwarped mountains.*

Differentiated Instruction

Challenge In small groups, have the students create a board game to help learn the major types of landforms found in the United States. Give them a blank outline map of the continental U.S., colored paper, and access to reference materials. Be sure they include detailed instructions with their game. Once their game is complete, ask groups to switch games and play a different game within their group. L3

Science Journal

Young Mountains Tell students that the Himalayas are still growing. Have students locate these mountains on a map and describe in their Science Journals why these mountains are still growing. Students should discover that the Himalayas are being pushed up as the Indian plate converges with the Asian plate. L2

12 ◆ G CHAPTER 1 Views of Earth

Figure 7 Mount Shasta is a volcanic mountain made up of layers of lava flows and ash.

Volcanic Mountains **Volcanic mountains**, like the one shown in **Figure 7,** begin to form when molten material reaches the surface through a weak area of the crust. The deposited materials pile up, layer upon layer, until a cone-shaped structure forms. Two volcanic mountains in the United States are Mount St. Helens in Washington and Mount Shasta in California. The Hawaiian Islands are the peaks of huge volcanoes that sit on the ocean floor. Measured from the base, Mauna Loa in Hawaii would be higher than Mount Everest.

Plains, plateaus, and mountains offer different kinds of landforms to explore. They range from low, coastal plains and high, desert plateaus to mountain ranges thousands of meters high.

section 1 review

Summary

Plains and Plateaus

- Plains are large, flat landforms that are usually found in the interior region of a continent.
- Plateaus are flat, raised landforms made of nearly horizontal, uplifted rocks.

Mountains

- Folded mountains form when horizontal rock layers are squeezed from opposite sides.
- Upwarped mountains form when blocks of Earth's crust are pushed up by forces inside Earth.
- Fault-block mountains form from huge, tilted blocks of rock that are separated by faults.
- Volcanic mountains form when molten rock forms cone-shaped structures at Earth's surface.

Self Check

1. **Describe** the eight major landform regions in the United States that are mentioned in this chapter.
2. **Compare and contrast** volcanic mountains, folded mountains, and upwarped mountains using a three-circle Venn diagram.
3. **Think Critically** If you wanted to know whether a particular mountain was formed by movement along a fault, what would you look for? Support your reasoning.

Applying Skills

5. **Concept Map** Make an events-chain concept map to explain how interior plains and coastal plains form.

Visual Learning

Figure 7 What visual clue tells you that Mount Shasta is a volcanic mountain? Mount Shasta has a cone shape.

3 Assess

DAILY INTERVENTION

Check for Understanding

Visual-Spatial Have students create labeled diagrams to illustrate how the processes that formed the Appalachian Mountains differed from those that formed the Black Hills. L2

Reteach

Fault-block Mountains Use wooden blocks or rectangular sponges cut at various angles to demonstrate the formation of fault-block mountains. L1 ELL LS **Visual-Spatial**

Assessment

Process Have students make a concept map that shows how volcanic mountains form. Use **Performance Assessment in the Science Classroom,** p. 161. L2

section 1 review

1. Pacific Mountain and Valley System: coastal ranges and valleys; Rocky Mountains: high mountain WA to NM; Superior Uplands: highlands near Lake Superior; Appalachian Highlands: old mountains; Coastal plains: low elevations near Gulf and Atlantic Ocean; Interior highlands: Ozark Plateau; Interior plains: expansive flatlands; Intermontane plateaus and basins: region between Pacific Ocean and Rocky Mountains.
2. folded: rock layers were squeezed laterally; upwarped: forces within Earth caused upward arching of rock layers; volcanic: molten material was deposited on Earth's surface
3. nearby faults and blocky-looking mountains
4. crust blocks pushed up → soil erodes → hard rocks exposed forming sharp peaks

section 2

Viewpoints

as you read

***What* You'll Learn**

- **Define** latitude and longitude.
- **Explain** how latitude and longitude are used to identify locations on Earth.
- **Determine** the time and date in different time zones.

***Why* It's Important**

Latitude and longitude allow you to locate places on Earth.

Review Vocabulary
pole: either end of an axis of a sphere

New Vocabulary
- equator
- latitude
- prime meridian
- longitude

Latitude and Longitude

During hurricane season, meteorologists track storms as they form in the Atlantic Ocean. To identify the exact location of a storm, latitude and longitude lines are used. These lines form an imaginary grid system that allows people to locate any place on Earth accurately.

Latitude Look at **Figure 8.** The **equator** is an imaginary line around Earth exactly halfway between the north and south poles. It separates Earth into two equal halves called the northern hemisphere and the southern hemisphere. Lines running parallel to the equator are called lines of **latitude,** or parallels. Latitude is the distance, measured in degrees, either north or south of the equator. Because they are parallel, lines of latitude do not intersect, or cross, one another.

The equator is at 0° latitude, and the poles are each at 90° latitude. Locations north and south of the equator are referred to by degrees north latitude and degrees south latitude, respectively. Each degree is further divided into segments called minutes and seconds. There are 60 minutes in one degree and 60 seconds in one minute.

Figure 8 Latitude and longitude are measurements that are used to indicate locations on Earth's surface.

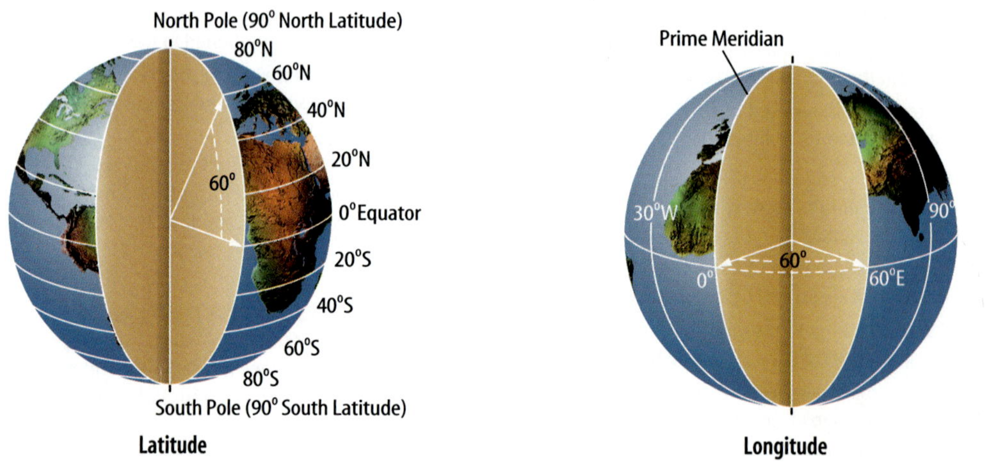

1 Motivate

Bellringer

INTERACTIVE CHALKBOARD PowerPoint® Presentations

Section Focus Transparencies also are available on the Interactive Chalkboard CD-ROM. L2 ELL

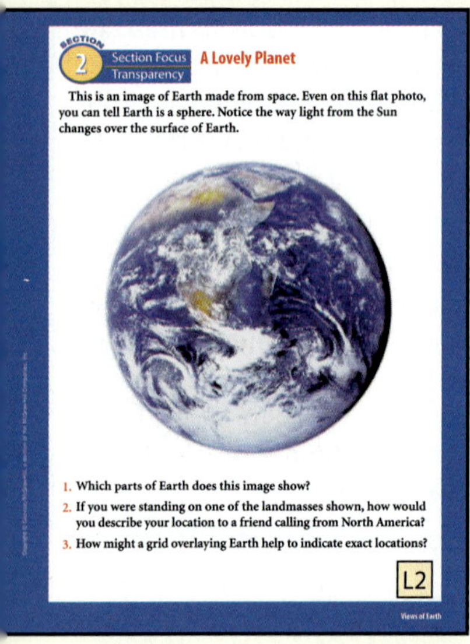
Section 2 Section Focus Transparency **A Lovely Planet**

This is an image of Earth made from space. Even on this flat photo, you can tell Earth is a sphere. Notice the way light from the Sun changes over the surface of Earth.

1. Which parts of Earth does this image show?
2. If you were standing on one of the landmasses shown, how would you describe your location to a friend calling from North America?
3. How might a grid overlaying Earth help to indicate exact locations?

L2

Views of Earth

Tie to Prior Knowledge

Map Lines Have students recall the appearance of maps and globes, and the lines that form a grid on them. Explain that in this section, they will learn how these lines help pinpoint locations on Earth's surface.

IDENTIFYING Misconceptions

Spherical Earth Some students may think that Earth is not a sphere, but flat. See page F at the beginning of this chapter for strategies that address this misconception.

Section 2 Resource Manager

Chapter *FAST FILE* Resources

Transparency Activity, p. 43
Directed Reading for Content Mastery, p. 19
MiniLAB, p. 4
Lab Activity, pp. 9–11, 13–14
Enrichment, p. 29
Reinforcement, p. 26

14 ◆ G CHAPTER 1 Views of Earth

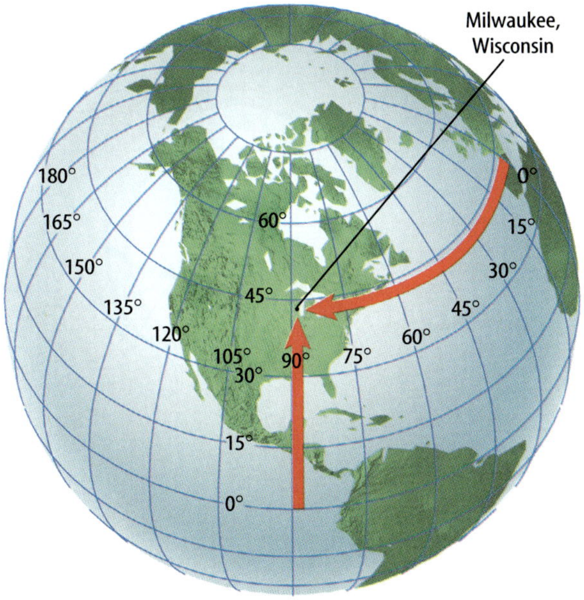

Figure 9 The city of Milwaukee, Wisconsin is located at about 43°N, 88°W. **Explain** *the difference between latitude and longitude.*

Longitude The vertical lines, seen in **Figure 8,** have two names—meridians and lines of longitude. Longitude lines are different from latitude lines in many important ways. Just as the equator is used as a reference point for lines of latitude, there's a reference point for lines of longitude—the **prime meridian.** This imaginary line represents 0° longitude. In 1884, astronomers decided the prime meridian should go through the Greenwich (GREN ihtch) Observatory near London, England. The prime meridian had to be agreed upon, because no natural point of reference exists.

Longitude refers to distances in degrees east or west of the prime meridian. Points west of the prime meridian have west longitude measured from 0° to 180°, and points east of the prime meridian have east longitude, measured similarly.

Prime Meridian The prime meridian does not circle Earth as the equator does. Rather, it runs from the north pole through Greenwich, England, to the south pole. The line of longitude on the opposite side of Earth from the prime meridian is the 180° meridian. East lines of longitude meet west lines of longitude at the 180° meridian. You can locate places accurately using latitude and longitude as shown in **Figure 9.** Note that latitude position always comes first when a location is given.

Reading Check *What line of longitude is found opposite the prime meridian?*

Mini LAB

Interpreting Latitude and Longitude

Procedure

1. Find the equator and prime meridian on a **world map.**
2. Move your finger to latitudes north of the equator, then south of the equator. Move your finger to longitudes west of the prime meridian, then east of the prime meridian.

Analysis

1. Identify the cities that have the following coordinates:
   a. 56°N, 38°E
   b. 34°S, 18°E
   c. 23°N, 82°W
2. Determine the latitude and longitude of the following cities:
   a. London, England
   b. Melbourne, Australia
   c. Buenos Aires, Argentina

2 Teach

Caption Answer

Figure 9 Both are part of a grid system for locating places on Earth; latitude refers to distances north and south of the equator; longitude refers to distances east and west of the prime meridian.

Mini LAB

Purpose Students learn to use latitude and longitude to locate places on Earth's surface. L2 ELL COOP LEARN IS **Visual-Spatial**

Materials world map, pencil, paper

Teaching Strategy Have students work in pairs. Encourage partners to compare their answers with those of other pairs.

Analysis

Allow a couple of degrees leeway in each answer, as students will be estimating in areas between marked longitude and latitude lines.

1. a) Moscow, Russia
   b) Cape Town, South Africa
   c) Havana, Cuba
2. a) 51°N, 1°W
   b) 38°S, 145°E
   c) 35°S, 58°W

Assessment

Performance Have students use a map to determine the latitude and longitude of their hometown. Use **Performance Assessment in the Science Classroom,** p. 99.

Reading Check

Answer 180° meridian

Differentiated Instruction

English-Language Learners Have students locate their native country or city on a world map. Ask them to talk with four other students to choose cities and countries around the world. Have English-Language Learners design a chart to display the name, longitude, latitude, and time for each location they have chosen. L2

Visually Impaired Have students work with a partner to outline continents and landforms on a large, paper, world map using white glue. Latitude and longitude lines and the prime meridian can be added by gluing on string or other craft supplies. L2

Visual Learning

Figure 10 What time is it where we live when it is 7:00 A.M. in Denver? Answers will vary depending on your location.

Caption Answer

Figure 10 6:00 P.M.

Quick Demo

Earth Time

Materials bright light, globe

Estimated Time 15 minutes

Procedure use a bright light to represent the Sun and a rotating globe. Point out that noon, midnight, 6 P.M., and 6 A.M. remain stationary with regard to the Sun, while positions on Earth's surface rotate toward or away from a specific time.

INTEGRATE Social Studies

Jet Lag Have students discuss what it felt like, how it affected their eating and sleeping routines, how long it took to recover, and whether they took special actions to get over jet lag. How is jet lag similar to the adjustments they might go through during Daylight Saving Time transition?

Reading Check

Answer Each day has 24 hours.

Discussion

Calendar Difference Tell the class that students living in Barrow, Alaska, want to watch live coverage of the finals of a men's volleyball championship that is scheduled for 9:00 AM on Tuesday, August 24, 2004, in Athens, Greece. What day and time will the students in Alaska be able to watch the game? Ask students to explain how it can be a certain day of the week in 1 city and a different day of the week in a different city.

Figure 10 The United States has six time zones.

A Washington, D.C., lies in the eastern time zone. Students there would be going to sleep at 9:00 P.M.

B But students in Seattle, Washington, which lies in the Pacific time zone, are eating dinner. **Determine** *what time it would be in Seattle when the students in Washington, D.C., are sleeping at 9:00 P.M.*

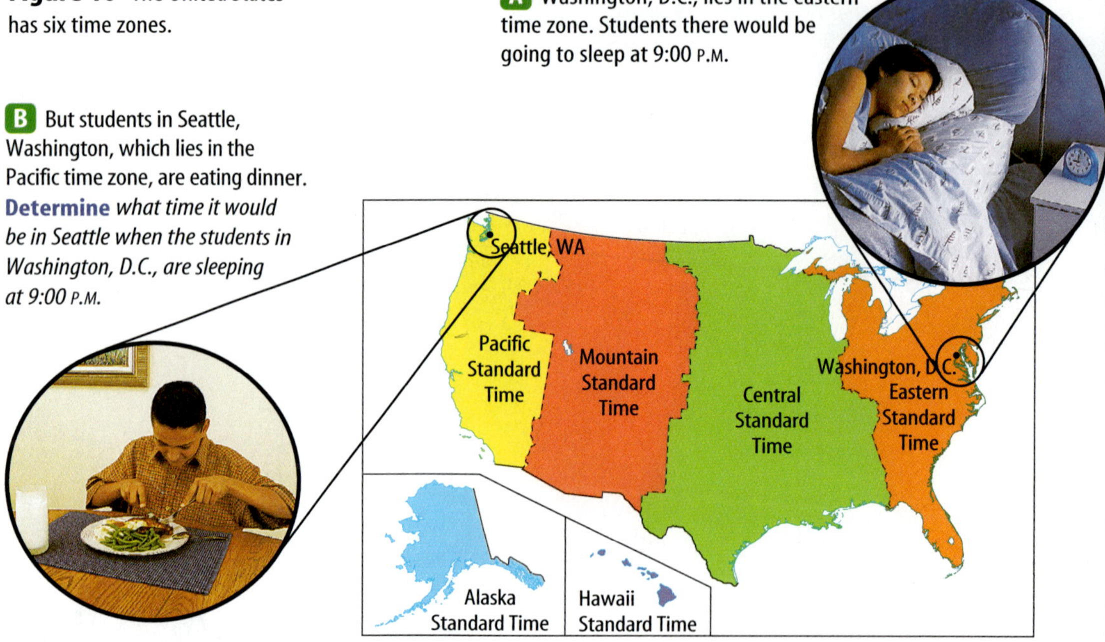

Time Zones

What time it is depends on where you are on Earth. Time is measured by tracking Earth's movement in relation to the Sun. Each day has 24 h, so Earth is divided into 24 time zones. Each time zone is about 15° of longitude wide and is 1 h different from the zones on each side of it. The United States has six different time zones. As you can see in **Figure 10,** people in different parts of the country don't experience dusk simultaneously. Because Earth rotates, the eastern states end a day while the western states are still in sunlight.

Reading Check *What is the basis for dividing Earth into 24 time zones?*

Time zones do not follow lines of longitude strictly. Time zone boundaries are adjusted in local areas. For example, if a city were split by a time zone boundary, the results would be confusing. In such a situation, the time zone boundary is moved outside the city.

Calendar Dates

In each time zone, one day ends and the next day begins at midnight. If it is 11:59 P.M. Tuesday, then 2 min later it will be 12:01 A.M. Wednesday in that particular time zone.

INTEGRATE Social Studies

International Travel If you travel east or west across three or more time zones, you could suffer from jet lag. Jet lag occurs when your internal time clock does not match the new time zone. Jet lag can disrupt the daily rhythms of sleeping and eating. Have you or any of your classmates ever traveled to a foreign country and suffered from jet lag?

Differentiated Instruction

Challenge Ask students to research Daylight Saving Time. Then ask them to look into the impact this has on the time zone differences within the United States. They should realize that not all locations set their clocks back in the fall. Challenge the students to create a fictional travel journey on a trip across the U.S., using what they have learned about Daylight Saving Time. L2

Teacher FYI

Time Zone Decision In 1884, an international committee agreed upon our current system of 24 time zones, 15° apart, starting with the Prime Meridian. Daylight Saving Time was instituted during World War I to conserve fuel for the war effort by extending evening daylight hours.

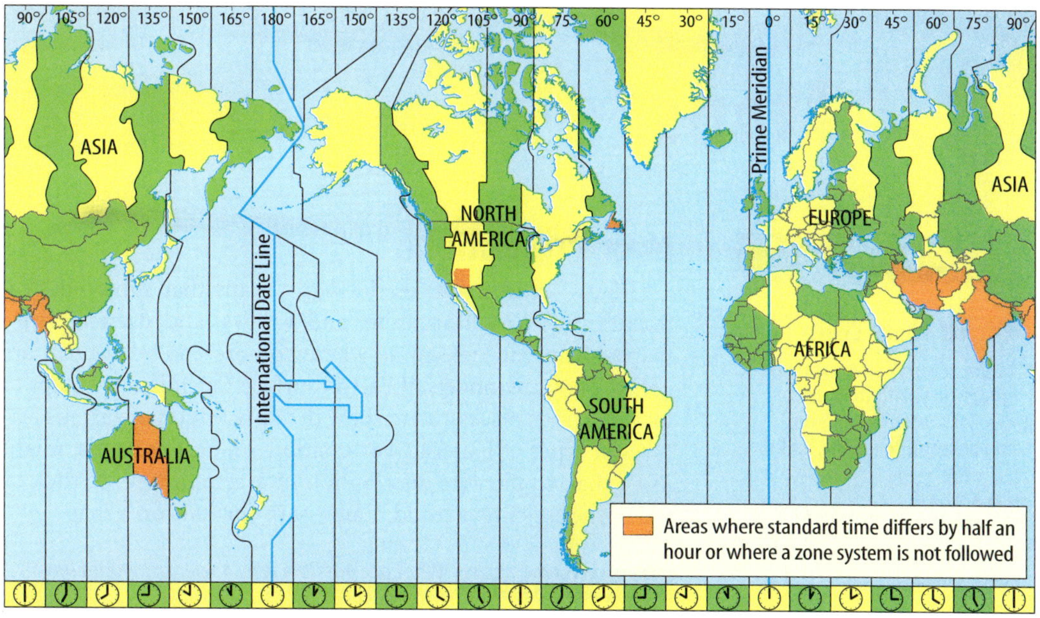

International Date Line You gain or lose time when you enter a new time zone. If you travel far enough, you can gain or lose a whole day. The International Date Line, shown on **Figure 11,** is the transition line for calendar days. If you were traveling west across the International Date Line, located near the 180° meridian, you would move your calendar forward one day. Traveling east, you would move your calendar back one day.

Figure 11 Lines of longitude roughly determine the locations of time zone boundaries. These boundaries are adjusted locally to avoid splitting cities and other political subdivisions, such as counties, into different time zones.

section 2 review

Summary

Latitude and Longitude

- The equator is the imaginary line that wraps around Earth at 0° latitude.
- Latitude is the distance in degrees north or south of the equator.
- The prime meridian is the imaginary line that represents 0° longitude and runs north to south through Greenwich, England.
- Longitude is the distance in degrees east or west of the prime meridian.

Time Zones and Calendar Dates

- Earth is divided into 24 one-hour time zones.
- The International Date Line is the transition line for calendar days.

Self Check

1. **Explain** how lines of latitude and longitude help people find locations on Earth.
2. **Determine** the latitude and longitude of New Orleans, Louisiana.
3. **Calculate** what time it would be in Los Angeles if it were 7:00 P.M. in New York City.
4. **Think Critically** How could you leave home on Monday to go sailing on the ocean, sail for 1 h on Sunday, and return home on Monday?

Applying Math

5. **Use Fractions** If you started at the prime meridian and traveled east one-fourth of the way around Earth, what line of longitude would you reach?

Science Online bookg.msscience.com/self_check_quiz

Activity

Time Difference Working with a partner, have students design an activity that will take place in one location at a specific time, and be viewed at a different time on a different day in another part of the world. Activities may include sporting events, flight information, phone calls, or birth announcements. L2

3 Assess

DAILY INTERVENTION

Check for Understanding

Logical-Mathematical Ask students to explain in writing how the spacing between longitude lines changes as you travel from the north pole, through the equator, and then to the south pole. The lines of longitude will be closer together on the smaller Arctic Circle. As you travel south to the equator, the lines get further apart, but still remain 15° apart on the larger equatorial circle. The lines again move closer together as you move toward the 0° south pole, staying at a constant 15° apart. L2

Reteach

Time Zones Obtain a globe that shows time zones. Using the current time in your area, have students determine the corresponding time in Quito, Equador; Paris, France; Beijing, China; and Cairo, Egypt. L2

Assessment

Process Assess students' abilities to interpret scientific illustrations by asking them to name the continent located between 11° and 39° south latitude and 113° and 153° east longitude. Australia Use **PASC**, p. 127. L2

section 2 review

1. They form an imaginary grid system that helps pinpoint locations.
2. About 30°N, 90°W
3. 4:00 P.M.
4. It's possible if you sail in waters through which the International Date Line passes. If you started west of the line on Monday, then sailed east, it would be Sunday. If you then sailed west over the line again, it would again be Monday.
5. 90°E

section 3

Maps

1 Motivate

INTERACTIVE CHALKBOARD PowerPoint® Presentations

Bellringer

Section Focus Transparencies also are available on the Interactive Chalkboard CD-ROM. L2 ELL

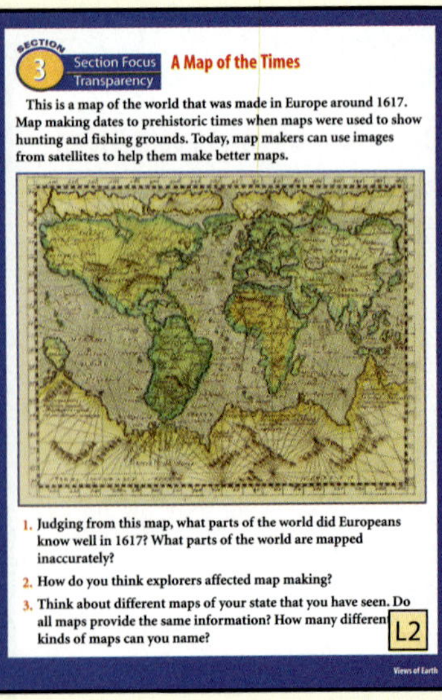
Section 3 Section Focus Transparency **A Map of the Times**

This is a map of the world that was made in Europe around 1617. Map making dates to prehistoric times when maps were used to show hunting and fishing grounds. Today, map makers can use images from satellites to help them make better maps.

1. Judging from this map, what parts of the world did Europeans know well in 1617? What parts of the world are mapped inaccurately?
2. How do you think explorers affected map making?
3. Think about different maps of your state that you have seen. Do all maps provide the same information? How many different kinds of maps can you name?

L2

Tie to Prior Knowledge

Map Projections Ask students if they've ever noticed that maps of the same place on Earth often look very different. To illustrate, show students two maps of the United States made with two different map projections. Have students observe how the maps differ.

Reading Check

Answer Scientists might use maps to find elevations, study Earth's surface, calculate distance or direction, or find specific locations.

Visual Learning

Figure 12 Libya, Egypt, Algeria, and Morocco—countries on the northern shore of Africa—are at the same latitude as what states in the United States? Florida, Texas, California (southern region of state)

as you read

What You'll Learn

- **Compare and contrast** map projections and their uses.
- **Analyze** information from topographic, geologic, and satellite maps.

Why It's Important

Maps help people navigate and understand Earth.

Review Vocabulary

globe: a spherical representation of Earth

New Vocabulary

- conic projection
- topographic map
- contour line
- map scale
- map legend

Map Projections

Maps—road maps, world maps, maps that show physical features such as mountains and valleys, and even treasure maps—help you determine where you are and where you are going. They are models of Earth's surface. Scientists use maps to locate various places and to show the distribution of various features or types of material. For example, an Earth scientist might use a map to plot the distribution of a certain type of rock or soil. Other scientists could draw ocean currents on a map.

Reading Check *What are possible uses a scientist would have for maps?*

Many maps are made as projections. A map projection is made when points and lines on a globe's surface are transferred onto paper, as shown in **Figure 12.** Map projections can be made in several different ways, but all types of projections distort the shapes of landmasses or their areas. Antarctica, for instance, might look smaller or larger than it is as a result of the projection that is used for a particular map.

Figure 12 Lines of longitude are drawn parallel to one another in Mercator projections.
Describe *what happens near the poles in Mercator projections.*

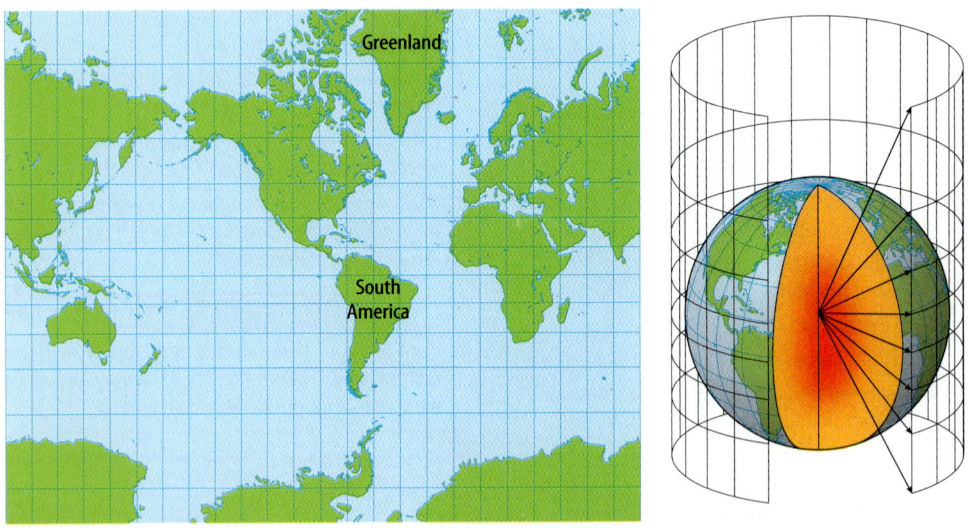

Section 3 Resource Manager

Chapter FAST FILE Resources
Transparency Activity, p. 44
Directed Reading for Content Mastery, pp. 19, 20
Enrichment, p. 30
Reinforcement, p. 27
Lab Worksheets, pp. 5–6, 7–8

Earth Science Critical Thinking/Problem Solving, p. 5
Cultural Diversity, pp. 27, 49
Reading and Writing Skill Activities, p. 21
Lab Management and Safety, p. 66

Figure 13 Robinson projections show little distortion in continent shapes and sizes.

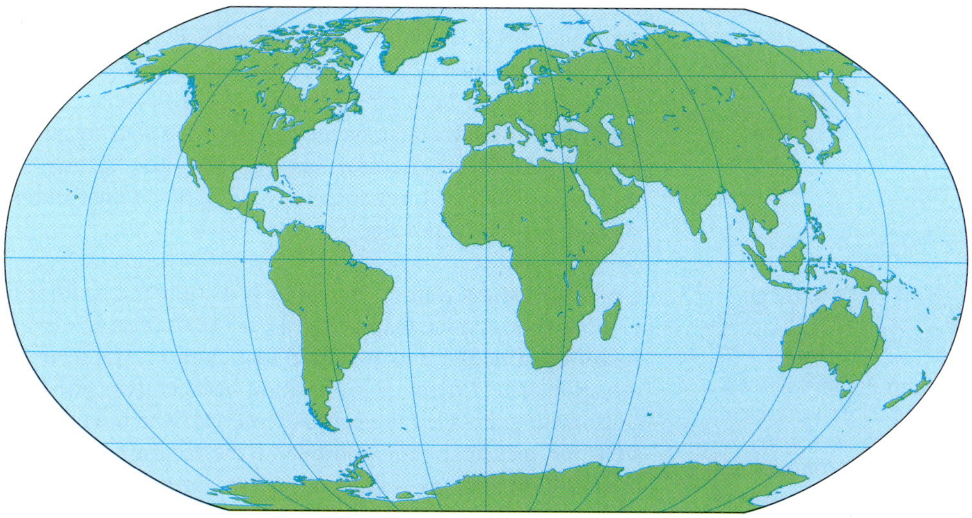

Mercator Projection Mercator (mer KAY ter) projections are used mainly on ships. They project correct shapes of continents, but the areas are distorted. Lines of longitude are projected onto the map parallel to each other. As you learned earlier, only latitude lines are parallel. Longitude lines meet at the poles. When longitude lines are projected as parallel, areas near the poles appear bigger than they are. Greenland, in the Mercator projection in **Figure 12,** appears to be larger than South America, but Greenland is actually smaller.

Robinson Projection A Robinson projection shows accurate continent shapes and more accurate land areas. As shown in **Figure 13,** lines of latitude remain parallel, and lines of longitude are curved as they are on a globe. This results in less distortion near the poles.

Conic Projection When you look at a road map or a weather map, you are using a conic (KAH nihk) projection. Conic projections, like the one shown in **Figure 14,** often are used to produce maps of small areas. These maps are well suited for middle latitude regions but are not as useful for mapping polar or equatorial regions. **Conic projections** are made by projecting points and lines from a globe onto a cone.

Reading Check *How are conic projections made?*

Figure 14 Small areas are mapped accurately using conic projections.

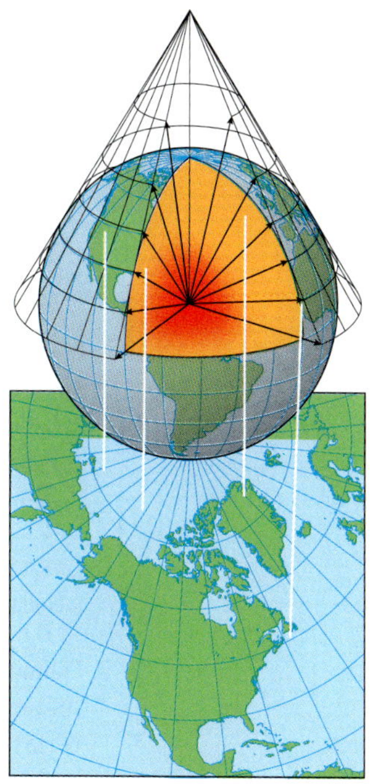

2 Teach

Caption Answer

Figure 12 Size is exaggerated.

Quick Demo

It's a Fruity World

Materials grapefruit or other citrus fruit, knife

Estimated Time 15 minutes

Procedure Draw a rough world map on a grapefruit. Then slice the grapefruit into eight identical wedges. Remove the fruit and press the wedges down side by side on a flat surface as they were on the fruit. Have students observe the way this "projection" shows the world. Discuss the pros and cons of such a projection. Pros: It doesn't distort the size of land areas. Cons: The features are interrupted, making the map hard to use.

LS **Visual-Spatial** L2

Reading Check

Answer by projecting points and lines from a globe onto a cone

LAB DEMONSTRATION

Purpose to make a Mercator projection grid

Materials globe, large sheet of tracing paper, marking pen, meterstick

Procedure Form a paper cylinder around the globe as shown in **Figure 12**. Trace latitude lines by bending the paper so that it is touching the globe. Mark where longitude lines intersect the equator. Remove the paper from the globe and use a meterstick to draw the longitude lines perpendicular to the equator.

Expected Outcome A Mercator projection grid is created.

Assessment

Where is the greatest distortion on the projection? near the poles Which area has the least distortion? areas near the equator

SECTION 3 Maps G ◆ 19

Teacher FYI

English Units Most United States Geological Survey (USGS) topographic maps use the English system of units rather than SI. Therefore, contour intervals are generally given in feet rather than meters.

Text Question Answer

It avoids having contour lines too close together or too far apart. If a small interval were used for a mountainous area, the closely-spaced contours would be impossible to read. If a large contour interval were used for a flat area, the lines would be so widely spaced that it would be difficult to interpret elevation.

Activity

Compare Topography Obtain two topographic maps of your area, one up-to-date and the other as old as possible. Have students compare and contrast the two maps. Students may find that one map contains features such as buildings or roads not found on the other map. Have students use their observations of these maps to explain why maps should be regularly updated. L2 ELL IS Visual-Spatial

INTEGRATE Physics

Mapping Planets The data collected by these probes provides scientists with information on elevation at various points on the surface. Contour maps of the planets' surfaces can be made using these data.

Research Have students investigate topographic maps of the Moon, how they are made, what tools were used, and how topographic maps are useful to astronomers, astrogeologists, and earth scientists.

Topographic Maps

For nature hiking, a conic map projection can be helpful by directing you to the location where you will start your hike. On your hike, however, you would need a detailed map identifying the hills and valleys of that specific area. A **topographic map**, shown in **Figure 15,** models the changes in elevation of Earth's surface. With such a map, you can determine your location relative to identifiable natural features. Topographic maps also indicate cultural features such as roads, cities, dams, and other structures built by people.

INTEGRATE Physics

Mapping Planets Satellites are used to map the surface of Earth and other planets. Space probes have made topographic maps of Venus and Mars. Satellites and probes send a radar beam or laser pulses to the surface and measure how long it takes for the beam or pulses to return to the space vehicle.

Contour Lines Before your hike, you study the contour lines on your topographic map to see the trail's changes in elevation. A **contour line** is a line on a map that connects points of equal elevation. The difference in elevation between two side-by-side contour lines is called the contour interval, which remains constant for each map. For example, if the contour interval on a map is 10 m and you walk between two lines anywhere on that map, you will have walked up or down 10 m.

In mountainous areas, the contour lines are close together. This situation models a steep slope. However, if the change in elevation is slight, the contour lines will be far apart. Often large contour intervals are used for mountainous terrain, and small contour intervals are used for fairly flat areas. Why? **Table 1** gives additional tips for examining contour lines.

Index Contours Some contour lines, called index contours, are marked with their elevation. If the contour interval is 5 m, you can determine the elevation of other lines around the index contour by adding or subtracting 5 m from the elevation shown on the index contour.

Table 1 Contour Rules

1. **Contour lines close around hills and basins.** To decide whether you're looking at a hill or basin, you can read the elevation numbers or look for hachures (ha SHOORZ). These are short lines drawn at right angles to the contour line. They show depressions by pointing toward lower elevations.
2. **Contour lines never cross.** If they did, it would mean that the spot where they cross would have two different elevations.
3. **Contour lines form Vs that point upstream when they cross streams.** This is because streams flow in depressions that are beneath the elevation of the surrounding land surface. When the contour lines cross the depression, they appear as Vs pointing upstream on the map.

Differentiated Instruction

Challenge Have students interview a cartographer, and then write a report detailing the steps involved in the map-making process, from obtaining raw data to making the actual map. Encourage students to illustrate each step. L3 P

Active Reading

Buddy Interviews This strategy helps students understand and clarify the reading. Have students interview one another to find out what helps them understand what they are reading, how they find answers, and how they assimilate new vocabulary terms. Have students use Buddy Interviews to help them master the concept of topographic maps. L2

NATIONAL GEOGRAPHIC VISUALIZING TOPOGRAPHIC MAPS

Figure 15

Planning a hike? A topographic map will show you changes in elevation. With such a map, you can see at a glance how steep a mountain trail is, as well as its location relative to rivers, lakes, roads, and cities nearby. The steps in creating a topographic map are shown here.

Ⓐ To create a topographic map of Old Rag Mountain in Shenandoah National Park, Virginia, mapmakers first measure the elevation of the mountain at various points.

Ⓑ These points are then projected onto paper. Points at the same elevation are connected, forming contour lines that encircle the mountain.

Ⓒ Where contour lines on a topographic map are close together, elevation is changing rapidly—and the trail is very steep!

NATIONAL GEOGRAPHIC

Visualizing Topographic Maps

Have students examine the pictures and read the captions. Then ask the following questions.

Besides hiking, what are some other reasons people would use a topographic map? Possible answers: For planning highways, bridges, and tunnels

How would a topographic map help someone planning to build a house? These maps could be used to show the differences in elevation, the steepness of slopes, the location of streams, and the proximity to roads.

Activity

Landscape Observation Take students to an area near school where they can observe a prominent landscape feature. Take along a topographic map and have students explain how the feature compares to its representation on the map. L2

Fun Fact

A circular depression looks very similar to a hill on a topographic map. The only difference is that the contour lines on a depression have small hatch marks on the inside, indicating that the contour interval should decrease.

SECTION 3 Maps G ◆ 21

Inquiry Lab

Schoolyard Topo Map

Purpose to introduce students to the process of making their own topographic map of a feature found in the schoolyard; students should also become accustomed to using a variety of measuring tools. L2

Possible Materials measuring sticks, string, compass, level, small craft sticks

Estimated Time 1 class session

Teaching Strategies

- Students should identify a small feature in the schoolyard to create a topographic map of. A small hill would work well.
- Using several meter sticks taped together, they should identify a contour interval that would be appropriate to show the details of the feature.
- Students should measure each interval around the feature and place a line of string along that contour line. They should continue this until the entire feature has been mapped.
- Encourage students to draw their contour map on a sheet of paper at a scale that is appropriate for a smaller representation.

For additional inquiry activities, see *Science Inquiry Labs*.

Discussion

Contour Scale What is the contour interval of a map if two side-by-side contour lines represent elevations of 200 m and 300 m? 100 m Suppose a student draws a map using the scale 1 cm: 100 miles. Is this a proper scale? Explain. No; units of measurement on each side of the ratio must be the same.

Text Question Answer

Drill into rock or soil to get core samples or other evidence.

Science Online

Topic: Map Technology
Visit bookg.msscience.com for Web links to information about modern cartography and technology.

Activity Become a cartographer by developing a plan that uses map technology. For example, plan a traveling vacation or design a home or recreation area. Make a map of your trip or new environment.

Map Scale When planning your hike, you'll want to determine the distance to your destination before you leave. Because maps are small models of Earth's surface, distances and sizes of things shown on a map are proportional to the real thing on Earth. Therefore, real distances can be found by using a scale.

The **map scale** is the relationship between the distances on the map and distances on Earth's surface. Scale often is represented as a ratio. For example, a topographic map of the Grand Canyon might have a scale that reads 1:80,000. This means that one unit on the map represents 80,000 units on land. If the unit you wanted to use was a centimeter, then 1 cm on the map would equal 80,000 cm on land. The unit of distance could be feet or millimeters or any other measure of distance. However, the units of measure on each side of the ratio must always be the same. A map scale also can be shown in the form of a small bar that is divided into sections and scaled down to match real distances on Earth.

Map Legend Topographic maps and most other maps have a legend. A **map legend** explains what the symbols used on the map mean. Some frequently used symbols for topographic maps are shown in the appendix at the back of the book.

Map Series Topographic maps are made to cover different amounts of Earth's surface. A map series includes maps that have the same dimensions of latitude and longitude. For example, one map series includes maps that are 7.5 minutes of latitude by 7.5 minutes of longitude. Other map series include maps covering larger areas of Earth's surface.

Geologic Maps

One of the more important tools to Earth scientists is the geologic map. Geologic maps show the arrangement and types of rocks at Earth's surface. Using geologic maps and data collected from rock exposures, a geologist can infer how rock layers might look below Earth's surface. The block diagram in **Figure 16** is a 3-D model that illustrates a solid section of Earth. The top surface of the block is the geologic map. Side views of the block are called cross sections, which are derived from the surface map. Developing geologic maps and cross sections is extremely important for the exploration and extraction of natural resources. What can a scientist do to determine whether a cross section accurately represents the underground features?

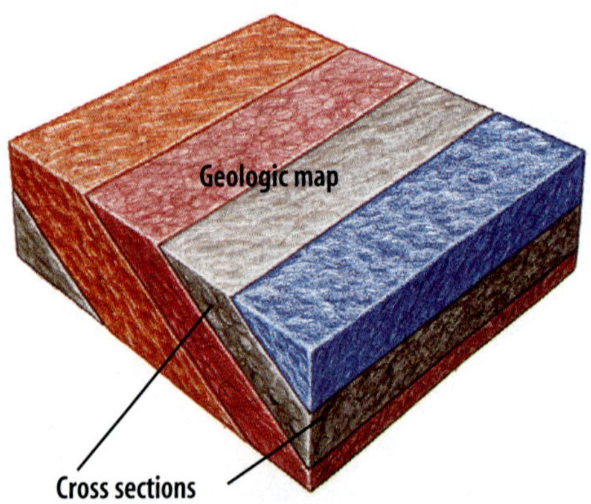

Figure 16 Geologists use block diagrams to understand Earth's subsurface. The different colors represent different rock layers.

Differentiated Instruction

Learning Disabled Have students draw a contour map with five concentric contours. Have them cut along the outermost contour line, then trace the pattern on a piece of cardboard. Direct students to cut out the cardboard along the traced line. Have them repeat this process with each of the other four contours, moving inward. When all the contours have been cut out, have students paste the cardboard cutouts together, one on top of the other, lining up the contour patterns. The stair-step result is a model of a hill. LS **Kinesthetic** L2

Three-Dimensional Maps Topographic maps and geologic maps are two-dimensional models that are used to study features of Earth's surface. To visualize Earth three dimensionally, scientists often rely on computers. Using computers, information is digitized to create a three-dimensional view of features such as rock layers or river systems. Digitizing is a process by which points are plotted on a coordinate grid.

Map Uses As you have learned, Earth can be viewed in many different ways. Maps are chosen depending upon the situation. If you wanted to determine New Zealand's location relative to Canada and you didn't have a globe, you probably would examine a Mercator projection. In your search, you would use lines of latitude and longitude, and a map scale. If you wanted to travel across the country, you would rely on a road map, or conic projection. You also would use a map legend to help locate features along the way. To climb the highest peak in your region, you would take along a topographic map.

Applying Science

How can you create a cross section from a geologic map?

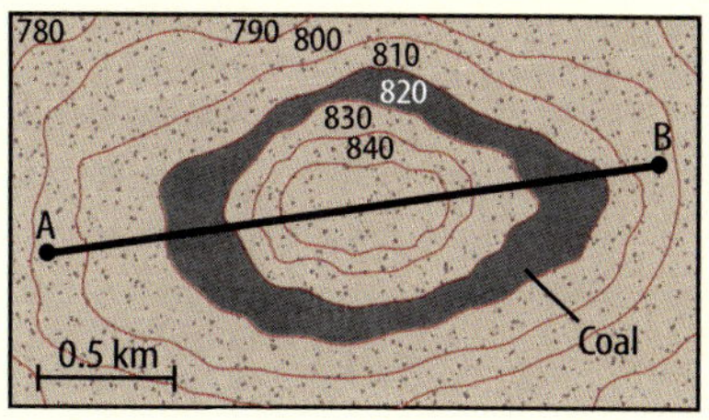

Earth scientists are interested in knowing the types of rocks and their configurations underground. To help them visualize this, they use geologic maps. Geologic maps offer a two-dimensional view of the three-dimensional situation found under Earth's surface. You don't have to be a professional geologist to understand a geologic map. Use your ability to create graphs to interpret this geologic map.

Identifying the Problem

Above is a simple geologic map showing where a coal seam is found on Earth's surface. Place a straight edge of paper along the line marked A–B and mark the points where it meets a contour. Make a different color mark where it meets the exposure of coal. Make a graph on which the various elevations (in meters) are marked on the y-axis. Lay your marked edge of paper along the x-axis and transfer the points directly above onto the proper elevation line. Now connect the dots to draw in the land's surface and connect the marks you made for the coal seam separately.

Solving the Problem

1. What type of topography does the map represent?
2. At what elevation is the coal seam?
3. Does this seam tilt, or is it horizontal? Explain how you know.

Use Science Words

Word Origin Have students use dictionaries to find the meaning and origin of the word *cartography*. Students will discover that cartography is the science of making maps. The word has two roots: *carte* comes from the French word meaning "card" or "map"; *-ography* comes from the Latin *-graphia*, meaning "writing." Ask students the technical name for a mapmaker. cartographer

Applying Science

Answers

1. a hill
2. 810 to 820 m
3. Horizontal; in the profile, the lines representing the top and bottom of the coal seam are horizontal.

Teacher FYI

Map Types Two common types of maps are thematic maps and inventory maps. A thematic map shows the distribution of a particular feature over a certain area, for example, the difference in population density or amount of rainfall over a state or country. An inventory map shows the location of a certain type of item over an area, for example, the location of nuclear power plants in a state.

SECTION 3 Maps G ◆ 23

3 Assess

DAILY INTERVENTION

Check for Understanding

Kinesthetic Ask student to determine their own pace (stride length). They can make their own maps of the classroom including a map scale. L2

Reteach

Community Map Have pairs of students prepare one type of map that gives the reader information about your community. The map can be a topographic map of a neighborhood, a road map, an inventory map showing the location of schools, or some other type. Have students present their maps to the class and explain how they were made. L2

COOP LEARN IS Interpersonal

Assessment

Process Have students prepare charts that compare and contrast topographic and geologic maps. L2

Figure 17 Hurricane Isabel's wind lashed the North Carolina and Virginia coasts on September 18, 2003.

Analyze *this satellite photo of Hurricane Isabel approaching the North Carolina Outer Banks. How many states do you think might be affected by this weather system?*

Remote Sensing

Scientists use remote-sensing techniques to collect much of the data used for making maps. Remote sensing is a way of collecting information about Earth from a distance, often using satellites.

Landsat One way that Earth's surface has been studied is with data collected from Landsat satellites, as shown in **Figure 17.** These satellites take pictures of Earth's surface using different wavelengths of light. The images can be used to make maps of snow cover over the United States or to evaluate the impact of forest fires, such as those that occurred in the western United States during the summer of 2000. The newest Landsat satellite, *Landsat 7,* can acquire detailed images by detecting light reflected off landforms on Earth.

Global Positioning System The Global Positioning System, or GPS, is a satellite-based, radio-navigation system that allows users to determine their exact position anywhere on Earth. Twenty-four satellites orbit 20,200 km above the planet. Each satellite sends a position signal and a time signal. The satellites are arranged in their orbits so that signals from at least six can be picked up at any given moment by someone using a GPS receiver. By processing the signals, the receiver calculates the user's exact location. GPS technology is used to navigate, to create detailed maps, and to track wildlife.

section 3 review

Summary

Map Projections

- A map projection is the projection of points and lines of a globe's surface onto paper.

Topographic Maps

- Topographic maps show the changes in elevation of Earth's surface by using contour lines.

Geologic Maps

- Geologic maps show the arrangement and types of rocks at Earth's surface.

Remote Sensing

- Remote sensing is a way of collecting information about Earth from a distance, often by using satellites.
- Distant planets can be mapped using satellites.

Self Check

1. **Compare and contrast** Mercator and conic projections.
2. **Explain** why Greenland appears larger on a Mercator projection than it does on a Robinson projection.
3. **Describe** why contour lines never cross.
4. **Explain** whether a topographic map or a geologic map would be most useful for drilling a water well.
5. **Think Critically** Review the satellite photograph at the beginning of this chapter. Is most of the city near or far from the water? Why is it located there?

Applying Skills

6. **Make Models** Architects make detailed maps called scale drawings to help them plan their work. Make a scale drawing of your classroom.

Science Online bookg.msscience.com/self_check_quiz

section 3 review

1. Areas of distortion differ, so they are best for different purposes. Mercator has great distortion in polar areas. Robinson has less distortion at the poles. Conic projections are accurate for relatively small areas of Earth's surface.
2. Lines of longitude are parallel on a Mercator, stretching any landmasses in polar areas. There is less distortion in polar areas on a Robinson projection, so Greenland isn't as stretched out.
3. One location cannot have two different elevations.
4. drilling a well—geologic map; hiking trip—topographic map
5. Most of the uncovered land is far from the water. Population needs water to drink, transport, and for daily use. The cities would grow up along the waterway.
6. Encourage students to start by making a grid system for the class and determining a scale, such as 1:100.

LAB

Making a Topographic Map

Have you ever wondered how topographic maps are made? Today, radar and remote-sensing devices aboard satellites collect data, and computers and graphic systems make the maps. In the past, surveyors and aerial photographers collected data. Then, maps were hand drawn by cartographers, or mapmakers. In this lab, you can practice cartography.

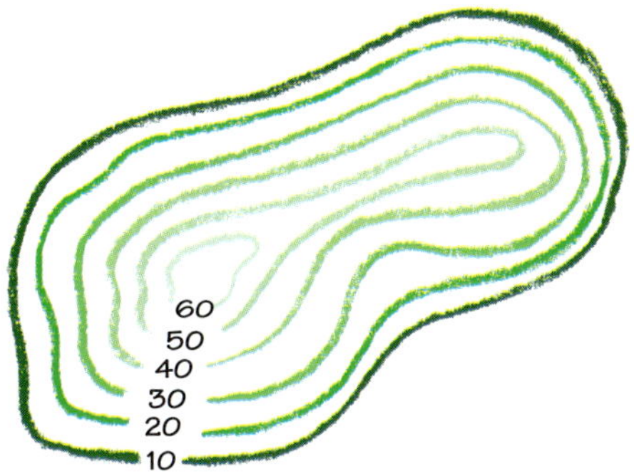

Real-World Question

How is a topographic map made?

Goals

- **Draw** a topographic map.
- **Compare and contrast** contour intervals.

Materials

plastic model of a landform
water tinted with food coloring
transparency
clear-plastic storage box with lid
beaker
metric ruler
tape
transparency marker

Procedure

1. Using the ruler and the transparency marker, make marks up the side of the storage box that are 2 cm apart.
2. Secure the transparency to the outside of the box lid with tape.
3. Place the plastic model in the box. The bottom of the box will be zero elevation.
4. Using the beaker, pour water into the box to a height of 2 cm. Place the lid on the box.
5. Use the transparency marker to trace the top of the water line on the transparency.
6. Using the scale 2 cm = 10 m, mark the elevation on the line.
7. Repeat the process of adding 2 cm of water and tracing until the landform is mapped.
8. Transfer the tracing of the landform onto a sheet of white paper.

Conclude and Apply

1. **Identify** the contour interval of this topographic map.
2. **Evaluate** how the distance between contour lines on the map shows the steepness of the slope on the landform model.
3. **Determine** the total elevation of the landform you have selected.
4. **Describe** how elevation was represented on your map.
5. **Explain** how elevations are shown on topographic maps.
6. Must all topographic maps have a contour line that represents 0 m of elevation? Explain.

BENCH TESTED

Real-World Question

Purpose Students make a topographic map from a landform model. L2 ELL COOP LEARN LS **Kinesthetic**

Process Skills measure in SI, make models, interpret data, compare and contrast

Time Required 45–50 minutes

Procedure

Teaching Strategy Suggest that students stand directly over the model with one eye closed to make each tracing, so that each will be done from the same position.

Troubleshooting Some students may need extra help in understanding the scale to be used.

Conclude and Apply

1. 10 m
2. The closer the contour lines, the steeper the slope.
3. Answers will vary, but they should equal the height of the model in centimeters multiplied by 500. (2 cm = 10 m; therefore 1 cm = 5 m or 500 cm)
4. a series of lines drawn on the transparency
5. contour lines
6. No, the 0-m contour line is only on maps with elevations at sea level.

Assessment

Performance To further assess students' understanding of making topographic maps from landform models, have them repeat the activity using a second landform model. Have them compare and contrast their two completed topographic maps. Use **Performance Assessment in the Science Classroom**, p. 123. L2

Communicating Your Data

Have students compare the topographic maps produced by other groups with their own.

BENCH TESTED

Real-World Question

Purpose Students will show the relationship between topographic maps and landforms by making a 3-D model from a topographic map. L2 ELL

COOP LEARN LS Kinesthetic P

Process Skills interpret scientific illustrations, make models, use numbers, observe and infer

Time Required 45 minutes

Make a Model

Discussion Ask students to discuss how they would walk across a sloping field while keeping their eyes level. Students will walk on a line of equal elevation, similar to a contour line on a topographic map. Mention to students that steep slopes are shown on a topographic map as areas where contour lines are very close.

Possible Materials Obtain 7.5-minute quadrangle maps from the U.S. Geological Survey and the book, *100 Topographic Maps*, by Richard Debruin, available from Hubbard. Obtain sandbox sand from a local hardware store. Ask maintenance for rolls of brown paper towels and empty spray bottles.

LAB Model and Invent

Constructing Landforms

Goals

- **Research** how contour lines show relief on a topographic map.
- **Determine** what scale you can best use to model a landscape of your choice.
- Working cooperatively with your classmates, model a landscape in three dimensions from the information given on a topographic map.

Possible Materials

U.S. Geological Survey 7.5-minute quadrangle maps
sandbox sand
rolls of brown paper towels
spray bottle filled with water
ruler

Real-World Question

Most maps perform well in helping you get from place to place. A road map, for example, will allow you to choose the shortest route from one place to another. If you are hiking, though, distance might not be so important. You might want to choose a route that avoids steep terrain. In this case you need a map that shows the highs and lows of Earth's surface, called relief. Topographic maps use contour lines to show the landscape in three dimensions. Among their many uses, such maps allow hikers to choose routes that maximize the scenery and minimize the physical exertion. What does a landscape depicted on a two-dimensional topographic map look like in three dimensions? How can you model a landscape?

Make a Model

1. **Choose** a topographic map showing a landscape easily modeled using sand. Check to see what contour interval is used on the map. Use the index contours to find the difference between the lowest and the highest elevations shown on the landscape. Check the distance scale to determine how much area the landscape covers.
2. **Determine** the scale you will use to convert the elevations shown on your map to heights on your model. Make sure the scale is proportional to the distances on your map.
3. **Plan** a model of the landscape in sand by sketching the main features and their scaled heights onto paper. Note the degree of steepness found on all sides of the features.

Differentiated Instruction

Physically Challenged Place your physically challenged students with other students who will help them make their models. Provide a large flat board on which lab groups can work on their models. This will enable them to move the project from place to place, making it easier for physically challenged students to participate. L2

Alternative Inquiry Lab

Topographic Golf Ask students to design a 1:220 scale topographic map of a fairway and green on a golf course. Creative plans may include depressions, elevations, and flat areas representing sand, water, and other hazards, as well as hills and greens. Students may make a scale model of their hole to join with other students' to form a class golf course display. L2

Using Scientific Methods

4. **Prepare** a document that shows the scale you plan to use for your model and the calculations you used to derive that scale. Remember to use the same scale for distance as you use for height. If your landscape is fairly flat, you can exaggerate the vertical scale by a factor of two or three. Be sure your paper is neat, is easy to follow, and includes all units. Present the document to your teacher for approval.

Test Your Model

1. Using the sand, spray bottle, and ruler, create a scale model of your landscape on the brown paper towels.
2. **Check** your topographic map to be sure your model includes the landscape features at their proper heights and proper degrees of steepness.

Analyze Your Data

1. **Determine** if your model accurately represents the landscape depicted on your topographic map. Discuss the strengths and weaknesses of your model.
2. **Explain** why it was important to use the same scale for height and distance. If you exaggerated the height, why was it important to indicate the exaggeration on your model?

Conclude and Apply

1. **Infer** why the mapmakers chose the contour interval used on your topographic map?
2. **Predict** the contour intervals mapmakers might choose for topographic maps of the world's tallest mountains—the Himalaya—and for topographic maps of Kansas, which is fairly flat.

Communicating Your Data

Prepare a vacation getaway commercial to advertise the topographical features of your model landscape. Be sure to discuss the landscape elevation and features, scale, and similarities to actual landforms.

Teaching Strategies

- Have students use a topographic map that has a feature on which at least four contour lines have been drawn.
- Some students will need help in making their sand models of the topographic map. They should begin with the highest and lowest elevations and shape the sand in between to the proper contours.

Test Your Model

Expected Outcome The model will look like a hill, a slope, or a valley as presented on the topographic map. The model should show a gradual increase in elevation from one contour line to the next.

Analyze Your Data

1. Students should note that their models approximate the shape of landforms on their maps. One strength is the fact that the sand models actual land features. One weakness is that it is difficult to keep the sand from shifting.
2. If the same scale is not used, the shape of the landform will be distorted. Indicating how the model was exaggerated will help explain distortions in the model.

Conclude and Apply

1. The contour interval is based on how great the change in elevation is on the map.
2. 100 meters; 5 meters

Assessment

Performance To further assess students' understanding of making a 3-D landform model from a topographic map, have them make a model using the topographic map they made in the activity. Have them compare their completed model with the plastic model used in that activity. Use **Performance Assessment in the Science Classroom**, p. 123. L2

Communicating Your Data

In their presentations, students should explain how their models are similar to the actual landforms and the method used to make the model. Students should also explain any exaggerations used in the model.

TIME

Content Background

Cultural geography is the study of the relationship between people and their environment. A major focus of cultural geographers is how culture and the surrounding environment fit together.

When studying the interactions between these people and their environment, geographers try to answer certain questions. How did the people living in an area get there? What aspects of the environment enticed people to settle in the area? How have the people influenced their environment? If culture or the environment changes, will the people be able to keep up with the changes?

The general consensus among geographers is that the environment does not solely determine culture nor does culture solely control the environment. Rather they interact and influence each other, thus shaping the lives of the people living in an area.

Discussion

Settle where? Facilitate a brainstorming session in which students list the factors that might affect where people settle. Have students discuss how each factor might affect the establishment and success of a community.

Historical Significance

Throughout history, civilizations have established themselves in areas that supported their daily needs. Settlements were established where other natural resources, fresh water, and minerals also were abundant. In general, populations flourished in areas of mild climate where the terrain was level and the soil was fertile.

TIME **SCIENCE AND HISTORY** SCIENCE CAN CHANGE THE COURSE OF HISTORY!

LOCATION, LOCATION

Why is New York City at the mouth of the Hudson River and not 300 km inland? Why are there more farms in Iowa than in Alaska? What's the reason for growing lots of peaches in Georgia but not in California's Death Valley? It's all about location. The landforms, climate, soil, and resources in an area determine where cities and farms grow and what people connected with them do.

LANDFORMS ARE KEY

When many American cities were founded hundreds of years ago, waterways were the best means of transportation. Old cities such as New York City and Boston are located on deep harbors where ships could land with people and goods. Rivers also were major highways centuries ago. They still are.

Topography and soil also play a role in where activities such as farming take root. States such as Iowa and Illinois have many farms because they have flat land and fertile soil. Growing crops is more difficult in mountainous areas or where soil is stony and poor.

CLIMATE AND SOIL

Climate limits the locations of cities and farms, as well. The fertile soil and warm, moist climate of Georgia make it a perfect place to grow peaches. California's Death Valley can't support such crops because it's a hot, dry desert.

RESOURCES RULE

The location of an important natural resource can change the rules. A gold deposit or an oil field can cause a town to grow in a place where the topography, soil, and climate are not favorable. For example, thousands of people now live in parts of Alaska only because of the great supply of oil there. Maine has a harsh climate and poor soil. But people settled along its coast because they could catch lobsters and fish in the nearby North Atlantic.

The rules that govern where towns grow and where people live are different now than they used to be. Often information, not goods, moves from place to place on computers that can be anywhere. But as long as people farm, use minerals, and transport goods from place to place, the natural environment and natural resources will always help determine where people are and what they do.

Research Why was your community built where it is? Research its history. What types of economic activity were important when it was founded? Did topography, climate, or resources determine its location? Design a Moment in History to share your information.

Science Online
For more information, visit bookg.msscience.com/time

Research In addition to using library or online resources, suggest that students research the questions posed in this activity by interviewing parents, grandparents, or other people in their community. Have students write a short newspaper article to report their findings. L2

Resources for Teachers and Students

Cultural Geography: A Critical Introduction, by Don Mitchell. Blackwell Publishers, Malden, MA, 2000

People, Land and Time, by Peter Atkins, Ian Simmons, and Brian Roberts. John Wiley & Sons, New York, 1998

Reviewing Main Ideas

Section 1 Landforms

1. The three main types of landforms are plains, plateaus, and mountains.
2. Plains are large, flat areas. Plateaus are relatively flat, raised areas of land made up of nearly horizontal rocks that have been uplifted. Mountains rise high above the surrounding land.

Section 2 Viewpoints

1. Latitude and longitude form an imaginary grid system that enables points on Earth to be located exactly.
2. Latitude is the distance in degrees north or south of the equator. Longitude is the distance in degrees east or west of the prime meridian.
3. Earth is divided into 24 time zones. Each time zone represents a 1-h difference. The International Date Line separates different calendar days.

Section 3 Maps

1. Mercator, Robinson, and conic projections are made by transferring points and lines on a globe's surface onto paper.
2. Topographic maps show the elevation of Earth's surface. Geologic maps show the types of rocks that make up Earth's surface.
3. Remote sensing is a way of collecting information from a distance. Satellites are important remote-sensing devices.

Visualizing Main Ideas

Copy and complete the following concept map on landforms.

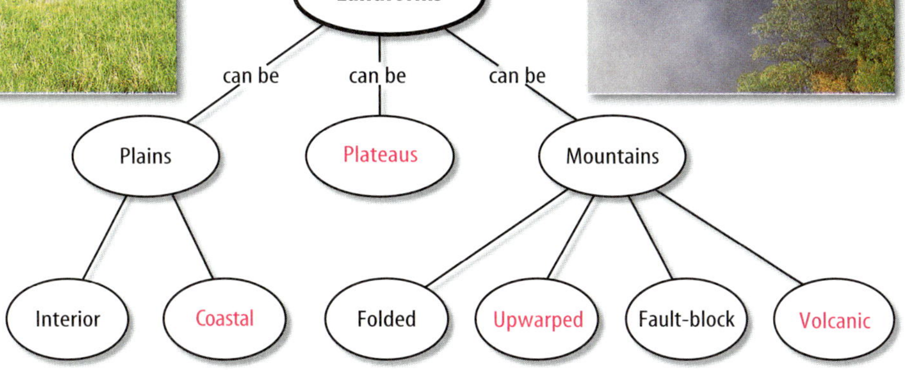

Identifying Misconceptions

Assess

Use this activity as follow-up to page F at the beginning of this chapter.

Activity Organize the class into small groups. Challenge each group to devise a skit that they can perform for the class. The skit should illustrate what might happen if Earth were flat. For example, they might act out a ship sailing off the edge of the world or Earth's shadow on the Moon as having a straight edge. After each group acts out its skit, have the class explain what actually happens on Earth, since it is a sphere.

Expected Outcome Students should understand by this point that Earth is a sphere, not flat.

chapter 1 Study Guide

Reviewing Main Ideas

Summary statements can be used by students to review the major concepts of the chapter.

Visualizing Main Ideas

See student page.

Science Online

Visit bookg.msscience.com
/self_check_quiz
/interactive_tutor
/vocabulary_puzzlemaker
/chapter_review
/standardized_test

Assessment Transparency

For additional assessment questions, use the *Assessment Transparency* located in the transparency book.

Assessment

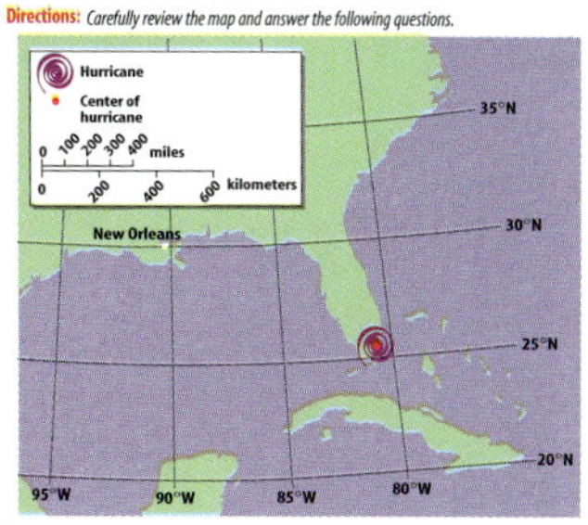

Assessment Transparency **Views of Earth**

Directions: *Carefully review the map and answer the following questions.*

1. What is the approximate latitude and longitude of New Orleans?
   A 30°N, 91°W
   B 29°N, 89°W
   C 91°N, 31°W
   D 89°N, 29°W
2. Approximately how far is New Orleans from the center of the hurricane?
   F 300 kilometers
   G 500 kilometers
   H 700 kilometers
   J 800 kilometers
3. What is the approximate latitude and longitude of the center of the hurricane?
   A 26°N, 81°W
   B 34°N, 78°W
   C 81°N, 26°W
   D 78°N, 34°W

L2

Views of Earth

Using Vocabulary

1. The equator is not a type of mountain.
2. The prime meridian is not a landform.
3. A volcanic mountain is not part of a topographic map.
4. A folded mountain is not a reference point on a map grid.
5. A plateau is not a type of mountain.

Checking Concepts

6. B	11. D
7. D	12. B
8. C	13. B
9. A	14. C
10. D	15. C

chapter 1 Review

Using Vocabulary

conic projection p. 19	map scale p. 22
contour line p. 20	plain p. 8
equator p. 14	plateau p. 10
fault-block mountain p. 12	prime meridian p. 15
folded mountain p. 11	topographic map p. 20
latitude p. 14	upwarped mountain p. 12
longitude p. 15	volcanic mountain p. 13
map legend p. 22	

For each set of terms below, choose the one term that does not belong and explain why it does not belong.

1. upwarped mountain—equator—volcanic mountain
2. plain—plateau—prime meridian
3. topographic map—contour line—volcanic mountain
4. prime meridian—equator—folded mountain
5. fault-block mountain—upwarped mountain—plateau

Checking Concepts

Choose the word or phrase that best answers the question.

6. What makes up about 50 percent of all land areas in the United States?
   A) plateaus **C)** mountains
   B) plains **D)** volcanoes

7. Which type of map shows changes in elevation at Earth's surface?
   A) conic **C)** Robinson
   B) topographic **D)** Mercator

8. How many degrees apart are the 24 time zones?
   A) 10° **C)** 15°
   B) 34° **D)** 25°

Use the photo below to answer question 9.

9. What kind of mountains are the Grand Tetons of Wyoming?
   A) fault-block **C)** upwarped
   B) volcanic **D)** folded

10. Landsat satellites collect data by using
    A) sonar. **C)** sound waves.
    B) echolocation. **D)** light waves.

11. Which type of map is most distorted at the poles?
    A) conic **C)** Robinson
    B) topographic **D)** Mercator

12. Where is the north pole located?
    A) 0°N **C)** 50°N
    B) 180°N **D)** 90°N

13. What is measured with respect to sea level?
    A) contour interval
    B) elevation
    C) conic projection
    D) sonar

14. What kind of map shows rock types making up Earth's surface?
    A) topographic **C)** geologic
    B) Robinson **D)** Mercator

15. Which major U.S. landform includes the Grand Canyon?
    A) Great Plains **C)** Colorado Plateau
    B) Great Basin **D)** Gulf Coastal Plain

Science Online bookg.msscience.com/vocabulary_puzzlemaker

Use the Exam*View*® Pro Testmaker CD-ROM to:

- create multiple versions of tests
- create modified tests with one mouse click for inclusion students
- edit existing questions and add your own questions
- build tests aligned with state standards using built-in State Curriculum Tags
- change English tests to Spanish with one mouse click and vice versa

Thinking Critically

16. **Explain** how a topographic map of the Atlantic Coastal Plain differs from a topographic map of the Rocky Mountains.

17. **Determine** If you left Korea early Wednesday morning and flew to Hawaii, on what day of the week would you arrive?

Use the illustration below to answer question 18.

18. **Determine** Using the map above, arrange these cities in order from the city with the earliest time to the one with the latest time on a given day: Anchorage, Alaska; San Francisco, California; Bangor, Maine; Denver, Colorado; Houston, Texas.

19. **Describe** how a map with a scale of 1:50,000 is different from a map with a scale of 1:24,000.

20. **Compare and contrast** Mercator, Robinson, and conic map projections.

21. **Form Hypotheses** You are visiting a mountain in the northwest part of the United States. The mountain has steep sides and is not part of a mountain range. A crater can be seen at the top of the mountain. Hypothesize about what type of mountain you are visiting.

22. **Concept Map** Copy and complete the following concept map about parts of a topographic map.

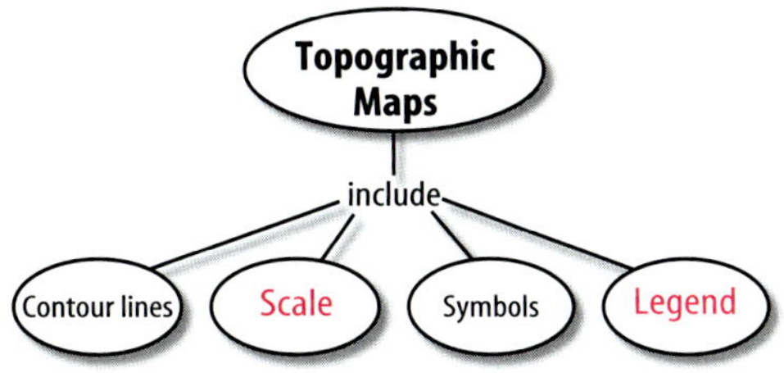

Performance Activities

23. **Poem** Create a poem about one type of landform. Include characteristics of the landform in your poem. How can the shape of your poem add meaning to your poem? Display your poem with those of your classmates.

Applying Math

24. **Calculate** If you were flying directly south from the north pole and reached 70° north latitude, how many more degrees of latitude would you pass over before reaching the south pole? Illustrate and label a diagram to support your answer.

Use the map below to answer question 25.

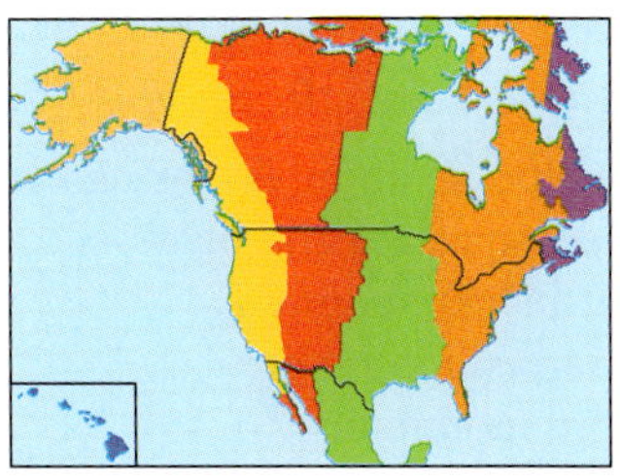

25. **Calculate** If it is 2:00 P.M. in Orlando, Florida, what time is it in Los Angeles, California? In Anchorage, Alaska? In your hometown?

Science Online bookg.msscience.com/chapter_review

chapter 1 Review

Thinking Critically

16. The map of the Atlantic Coastal Plain would show little change in elevation, and the contour interval would be small. The map of the Rockies would show many closed contours indicating mountain peaks, and the contour interval would be large.
17. Tuesday
18. Anchorage, San Francisco, Denver, Houston, Bangor
19. The map with a scale of 1:50,000 would show less detail than the map with a scale of 1:24,000.
20. Each depicts a round Earth on a flat surface. Mercator projections exaggerate the areas near the poles and are used mainly for navigation. Robinson projections are good for depicting accurate continent shapes and have less distortion near the poles. Conic projections are good for mapping small areas and are used for road maps.
21. volcanic mountain
22. See student page.

Performance Activities

23. Poems may be simple as long they describe different types of landforms. Encourage students to be creative. Use **PASC,** p. 151.

Applying Math

National Math Standards

1, 2, 3, 4, 9, 10

24. 160°
25. Los Angeles: 11:00 P.M. Anchorage: 10 A.M.

Assessment Resources

Reproducible Masters

Chapter *Fast File* Resources
- Chapter Review, pp. 35–36
- Chapter Tests, pp. 37–40
- Assessment Transparency Activity, p. 47

Glencoe Science Web site
- Chapter Review Test
- Standardized Test Practice

Glencoe Technology

- Assessment Transparency
- Exam*View*® Pro Testmaker
- MindJogger Videoquiz
- Interactive Chalkboard

FAST FILE

Answer Sheet A practice answer sheet can be found at bookg.msscience.com/answer_sheet.

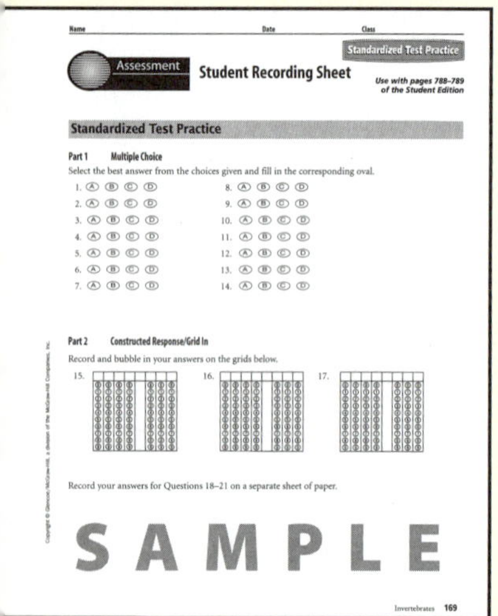

Part 1 Multiple Choice

1. D
2. D
3. B
4. D
5. C
6. D
7. C
8. B
9. B
10. A

Part 2 Short Response

11. Volcanic mountains form as lava and tephra build a cone-shaped structure around the vent.
12. Each day has 24 h, so Earth is divided into 24 time zones. Each time zone is 15° of longitude wide. ($360° \div 24 = 15°$)
13. geologic map; These maps show the types of rock that occur at the surface.
14. Both are fairly flat landforms. Plateaus have much higher elevation than plains.
15. because all lines of latitude are parallel
16. the equator, the north pole, the south pole
17. 100,000 cm
18. 1,000 m
19. 1 km

chapter 1 Standardized Test Practice

Part 1 Multiple Choice

Record your answers on the answer sheet provided by your teacher or on a sheet of paper.

Use the map below to answer question 1.

1. Which of the following is shown above?
   A. cross section
   B. geologic map
   C. topographic map
   D. road map

2. Which landform is a relatively flat area that has high elevation?
   A. mountain
   B. interior plain
   C. coastal plain
   D. plateau

3. Which of the following can provide detailed information about your position on Earth's surface?
   A. prime meridian
   B. global positioning system
   C. International Date Line
   D. LandSat 7

4. What connects points of equal elevation on a map?
   A. legend
   B. series
   C. scale
   D. contour line

5. Which type of mountain forms when rock layers are squeezed and bent?
   A. fault-block mountains
   B. upwarped mountains
   C. folded mountains
   D. volcanic mountains

Test-Taking Tip

Read Carefully Read all choices before answering the questions.

Use the illustration below to answer questions 6–8. The numbers on the drawing represent meters above sea level.

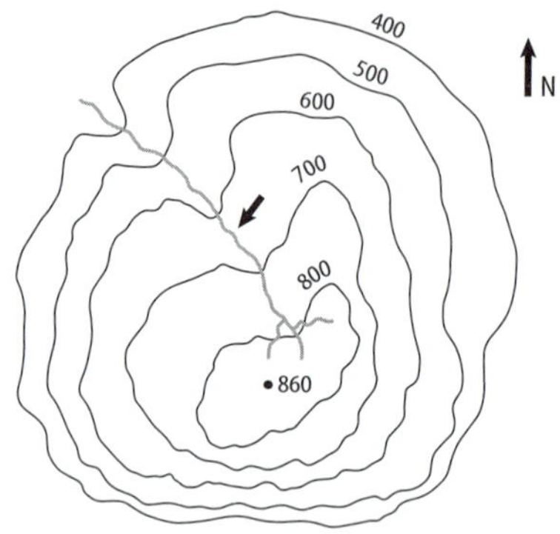

6. Which side of the feature has the steepest slope?
   A. north side
   B. east side
   C. west side
   D. south side

7. What is the highest elevation on the feature?
   A. 800 meters
   B. 960 meters
   C. 860 meters
   D. 700 meters

8. What does the line that is marked by the arrow represent?
   A. a contour line
   B. a stream
   C. a high ridge
   D. a glacier

9. Which type of map is made by projecting points and lines from a globe onto a cone?
   A. Mercator projection
   B. conic projection
   C. Robinson projection
   D. geologic map

10. Which are useful for measuring position north or south of the equator?
    A. lines of latitude
    B. lines of longitude
    C. index contours
    D. map legends

Part 3 Open Ended

20. In the west, the Rocky Mountains are higher and more rugged; formed as upwarped mountains. The Appalachians are lower and less rugged; formed as folded mountains. The central and coastal plains are flat, formed from sediment. Much of the Colorado Plateau is fairly flat; formed by uplifting.
21. In general, the western United States consists of mountains, valleys, basins, and plateaus. The central part of the country is flat interior plains. The eastern portion of the country includes the Appalachian range and the low, flat coastal plain.

32 ◆ G STANDARDIZED TEST PRACTICE

Part 2 Short Response/Grid In

Record your answers on the answer sheet provided by your teacher or on a separate sheet of paper.

11. How do volcanic mountains form?

12. What is a time zone? How are time zones determined around the world? Why are they needed?

13. Which type of map would you use to find the location of a layer of coal at Earth's surface? Why?

14. How are plateaus similar to plains? How are they different?

15. Why are lines of latitude sometimes called parallels?

16. List the locations on Earth that represent 0° latitude, 90°N latitude, and 90°S latitude.

Use the scale at the bottom of the map to answer questions 17–19.

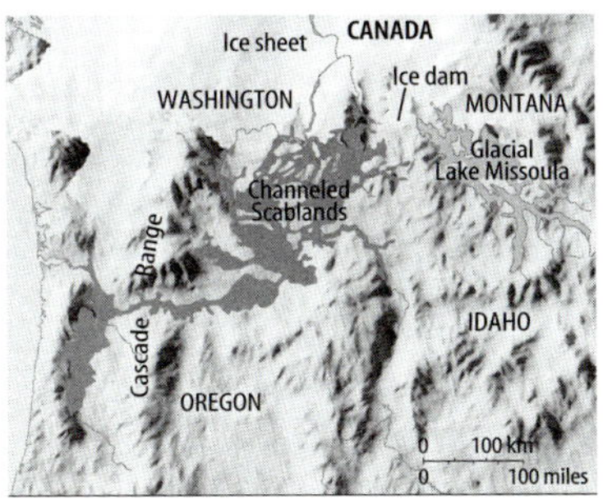

17. How many centimeters on Earth's surface are represented by one centimeter on the map?

18. How many meters on Earth's surface does one centimeter represent?

19. How many kilometers on Earth's surface does one centimeter represent?

Part 3 Open Ended

Record your answers on a sheet of paper.

Use the map below to answer questions 20–21.

20. Where do mountains, plateaus, and plains occur in the United States? Describe these regions and how they were formed.

21. Take the role of a fur trader, pioneer, or explorer. Write three journal entries that give a general description of how the landforms change across the United States.

22. Compare and contrast interior plains and coastal plains.

23. Why is remote sensing important to society? What types of information are obtained?

24. Why are computers often used to make maps?

25. Why might you experience jet lag if you travel across the United States on a plane? Predict which direction it would be more difficult to adjust to jet lag—when traveling from Hawaii to New York or New York to Hawaii. Support your answer with an example of each situation.

Science Online bookg.msscience.com/standardized_test

22. Both are fairly flat features. Interior plains occur in the interior of continents, whereas coastal plains occur along the coast. Coastal plains usually have lower elevation

23. Remote sensing allows data from large land regions to be acquired in a short time. It provides much useful information, such as atmospheric conditions, information about vegetative cover, elevations of land and sea, and much more.

24. Once information is digitized, the information can be quickly displayed in many ways. Three-dimensional mapping using computers allows scientists to examine surface and subsurface features in all three spatial dimensions.

25. Traveling across several time zones often causes one to experience jet lag because the body's internal clock requires some time to adapt to the different times of sunrise and sunset.

chapter 1 Standardized Test Practice

Rubrics

The following rubrics are sample scoring devices for short response and open-ended questions.

Short Response

Points	Description
2	The student demonstrates a thorough understanding of the science of the task. The response may contain minor flaws that do not detract from the demonstration of a thorough understanding.
1	The student has provided a response that is only partially correct.
0	The student has provided a completely incorrect solution or no response at all.

Open Ended

Points	Description
4	The student demonstrates a thorough understanding of the science of the task. The response may contain minor flaws that do not detract from the demonstration of a thorough understanding.
3	The student demonstrates an understanding of the science of the task. The response is essentially correct and demonstrates an essential but less than thorough understanding of the science.
2	The student demonstrates only a partial understanding of the science of the task. Although the student may have used the correct approach to a solution or may have provided a correct solution, the work lacks an essential understanding of the underlying science concepts.
1	The student demonstrates a very limited understanding of the science of the task. The response is incomplete and exhibits many flaws.
0	The student provides a completely incorrect solution or no response at all.

chapter 2 Organizer

Section/Objectives	Standards		Labs/Features
	National	State/Local	
Chapter Opener	See pp. 9T–10T for a Key to Standards.		**Launch Lab:** Stalactites and Stalagmites, p. 35 **Foldables,** p. 35
Section 1 Weathering 1 sessions 0.5 block 1. **Explain** how mechanical weathering and chemical weathering differ. 2. **Describe** how weathering affects Earth's surface. 3. **Explain** how climate affects weathering.	National Content Standards: UCP.1–3, UCP.5, A.1, A.2, B.1, D.1, D.2, F.4		**Science Online,** p. 39 **MiniLAB:** Observing the Formation of Rust, p. 40
Section 2 The Nature of Soil 2 sessions 1 block 4. **Explain** how soil forms. 5. **Describe** soil characteristics. 6. **Describe** factors that affect the development of soil.	National Content Standards: UCP.1–3, UCP.5, A.1, A.2, B.2, D.1, F.1		**Visualizing Soil Formation,** p. 43 **MiniLAB:** Comparing Components of Soil, p. 44 **Integrate Chemistry,** p. 45 **Applying Math:** Soil Texture, p. 46 **Lab:** Soil Texture, p. 49
Section 3 Soil Erosion 3 sessions 1.5 blocks 7. **Explain** why soil is important. 8. **Evaluate** ways that human activity has affected Earth's soil. 9. **Describe** ways to reduce soil erosion.	National Content Standards: UCP.1–3, UCP.5, A.1, A.2, B.2, D.2		**Integrate Career,** p. 51 **Science Online,** p. 51 **Lab:** Weathering Chalk, p. 54 **Science and Language Arts:** Landscape, History, and the Pueblo Imagination, p. 56

Glencoe Exclusive! TeacherWorks™ All-In-One Planner and Resource Center

Lab Materials	Reproducible Resources	Section Assessment	Technology
Launch Lab: water, beakers (2), large piece of cardboard, Epsom salt, yellow food coloring, cotton string, large steel nuts (2)	**Chapter *FAST FILE* Resources** Foldables Worksheet, p. 13 Directed Reading Overview, p. 15 Note-taking Worksheets, pp. 29–31	GLENCOE'S ASSESSMENT ADVANTAGE	TeacherWorks includes: • Interactive Teacher Edition • Lesson Planner with calendar • Access to all program blacklines • Correlations to standards • Web links
MiniLAB: steel wool, glass jar, water *Need materials?* Contact Science Kit at 1-800-828-7777 or www.sciencekit.com on the Internet.	**Chapter *FAST FILE* Resources** Transparency Activity, p. 40 MiniLAB, p. 3 Enrichment, p. 26 Reinforcement, p. 23 Directed Reading, p. 16 Lab Activity, pp. 9–10 **Earth Science Critical Thinking/ Problem Solving,** p. 4 **Physical Science Critical Thinking/ Problem Solving,** p. 16	**Portfolio** Science Journal, p. 37 Curriculum Connection, p. 40 **Performance** MiniLAB, p. 40 Applying Skills, p. 41 **Content** Section Review, p. 41	Section Focus Transparency Guided Reading Audio Program Interactive Chalkboard CD-ROM
MiniLAB: soil sample, magnifying lens **Lab:** soil sample, water bottle	**Chapter *FAST FILE* Resources** Transparency Activity, p. 41 MiniLAB, p. 4 Enrichment, p. 27 Reinforcement, p. 24 Transparency Activity, pp. 43–44 Directed Reading, p. 17 Lab Worksheets, pp. 5–6 Lab Activity, pp. 11–12 **Earth Science Critical Thinking/ Problem Solving,** p. 22 **Performance Assessment in the Science Classroom,** p. 44	**Portfolio** Science Journal, p. 44 **Performance** MiniLAB, p. 44 Applying Math, p. 46 Applying Skills, p. 48 **Content** Section Review, p. 48	Section Focus Transparency Teaching Transparency Virtual Labs CD-ROM Guided Reading Audio Program Interactive Chalkboard CD-ROM
Lab: equal-sized pieces of chalk (6), small beakers (2), metric ruler, water, white vinegar (100 mL), hot plate, 250-mL graduated cylinder, computer probe for temperature or thermometer	**Chapter *FAST FILE* Resources** Transparency Activity, p. 42 Enrichment, p. 28 Reinforcement, p. 25 Directed Reading, pp. 17, 18 Lab Worksheets, pp. 7–8 **Lab Management and Safety,** p. 65	**Portfolio** Assessment, p. 53 **Performance** Applying Skills, p. 53 **Content** Section Review, p. 53	Section Focus Transparency Guided Reading Audio Program Interactive Chalkboard CD-ROM Video Lab

End of Chapter Assessment

GLENCOE'S ASSESSMENT ADVANTAGE

Blackline Masters	Technology	Professional Series
Chapter *FAST FILE* Resources Chapter Review, pp. 33–34 Chapter Tests, pp. 35–38 **Standardized Test Practice,** pp. 11–14	MindJogger Videoquiz Virtual Labs CD-ROM Exam*View*® Pro Testmaker TeacherWorks CD-ROM Interactive Chalkboard CD-ROM	**Performance Assessment in the Science Classroom (PASC)**

chapter 2 Weathering and Soil

Transparencies

Section Focus

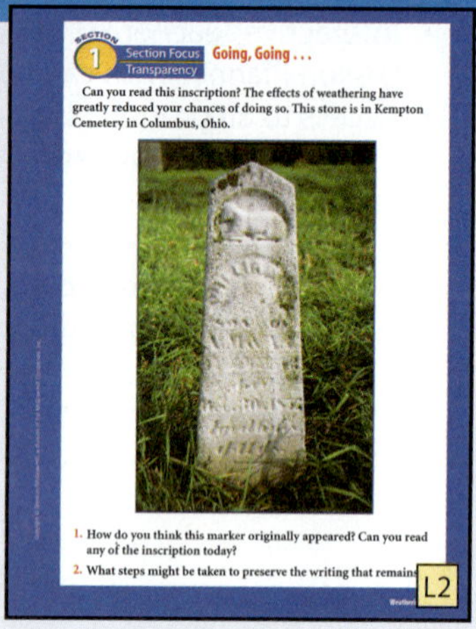
Section 1 Section Focus Transparency **Going, Going . . .**

Can you read this inscription? The effects of weathering have greatly reduced your chances of doing so. This stone is in Kempton Cemetery in Columbus, Ohio.

1. How do you think this marker originally appeared? Can you read any of the inscription today?
2. What steps might be taken to preserve the writing that remains?

L2

Section 2 Section Focus Transparency **More Than Just Dirt**

You probably see the material shown here every day, but you may not even notice it. You may only think about it when it makes your room or clothes dirty, but soil is also a valuable resource.

1. What components of soil can you identify in this picture?
2. How is the worm involved with the soil?
3. Why isn't soil the same everywhere?

L2

Section 3 Section Focus Transparency **Blowing in the Wind**

During the 1930s, catastrophic "black blizzards" were experienced in many states of the southern Great Plains. One storm carried over 300 million tons of dirt all the way to the east coast.

1. What is occurring in this picture?
2. Why is the loss of topsoil harmful?
3. What steps have been taken to prevent such serious dust storms from happening today?

L2

This is a representation of key blackline masters available in the Teacher Classroom Resources. See Resource Manager boxes within the chapter for additional information.

Assessment

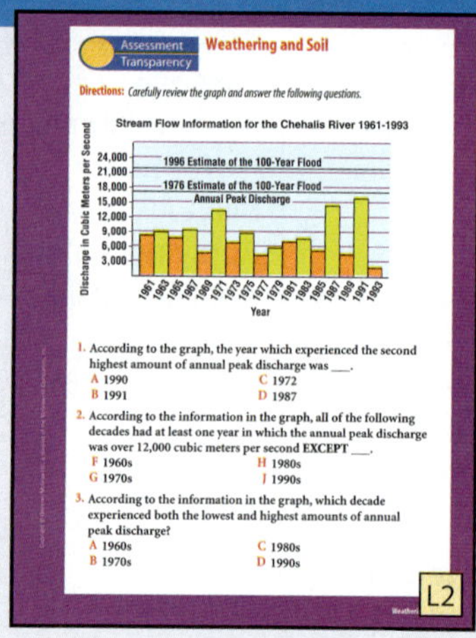
Assessment Transparency **Weathering and Soil**

Directions: Carefully review the graph and answer the following questions.

1. According to the graph, the year which experienced the second highest amount of annual peak discharge was ___.
   A 1990 C 1972
   B 1991 D 1987
2. According to the information in the graph, all of the following decades had at least one year in which the annual peak discharge was over 12,000 cubic meters per second EXCEPT ___.
   F 1960s H 1980s
   G 1970s J 1990s
3. According to the information in the graph, which decade experienced both the lowest and highest amounts of annual peak discharge?
   A 1960s C 1980s
   B 1970s D 1990s

L2

Teaching

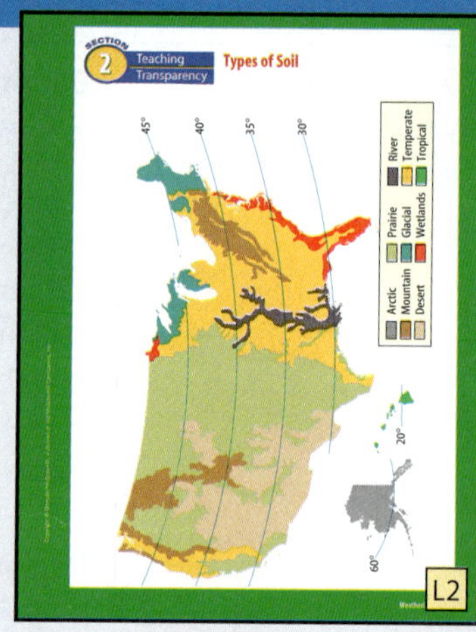
Section 2 Teaching Transparency **Types of Soil**

L2

Key to Teaching Strategies

The following designations will help you decide which activities are appropriate for your students.

- **L1** Level 1 activities should be appropriate for students with learning difficulties.
- **L2** Level 2 activities should be within the ability range of all students.
- **L3** Level 3 activities are designed for above-average students.
- **ELL** ELL activities should be within the ability range of English Language Learners.
- **COOP LEARN** Cooperative Learning activities are designed for small group work.
- **LS** Multiple Learning Styles logos, as described on page 6T, are used throughout to indicate strategies that address different learning styles.
- **P** These strategies represent student products that can be placed into a best-work portfolio.
- **PBL** Problem-Based Learning activities apply real-world situations to learning.

Hands-on Activities

Student Text Lab Worksheet

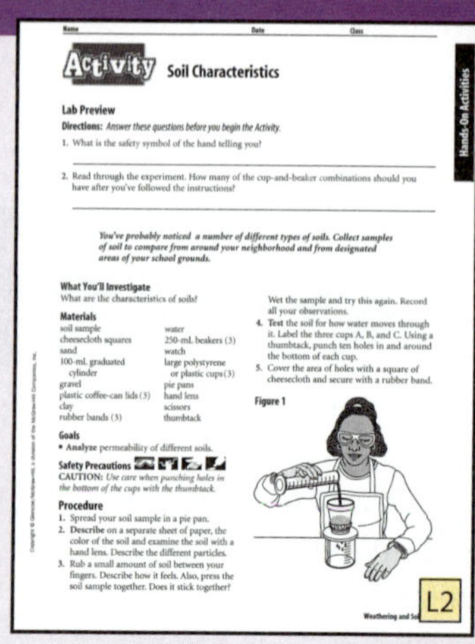
Activity **Soil Characteristics**

Hands-On Activities

Lab Preview

Directions: *Answer these questions before you begin the Activity.*

1. What is the safety symbol of the hand telling you?
2. Read through the experiment. How many of the cup-and-beaker combinations should you have after you've followed the instructions?

L2

Laboratory Activities

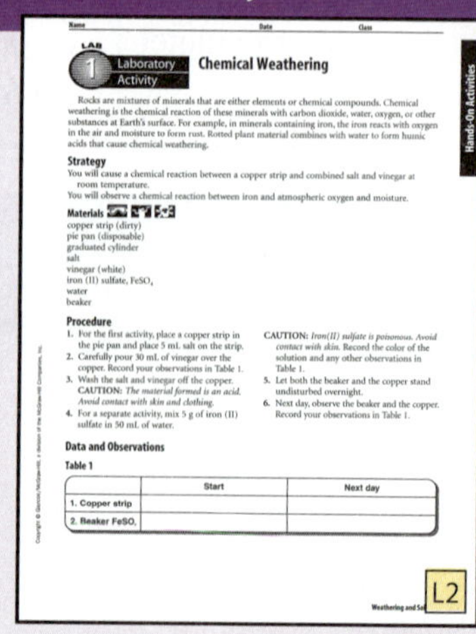
Lab 1 Laboratory Activity **Chemical Weathering**

Hands-On Activities

Strategy

Materials

Procedure

Data and Observations

Table 1

	Start	Next day
1. Copper strip		
2. Beaker $FeSO_4$		

L2

Meeting Different Ability Levels

Content Outline

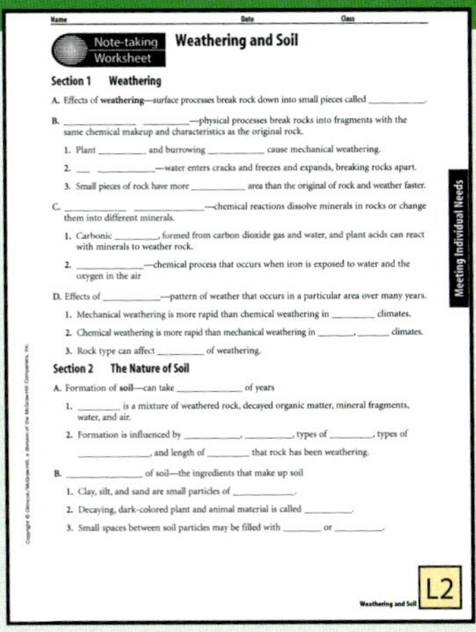

Reinforcement

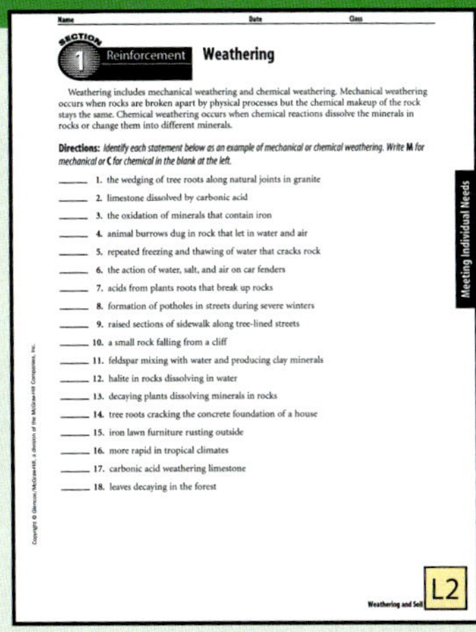

Enrichment

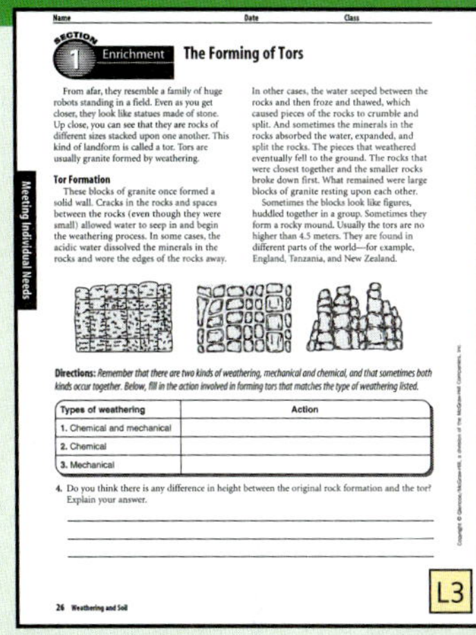

Directed Reading (English/Spanish)

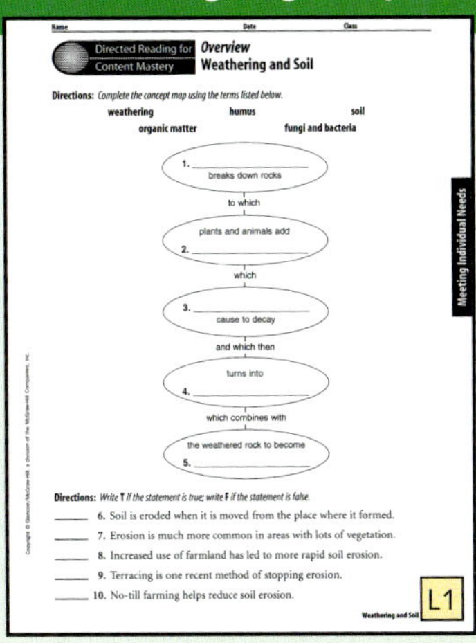

Study Guide

Reading Essentials

Assessment

Test Practice Workbook

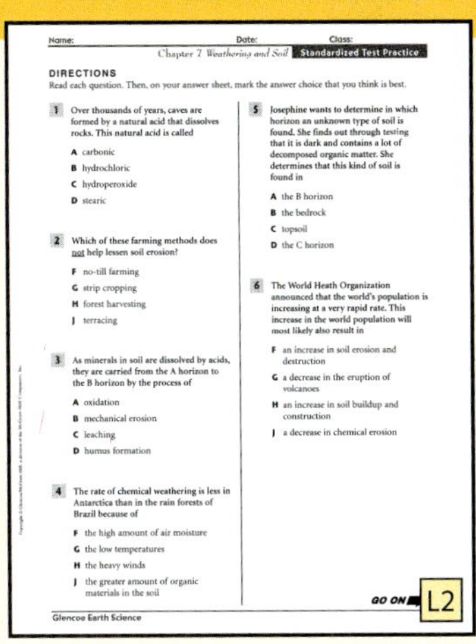

Chapter Review

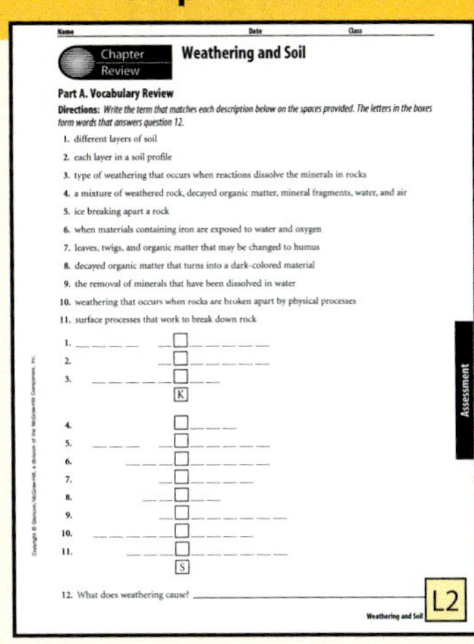

Chapter Tests

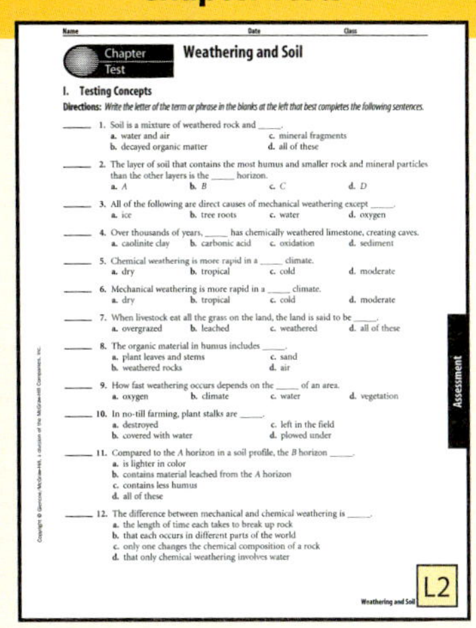

Science Content Background

section 1 Weathering

Evidence of Weathering

Potholes and sidewalk cracks, crumbling rocks, discolored paint, rusted automobiles, and eroded statues all show evidence of different types of weathering.

Chemical Weathering

When oxygen, water, or carbonic acid combine with minerals in rocks, new compounds form. These processes of chemical weathering are called oxidation, hydration, and carbonation, respectively.

Runk/Schoenberger/Grant Heilman Photography, Inc.

Nitrogen oxide and sulfur dioxide, two compounds released into the atmosphere when fossil fuels are burned, also can cause chemical weathering. When mixed with water, these compounds form acid rain, which can dissolve rocks, statues, and buildings.

section 2 The Nature of Soil

Soil Profiles

Leaching and redeposition of minerals can result in unique soil profiles that extend from the topsoil to the other layers in the profile and finally to the bedrock. Some soil horizons can be white because of the removal of soluble elements by leaching. These elements can precipitate as oxides in other horizons and give them a dark color.

section 3 Soil Erosion

Preventing Soil Erosion

Methods of reducing soil loss include planting shelter belts of trees to break the force of the wind and planting cover crops after harvest to hold soil in place. Water erosion can be reduced by terracing steep slopes or removing these slopes from cultivation. On gentle slopes, plowing along the natural contours of the land and planting crops in strips help reduce erosion by water. In strip cropping, a crop that covers the ground is alternated with a crop such as corn that exposes a considerable amount of soil. In recent years, many farmers have begun no-till farming, in which new plants are grown among the stalks of last year's crop.

chapter content resources

Internet Resources

For additional content background, visit bookg.msscience.com to:

- access your book online
- find references to related articles in popular science magazines
- access Web links with related content background
- access current events with science journal topics

Print Resources

The Dynamic Earth: An Introduction to Physical Geology by Brian J. Skinner and Stephen C. Porter, John Wiley & Sons, 2003

Process Geomorphology by Dale F. Ritter, R. Craig Kochel, and Jerry R. Miller, McGraw-Hill Higher Education, 2002

Fundamentals of Soil Science by Henry D. Foth, John Wiley & Sons, 1990

IDENTIFYING Misconceptions

Find Out What Students Think

Students may think that . . .

Climate does not affect weathering.

Students often think that mechanical and chemical weathering occur at the same rate regardless of climate. Students develop this misconception because they incorrectly infer that what can be experienced and observed locally applies globally.

Discussion

Organize the class into small groups. Have students in each group use a globe to identify the locations of a variety of different climates on Earth. Then have them discuss which agents might have the greatest effect on weathering in each different area. Some examples are listed below.

- Tundra (arctic areas and ice caps): mechanical weathering caused by freezing and thawing
- Tropical rain forests: mostly chemical weathering caused by high amounts of rainfall and high year-round temperatures

As an extension, have students identify which areas would have relatively slow weathering rates (desert and polar regions) and which would have relatively fast weathering rates (tropical regions). Have students make a Venn diagram comparing and contrasting the climates in these different areas.

Promote Understanding

Activity

Have small groups of students model leaching in a dry and a wet environment. Remind students that leaching is the removal of minerals that have been dissolved in water. Point out that the ions are freed by weathering and leaching moves them in a soil profile. Have students place two tea bags on folded paper towels. Instruct them to put 2 drops of water on one tea bag to model a dry environment, and to place 20 drops of water on the other tea bag to simulate a wet environment. After 2 minutes, have students remove the tea bags and compare the appearance of the paper towels. *They should see that the one beneath the tea bag with more water added had a bigger and darker spot.* Ask whether leaching would be more likely to occur in a desert or a tropical region. *In a tropical region because it receives a lot of rain which will free more ions as a result of weathering.*

Assess

After completing the chapter, see *Identifying Misconceptions* in the Study Guide at the end of the chapter.

chapter 2

Chapter Vocabulary

weathering, p. 36
mechanical weathering, p. 37
ice wedging, p. 38
chemical weathering, p. 39
oxidation, p. 40
climate, p. 40
soil, p. 42
humus, p. 44
horizon, p. 44
soil profile, p. 44
litter, p. 45
leaching, p. 45
no-till farming, p. 52
contour farming, p. 53
terracing, p. 53

Science Journal Possible answer:

Standing above the English moor;
Remnant of a landscape past.

The granite boulders of a tor;
Survivors of nature's wrath.

Subject of myth and lore;
All that remains is weathered core.

INTERACTIVE CHALKBOARD with Image Bank

PowerPoint® Presentations

This CD-ROM is an editable Microsoft® PowerPoint® presentation that includes:

- a pre-made presentation for every chapter
- interactive graphics
- animations
- audio clips
- image bank
- all new section and chapter questions
- Standardized Test Practice
- transparencies
- pre-lab questions for all labs
- Foldables directions
- links to bookg.msscience.com

chapter 2

Weathering and Soil

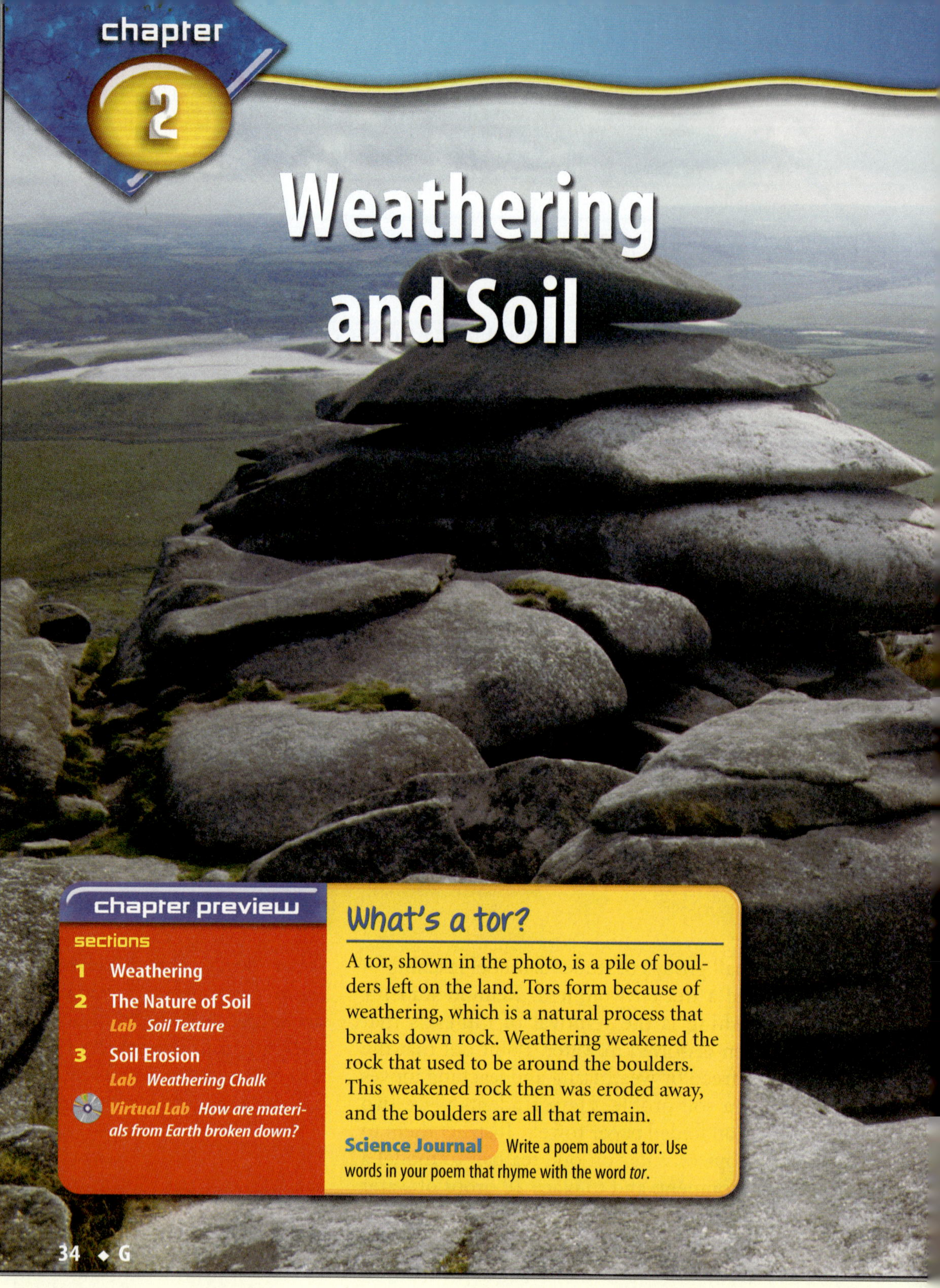

chapter preview

sections

What's a tor?

A tor, shown in the photo, is a pile of boulders left on the land. Tors form because of weathering, which is a natural process that breaks down rock. Weathering weakened the rock that used to be around the boulders. This weakened rock then was eroded away, and the boulders are all that remain.

Science Journal Write a poem about a tor. Use words in your poem that rhyme with the word *tor*.

34 ◆ G

Theme Connection

Stability and Change Students will learn that the rock and landscapes that often appear constant through the course of a human life are changing. Through geologic time, slow change has an enormous effect.

About the Photo

Chemical Weathering The tors shown in the photo occur at the summit of a large mass of fractured granite in Cornwall, England. Chemical weathering disintegrated rock along the fractures to produce the stacks of boulders.

Start-Up Activities

Launch LAB

Stalactites and Stalagmites

During weathering, minerals can be dissolved by acidic water. If this water seeps into a cave, minerals might precipitate. In this lab, you will model the formation of stalactites and stalagmites.

1. Pour 700 mL of water into two 1,000-mL beakers and place the beakers on a large piece of cardboard. Stir Epsom salt into each beaker until no more will dissolve.
2. Add two drops of yellow food coloring to each beaker and stir.
3. Measure and cut three 75-cm lengths of cotton string. Hold the three pieces of string in one hand and twist the ends of all three pieces to form a loose braid of string.
4. Tie each end of the braid to a large steel nut.
5. Soak the braid of string in one of the beakers until it is wet with the solution. Drop one nut into one beaker and the other nut into the second beaker. Allow the string to sag between the beakers. Observe for several days.
6. **Think Critically** Record your observations in your Science Journal. How does this activity model the formation of stalactites and stalagmites?

FOLDABLES™ Study Organizer

Weathering and Soil Make the following Foldable to help you understand the vocabulary terms in this chapter.

STEP 1 **Fold** a vertical sheet of notebook paper from side to side.

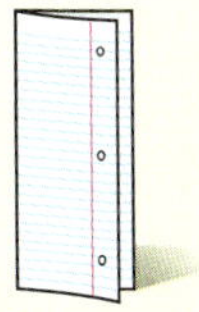

STEP 2 **Cut** along every third line of only the top layer to form tabs.

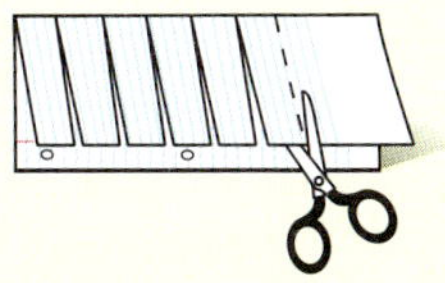

STEP 3 **Label** each tab.

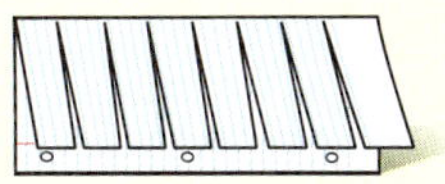

Build Vocabulary As you read the chapter, list the vocabulary words about weathering and soil on the tabs. As you learn the definitions, write them under the tab for each vocabulary word.

Science Online Preview this chapter's content and activities at bookg.msscience.com

Launch LAB

Purpose Students model the formation of stalactites and stalagmites. Students learn that minerals are deposited as water evaporates.

Preparation To reduce set-up time, you might want to prepare epsom-salt solution and cut lengths of string before class.

Materials three 75-cm lengths of cotton string, two large steel nuts, yellow food coloring, epsom salt (available at most drug stores), 1000-mL beakers (2), stirring rod, large piece of sturdy cardboard

Teaching Strategy Lead a discussion about stalactites, stalagmites, and caves before students conduct the activity.

Safety Precautions Students should wear safety goggles and aprons.

Think Critically

Students will observe "stalactites" forming on the string and "stalagmites" forming on the cardboard beneath the sag in the string. Capillary action draws solution through the string. As water evaporates, epsom salt precipitates. This is similar to how stalactites and stalagmites form in a cave.

Assessment

Process Give students photographs showing stalactites, stalagmites, columns, and other cave features. Ask them to speculate about how these features formed. Use **Performance Assessment in the Science Classroom,** p. 89.

FOLDABLES™ Study Organizer — Dinah Zike Study Fold

Student preparation materials for this Foldable are available in the Chapter *Fast File* Resources.

1 Motivate

INTERACTIVE CHALKBOARD PowerPoint® Presentations

Bellringer

Section Focus Transparencies also are available on the Interactive Chalkboard CD-ROM. L2 ELL

Section 1 Section Focus Transparency **Going, Going . . .**

Can you read this inscription? The effects of weathering have greatly reduced your chances of doing so. This stone is in Kempton Cemetery in Columbus, Ohio.

1. How do you think this marker originally appeared? Can you read any of the inscription today?
2. What steps might be taken to preserve the writing that rema

L2

Weathering and Soil

Tie to Prior Knowledge

Weathering Explain that in this section, students will learn about weathering—the breaking down of rock in nature. Where have you seen pieces of rock that could have been broken down by weathering? Possible answers: boulders, large pebbles, or sand on beaches; pebbles in streams; sediment in soil

section 1

Weathering

as you read

What You'll Learn

- **Explain** how mechanical weathering and chemical weathering differ.
- **Describe** how weathering affects Earth's surface.
- **Explain** how climate affects weathering.

Why It's Important

Through time, weathering turns mountains into sediment.

Review Vocabulary
surface area: the area of a rock or other object that is exposed to the surroundings

New Vocabulary
- weathering
- mechanical weathering
- ice wedging
- chemical weathering
- oxidation
- climate

Weathering and Its Effects

Can you believe that tiny moss plants, a burrowing vole shrew, and even oxygen in the air can affect solid rock? These things and many more weaken and break apart rock at Earth's surface. Together, surface processes that work to break down rock are called **weathering.**

Weathering breaks rock into smaller and smaller pieces, such as sand, silt, and clay. These smaller, loose pieces are called sediment. The terms *sand, silt,* and *clay* are used to describe specific sizes of sediment. Sand grains are larger than silt, and silt is larger than clay.

Over millions of years, weathering has changed Earth's surface. The process continues today. Weathering wears mountains down to hills, as shown in **Figure 1.** Rocks at the top of mountains are broken down by weathering, and the sediment is moved downhill by gravity, water, and ice. Weathering also produces strange rock formations like those shown at the beginning of this chapter. Two different types of weathering—mechanical weathering and chemical weathering—work together to shape Earth's surface.

Figure 1 Over long periods of time, weathering wears mountains down to rolling hills.
Explain *how this occurs.*

Section 1 Resource Manager

Chapter *FAST FILE* Resources
- Transparency Activity, p. 40
- Directed Reading for Content Mastery, pp. 15, 16
- Enrichment, p. 26
- Note-taking Worksheets, pp. 29–31
- MiniLAB, p. 3
- Reinforcement, p. 23

Physical Science Critical Thinking/Problem Solving, p. 16

Earth Science Critical Thinking/Problem Solving, p. 4

Figure 2 Growing tree roots can be agents of mechanical weathering.

Tree roots can crack a sidewalk.

Tree roots also can grow into cracks and break rock apart.

Mechanical Weathering

Mechanical weathering occurs when rocks are broken apart by physical processes. This means that the overall chemical makeup of the rock stays the same. Each fragment has characteristics similar to the original rock. Growing plants, burrowing animals, and expanding ice are some of the things that can mechanically weather rock. These physical processes produce enough force to break rocks into smaller pieces.

Reading Check *What can cause mechanical weathering?*

Figure 3 Small animals mechanically weather rock when they burrow by breaking apart sediment.

Plants and Animals Water and nutrients that collect in the cracks of rocks result in conditions in which plants can grow. As the roots grow, they enlarge the cracks. You've seen this kind of mechanical weathering if you've ever tripped on a crack in a sidewalk near a tree, as shown in **Figure 2.** Sometimes tree roots wedge rock apart, also shown in **Figure 2.**

Burrowing animals also cause mechanical weathering, as shown in **Figure 3.** As these animals burrow, they loosen sediment and push it to the surface. Once the sediment is brought to the surface, other weathering processes act on it.

2 Teach

Caption Answer

Figure 1 Rocks at the tops of mountains are broken down by weathering, and the sediment is moved downhill by gravity, water, and ice.

Discussion

Earth's Surface How would Earth's surface be different if weathering did not exist? Features such as mountains would be higher, and there would be a lot of bare rock in places where there is now soil. Accept reasonable answers.

Science Journal

Weathering and Mountains Have students draw a series of pictures that show how weathering changes a mountain's shape over time. Drawings should show a steep, rugged mountain slowly changing into a more rounded, less steep hill.

L2 ELL LS **Visual-Spatial** P

Visual Learning

Figure 2 How might this type of damage to concrete be avoided? Possible answers: Plant trees farther away from the sidewalk; remove large roots; build the sidewalk farther away from mature trees.

Reading Check

Answer growing plants, burrowing animals, expanding ice

Teacher FYI

Mechanical Weathering Additional types of mechanical weathering include stress release, insolation weathering, alternate wetting and drying, and salt weathering. Stress release occurs when overburden rock is eroded and the underlying rock expands and fractures. Insolation weathering causes rock to split after many cycles of heating and cooling. This type of weathering is common in deserts. Shales and mudrocks can fall apart as a result of repeated cycles of wetting and drying. Salt weathering occurs when salt crystals grow in rock pores and cause fragments of rock to break off.

Visual Learning

Figure 4 How might this cliff look several thousand years from now? Possible answer: Because of weathering, the face of the cliff would be farther back.

Discussion

Ice Wedging Why is ice wedging more likely to occur at higher elevations than at lower elevations? Temperatures are more extreme at higher elevations. Nighttime temperatures often are below freezing, whereas daytime temperatures are above freezing. This freezing and thawing cycle is most likely to cause ice wedging.

Fun Fact

According to the Federal Highway Administration, \$20–\$25 billion is spent each year on pothole repair for streets and highways, some of which is caused by ice wedging. This adds up to about \$14,000 per mile.

Use Science Words

Word Origin Have students research ways the Latin word *speleum* is used in cave terminology. *Speleum* means cave. People who explore caves for a hobby are called spelunkers. The scientific study of caves is called speleology; scientists who study caves are speleologists. Speleothems are cave formations. L2

Figure 4 When water enters cracks in rock and freezes, it expands, causing the cracks to enlarge and the rock to break apart.

Figure 5 As rock is broken apart by mechanical weathering, the amount of rock surface exposed to air and water increases. The background squares show the total number of surfaces exposed.

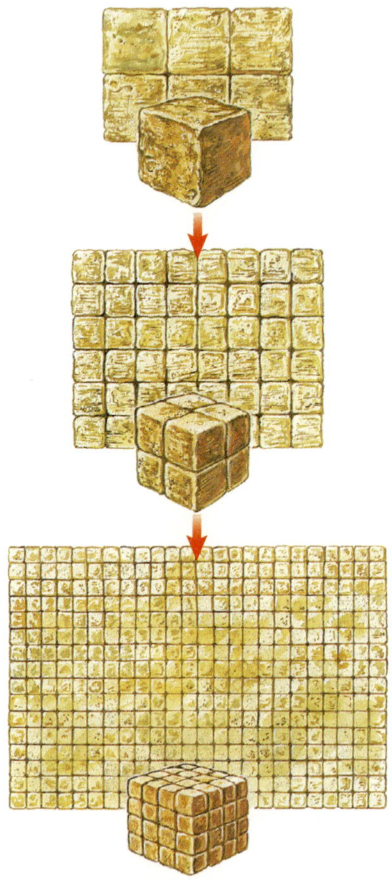

Ice Wedging A mechanical weathering process called ice wedging is shown in **Figure 4.** **Ice wedging** occurs in temperate and cold climates where water enters cracks in rocks and freezes. Because water expands when it turns to ice, pressure builds up in the cracks. This pressure can extend the cracks and break apart rock. The ice then melts, allowing more water to enter the crack, where it freezes and breaks the rock even more. Ice wedging is most noticeable in the mountains, where warm days and cold nights are common. It is one process that wears down mountain peaks. This cycle of freezing and thawing not only breaks up rocks, but also can break up roads and highways. When water enters cracks in road pavement and freezes, it forces the pavement apart. This causes potholes to form in roads.

Surface Area Mechanical weathering by plants, animals, and ice wedging reduces rocks to smaller pieces. These small pieces have more surface area than the original rock body, as shown in **Figure 5.** As the amount of surface area increases, more rock is exposed to water and oxygen. This speeds up a different type of weathering called chemical weathering.

Differentiated Instruction

Learning Disabled Have students collect and display photographs that exemplify weathering. Pictures might include cracks in rocks, animal burrows, or a thick soil profile. L1 ELL LS **Visual-Spatial**

Curriculum Connection

Art Direct students to draw a diagram, with labels, that illustrates one type of mechanical weathering. Students can choose from among growing plants, burrowing animals, or ice wedging. L2 ELL LS **Visual-Spatial**

38 ◆ G CHAPTER 2 Weathering and Soil

Chemical Weathering

INTEGRATE Chemistry The second type of weathering, **chemical weathering**, occurs when chemical reactions dissolve the minerals in rocks or change them into different minerals. This type of weathering changes the chemical composition of the rock, which can weaken the rock. Next, you will see how chemical weathering happens.

Natural Acids Naturally formed acids can weather rocks chemically. When water reacts with carbon dioxide gas in the air or soil, a weak acid, called carbonic acid, forms. This is the same weak acid that is found in many soft drinks. Carbonic acid reacts with minerals such as calcite, which is the main mineral that makes up limestone. This reaction causes the calcite to dissolve. Over many thousands of years, carbonic acid has weathered so much limestone that caves, such as the one shown in **Figure 6,** have formed.

Chemical weathering also occurs when naturally formed acids come in contact with other rocks. Over a long period of time, the mineral feldspar, which is found in granite, some types of sandstone, and other rocks, is broken down into a clay mineral called kaolinite (KAY oh luh nite). Kaolinite clay is common in some soils. Clay is an end product of weathering.

Reading Check *How does kaolinite clay form?*

Science Online

Topic: Chemical Weathering

Visit bookg.msscience.com for Web links to information about chemical weathering.

Activity List different types of chemical weathering. Next to each type, write an effect that you have observed.

Figure 6 Caves form when slightly acidic groundwater dissolves limestone.
Explain *why the groundwater is acidic.*

Use an Analogy

Dissolving Candy Give two student volunteers each a piece of hard candy. Have one student break the candy into pieces before allowing it to dissolve in the mouth. Have the other student let the candy dissolve without breaking it. Compare how long it takes for the candy to dissolve. Candy broken into pieces dissolves faster. The dissolving process is like chemical weathering, which occurs faster when more rock surface area is exposed. L1

Teacher FYI

Caverns Carlsbad Caverns in New Mexico is one of the world's largest caverns. Part of the caverns is a chamber called The Big Room. This room is as high as a 25-story building.

Identifying Misconceptions

Weathering Rates Some students may believe that mechanical and chemical weathering occur at the same rate regardless of climate. Refer to page F at the beginning of this chapter for teaching strategies that address this misconception.

Reading Check

Answer It forms when natural acids react with feldspar in granite and other rocks.

Caption Answer

Figure 6 As water seeps through soil, it reacts with carbon dioxide in soil air to form carbonic acid.

Differentiated Instruction

English-Language Learners Provide these students with a drawing or photo of a cave's interior. Have them add labels to identify and explain important features.

Challenge Have students research how stalactites form in caves. Students then should make sketches illustrating the steps in the process. Stalactites begin as narrow, hollow tubes called soda straws. Through time, the end of the straw plugs and mineral growth occurs around it. L3

Reading Check

Answer Oxygen reacts with some minerals to form new minerals.

Mini LAB

Purpose Students compare rust formation to chemical weathering. L2 ELL IN **Visual-Spatial**

Materials uncoated steel wool, shallow glass jar, water

Teaching Strategy Have students observe the steel wool each day.

Safety Precautions Use care when handling steel wool. Wash hands after handling it.

Analysis

1. Rust formed.
2. The steel wool oxidized in the presence of water.
3. Chemical changes break rock down, just as chemical changes in steel wool broke it down.

Assessment

Performance Have students design experiments to see how length of exposure to a weak acid affects the chemical weathering of various metals such as an iron nail or a copper penny. Use **Performance Assessment in the Science Classroom**, p. 95.

Try at Home

Caption Answer

Figure 7 In both, iron oxidizes.

Discussion

Desert Weathering Which type of weathering would be more effective in a desert? Explain. Mechanical; because it is dry.

Mini LAB

Observing the Formation of Rust

Procedure

1. Place some **steel wool** in a **glass jar** with 1 cm of **water.**
2. Observe for several days.

Analysis

1. What changes occurred?
2. What caused the changes?
3. How are these changes related to weathering?

Try at Home

Plant Acids Some roots and decaying plants give off acids that also dissolve minerals in rock. When these minerals dissolve, the rock is weakened. Eventually, the rock will break into smaller pieces. As the rock weathers, nutrients become available to plants.

Oxygen Oxygen causes chemical weathering. You've seen rusty swing sets, nails, and cars. Rust is caused by oxidation. **Oxidation** (ahk sih DAY shun) occurs when some materials are exposed to oxygen and water. For example, when minerals containing iron are exposed to water and the oxygen in air, the iron in the mineral reacts to form a new material that resembles rust. One common example of this type of weathering is the alteration of the iron-bearing mineral magnetite to a rustlike material called limonite, as shown in **Figure 7.** Oxidation of minerals gives some rock layers a red color.

Reading Check *How does oxygen cause weathering?*

Effects of Climate

Climate affects the rate of weathering. **Climate** is the pattern of weather that occurs in a particular area over many years. In cold climates, where freezing and thawing are frequent, mechanical weathering rapidly breaks down rock through the process of ice wedging. Chemical weathering is more rapid in warm, wet climates. Thus, chemical weathering occurs quickly in tropical areas such as the Amazon River region of South America. Lack of moisture in deserts and low temperatures in polar regions slow down chemical weathering. Which type of weathering do you think is more rapid where you live?

Figure 7 Iron-containing minerals like the magnetite shown here can weather to form a rustlike material called limonite.
Explain *how this is similar to rust forming on your bicycle chain.*

Curriculum Connection

History The Statue of Liberty has to be repaired periodically because of damage caused by weathering. Have students research the material of which the statue is made and the type of damage that weathering has caused to it over the years. Have them write reports on what has been done to repair and protect the statue from weathering. L2 P

40 ◆ G CHAPTER 2 Weathering and Soil

Marble statue

Granite statue

Figure 8 Different types of rock weather at different rates. In humid climates, marble statues weather rapidly and become discolored. Granite statues weather more slowly.

Effects of Rock Type Rock type also can affect the rate of weathering in a particular climate. In wet climates, for example, marble weathers more rapidly than granite, as shown in **Figure 8.**

Now you can understand how weathering affects rocks, caves, mountains, and even buildings and streets. Weathering is an important part of the rock cycle. When weathering breaks down rocks, it produces sediment that can form sedimentary rocks. Weathering also begins the process of forming soil from rock and sediment. This is discussed in the next section.

section 1 review

Summary

Weathering and Its Effects

- Weathering includes processes that break down rock.
- Weathering affects Earth's landforms.

Mechanical Weathering

- During mechanical weathering, rock is broken apart, but it is not changed chemically.
- Plant roots, burrowing animals, and expanding ice all weather rock.

Chemical Weathering

- During chemical weathering, minerals in rock dissolve or change to other minerals.
- Agents of chemical weathering include natural acids and oxygen.

Self Check

1. **Describe** how weathering reduces the height of mountains through millions of years.
2. **Explain** how both tree roots and prairie dogs mechanically weather rock.
3. **Summarize** the effects of carbonic acid on limestone.
4. **Describe** how climate affects weathering.
5. **Think Critically** Why does limestone often form cliffs in dry climates but rarely form cliffs in wet climates?

Applying Skills

6. **Venn Diagram** Make a Venn diagram to compare and contrast mechanical weathering and chemical weathering. Include the causes of mechanical and chemical weathering in your diagram.

Science Online bookg.msscience.com/self_check_quiz

Quick Demo

Effect of Acid

Materials piece of granite, piece of marble, small dropper bottle containing dilute (5%) HCl

Estimated Time five minutes

Procedure Place a drop of dilute HCl on the piece of marble and allow students to observe the acid fizzes. Do the same for the piece of granite the acid doesn't fizz. Ask students which rock is more vulnerable to attack by acid. the marble Why does marble weather more rapidly than granite in humid climates? Carbonic acid in rain weathers marble more rapidly than granite.

3 Assess

DAILY INTERVENTION

Check for Understanding

Interpersonal Have student groups list as many effects of weathering as possible. After the groups have finished, compile their responses to make a master list. L2

Reteach

Bulletin Board Have students make a bulletin board of local examples of weathering. Allow students to borrow an inexpensive camera (or use their own) to photograph chemical and mechanical weathering near their homes or in the schoolyard. After the pictures are posted, discuss what each shows. L1 ELL LS **Visual-Spatial**

Assessment

Process Have students draw an events chain concept map that describes the process of mechanical weathering by ice or chemical weathering through oxidation. Use **Performance Assessment in the Science Classroom,** p. 163. L2

section 1 review

1. Weathering breaks rock into pieces and dissolves rock. This material is moved downhill by water, gravity, or ice.
2. Tree roots grow into cracks in rock and enlarge them. Prairie dogs burrow into layers of sediment.
3. Carbonic acid reacts with the calcite in limestone, which causes the limestone to dissolve.
4. Chemical weathering is rapid in warm, moist climates. Mechanical weathering is more important in dry climates and in cold climates.
5. Chemical weathering has a greater effect on limestone in wet climates. This often produces more subdued landscapes.
6. Check to make certain that students understand how to use a Venn diagram. Their work should be accurate and logical.

1 Motivate

Bellringer

INTERACTIVE CHALKBOARD PowerPoint® Presentations

Section Focus Transparencies also are available on the Interactive Chalkboard CD-ROM. L2 ELL

SECTION 2 Section Focus Transparency **More Than Just Dirt**

You probably see the material shown here every day, but you may not even notice it. You may only think about it when it makes your room or clothes dirty, but soil is also a valuable resource.

1. What components of soil can you identify in this picture?
2. How is the worm involved with the soil?
3. Why isn't soil the same everywhere?

L2

Weathering and Soil

Tie to Prior Knowledge

Soil Ask students to describe different types of soil they have seen. Show photos or bring in different types of soil and have them describe the soils by color or particle size. Possible answers: coarse or fine soil, red soil, brown soil, black soil. L2

Caption Answer

Figure 9 Longer periods of time allow more weathering and the formation of smaller soil particles. Longer periods also allow more humus to form.

section 2

The Nature of Soil

as you read

***What* You'll Learn**

- **Explain** how soil forms.
- **Describe** soil characteristics.
- **Describe** factors that affect the development of soil.

***Why* It's Important**

Much of the food that you eat is grown in soil.

Review Vocabulary

profile: a vertical slice through rock, sediment, or soil

New Vocabulary

- soil
- humus
- horizon
- soil profile
- litter
- leaching

Formation of Soil

How often have you been told "Take off those dirty shoes before you come into this house"? Ever since you were a child, you've had experience with soil. Soil is found in many places—backyards, empty city lots, farm fields, gardens, and forests.

What is soil and where does it come from? A layer of rock and mineral fragments produced by weathering covers the surface of Earth. As you learned in Section 1, weathering gradually breaks rocks into smaller and smaller fragments. However, these fragments do not become high-quality soil until plants and animals live in them. Plants and animals add organic matter, the remains of once-living organisms, to the rock fragments. Organic matter can include leaves, twigs, roots, and dead worms and insects. **Soil** is a mixture of weathered rock, decayed organic matter, mineral fragments, water, and air.

Soil can take thousands of years to form and ranges from 60 m thick in some areas to just a few centimeters thick in others. Climate, slope, types of rock, types of vegetation, and length of time that rock has been weathering all affect the formation of soil, as shown in **Figure 9.** For example, different kinds of soils develop in tropical regions than in polar regions. Soils that develop on steep slopes are different from soils that develop on flat land. **Figure 10** illustrates how soil develops from rock.

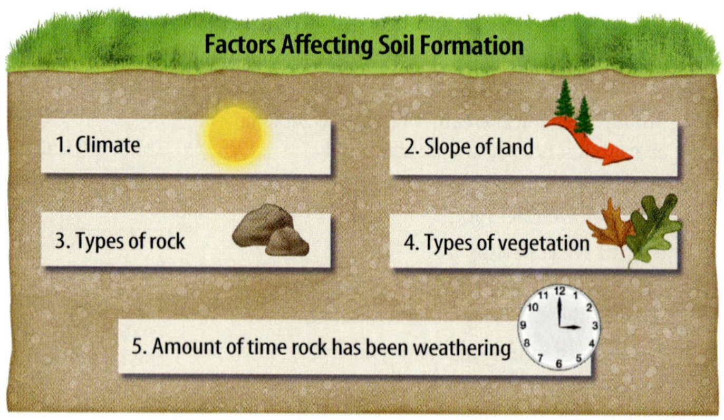

Figure 9 Five different factors affect soil formation.
Explain *how time influences the development of soils.*

Section 2 Resource Manager

Chapter *FAST FILE* Resources
- Transparency Activity, pp. 41, 43–44
- Directed Reading for Content Mastery, p. 17
- MiniLAB, p. 4
- Enrichment, p. 27
- Reinforcement, p. 24
- Lab Worksheets, pp. 5–6
- Lab Activity, pp. 11–12

Performance Assessment in the Science Classroom, p. 44

Earth Science Critical Thinking/Problem Solving, p. 22

Mathematics Skill Activities, p. 1

42 ◆ G CHAPTER 2 Weathering and Soil

NATIONAL GEOGRAPHIC **VISUALIZING SOIL FORMATION**

Figure 10

It may take thousands of years to form, but soil is constantly evolving from solid rock, as this series of illustrations shows. Soil is a mixture of weathered rock, mineral fragments, and organic material—the remains of dead plants and animals—along with water and air.

Ⓐ Natural acids in rainwater weather the surface of exposed bedrock. Water can also freeze in cracks, causing rocks to fracture and break apart. The inset photo shows weathered rock in the Tien Shan Mountains of Central Asia.

Ⓑ Plants take root in the cracks and among bits of weathered rock—shown in the inset photo above. As they grow, plants, along with other natural forces, continue the process of breaking down rocks, and a thin layer of soil begins to form.

Ⓒ Like the grub in the inset photo, insects, worms, and other living things take up residence among plant roots. Their wastes, along with dead plant material, add organic matter to the soil.

Ⓓ As organic matter increases and underlying bedrock continues to break down, the soil layer thickens. Rich topsoil supports trees and other plants with large root systems.

NATIONAL GEOGRAPHIC

Visualizing Soil Formation

Have students examine the pictures and read the captions. Then ask the following question.

What role does water play in each stage of soil formation? It begins the process by weathering the bedrock, then nourishes the plants and animals that continue to break up the rock. Finally, it supports the trees and other organisms that contribute to the thick organic layer of mature soil.

Activity

Circle Graphs Have students make four circle graphs showing the relative abundances of soil components—weathered rock/mineral grains, organic matter, water, and air—at each of the pictured stages of soil formation. Call on students to display their charts and discuss their reasoning.

Fun Fact

A deeply weathered tropical soil might take more than 50,000 years to develop.

Differentiated Instruction

Challenge Have students find out the composition of soils common in their state. Which regions have mature soil, and which regions have less mature soil? Have them speculate on the reasons for this and report their findings to the class. L3

2 Teach

Mini LAB

Purpose Students compare components of soil. L2 ELL LS Interpersonal

Materials soil samples (100 g), magnifying glass

Teaching Strategy To ensure variety, have students bring soil samples from home.

Safety Precautions Have students wash their hands after handling soil.

Analysis

1. Students might find small pieces of rock as well as organic matter. Remains of living organisms could also be present.
2. All samples should contain particles of weathered rock. Soils differ in particle size, color, and the amount and composition of organic matter.

Assessment

Performance Have students separate components of soil: rock fragments, humus, and remains of living organisms. Use **Performance Assessment in the Science Classroom,** p. 97.

Reading Check

Answer It provides nutrients for plants.

Caption Answer

Figure 11 It is darker because of large amounts of humus, and it contains fewer mineral and rock fragments.

Mini LAB

Comparing Components of Soil

Procedure

1. Collect a sample of soil.
2. Observe it closely with a magnifying lens.

Analysis

1. Describe the different particles found in your sample. Did you find any remains of once-living organisms?
2. Compare and contrast your sample with those other students have collected.

Composition of Soil

As you have seen already, soil is made up of rock and mineral fragments, organic matter, air, and water. The rock and mineral fragments found in soils come from rocks that have been weathered. Most of these fragments are small particles of sediment such as clay, silt, and sand. However, some larger pieces of rock also might be present.

Most organic matter in soil comes from plants. Plant leaves, stems, and roots all contribute organic matter to soil. Animals and microorganisms provide additional organic matter when they die. After plant and animal material gets into soil, fungi and bacteria cause it to decay. The decayed organic matter turns into a dark-colored material called **humus** (HYEW mus). Humus serves as a source of nutrients for plants. As worms, insects, and rodents burrow throughout soil, they mix the humus with the fragments of rock. Good-quality surface soil has approximately equal amounts of humus and weathered rock material.

Reading Check *What does humus do for soil?*

Soil has many small spaces between individual soil particles that are filled with water or air. When soil is moist, the spaces hold the water that plants need to grow. During a drought, the spaces are almost entirely filled with air.

Soil Profile

You have seen layers of soil if you've ever dug a deep hole or driven along a road that has been cut into a hillside. You probably observed that most plant roots grow in the top layer of soil. The top layer typically is darker than the soil layers below it. These different layers of soil are called **horizons.** All the horizons of a soil form a **soil profile.** Most soils have three horizons—labeled A, B, and C, as shown in **Figure 11.**

Figure 11 This soil, which developed beneath a grassy prairie, has three main horizons. **Describe** *how the A horizon is different from the other two horizons.*

Teacher FYI

Soil Quality In tropical areas, nutrients are held largely in vegetation. When forests are cut, heavy rain quickly washes away the thin layer of topsoil. In just a few years, the soil is often too poor to farm. To survive, people move, cut more forest, and continue the destructive cycle.

Science Journal

Prairie Soils In 1935, Laura Ingalls Wilder wrote *Little House on the Prairie.* In her book, she described life on the prairie in the late 1800s, when families were escaping from the crowded cities of the East by moving west. Ask students to research prairie topography and soil and write a paragraph on why settlers found it desirable. Fertile soils and flat lands were easy to farm. P L2

A Horizon The A horizon is the top layer of soil. In a forest, the A horizon might be covered with litter. **Litter** consists of leaves, twigs, and other organic material that eventually can be changed to humus by decomposing organisms. Litter helps prevent erosion and evaporation of water from the soil. The A horizon also is known as topsoil. Topsoil has more humus and fewer rock and mineral particles than the other layers in a soil profile. The A horizon generally is dark and fertile. The dark color of the soil is caused by the humus, which provides nutrients for plant growth.

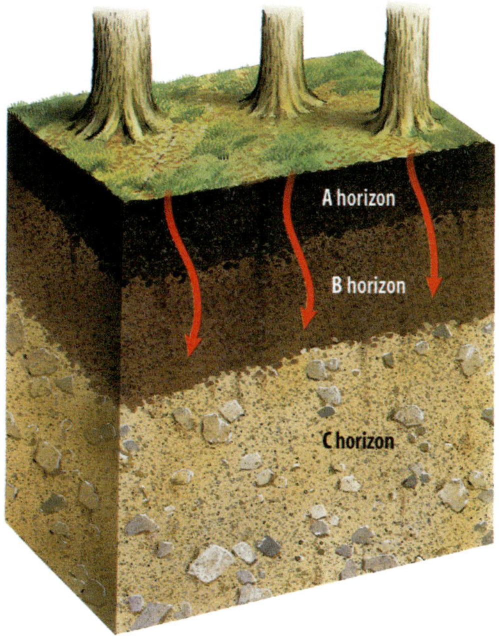

Figure 12 Leaching removes material from the upper layer of soil. Much of this material then is deposited in the B horizon.

B Horizon The layer below the A horizon is the B horizon. Because less organic matter is added to this horizon, it is lighter in color than the A horizon and contains less humus. As a result, the B horizon is less fertile. The B horizon contains material moved down from the A horizon by the process of leaching.

Leaching is the removal of minerals that have been dissolved in water. The process of leaching resembles making coffee in a drip coffeemaker. In a coffeemaker, water drips through ground coffee. In soil, water seeps through the A horizon. In a coffeemaker, water absorbs the flavor and color from the coffee and flows into a coffeepot below. In soil, water reacts with humus and carbon dioxide to form acid. The acid dissolves some of the minerals in the A horizon and carries the material into the B horizon, as shown in **Figure 12.**

Reading Check *How does leaching transport material from the A horizon to the B horizon?*

C Horizon The C horizon consists of partially weathered rock and is the bottom horizon in a soil profile. It is often the thickest soil horizon. This horizon does not contain much organic matter and is not strongly affected by leaching. It usually is composed of coarser sediment than the soil horizons above it. What would you find if you dug to the bottom of the C horizon? As you might have guessed, you would find rock—the rock that gave rise to the soil horizons above it. This rock is called the parent material of the soil. The C horizon is the soil layer that is most like the parent material.

INTEGRATE Chemistry

Soil Fertility Plants need a variety of nutrients for growth. They get these nutrients from the soil. Fertile soil supplies the nutrients that plants need in the proper amounts. Soil fertility usually is determined in a laboratory by a soil chemist. However, fertility sometimes can be inferred by looking at plants. Do research to discover some important plant nutrients.

Teacher FYI

Leaching The concept of leaching is important in constructing landfills. Leaching of toxic materials in hazardous wastes must be controlled so that dangerous chemicals do not enter soil or water outside the landfill. Using leachate collection techniques or landfill liners usually prevents this problem.

Quick Demo

Pore Space

Materials glass jar, marbles or beans

Estimated Time 10 minutes

Procedure Fill a glass jar with marbles or beans. Show students that space exists between the "grains." Explain that pore space also exists between the grains in soil.

Fun Fact

In 1 g of soil, there can be hundreds of millions of organisms.

Discussion

C Horizon If you could remove soil from the C horizon, would it be useful for growing plants? Explain. No; it consists of rock that is only partially weathered and contains very little humus.

Reading Check

Answer Water carries ions from the A horizon to the B horizon.

INTEGRATE Chemistry

Soil Fertility Students should discover major nutrients such as nitrogen, phosphorus, and potassium. They also might learn about micronutrients such as iron and manganese.

Virtual Labs

The Breakdown *How are materials from earth broken down?*

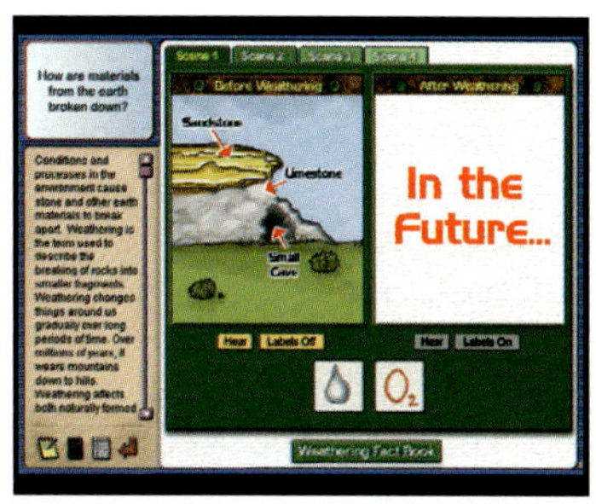

Text Question Answer

The soil developed from glacial till, a poorly sorted sediment deposited directly from ice.

Teacher FYI

Till The glacial sediment described on the student page is called till. Till can be deposited in two main ways: 1) Ice at the base of a glacier can melt and release sediment as the glacier slides over its substrate. 2) Glacier ice can stop flowing and melt in place. The sediment that the ice contains is gradually lowered onto the substrate.

Applying Math

National Math Standards

Correlation to Mathematics Objectives

1, 5, 6

Answers to Practice Problems

1. 20 g sand/ 50 g sample × 100 = 40% sand in the sample
2. 15 g silt/ 50 g sample × 100 = 30% silt in the sample
3. 40% sand, 30% silt, 30% clay; 40% + 30% + 30% = 100%

Glacial Deposits At many places on Earth, the land is covered by a thick layer of sediment that was deposited by glaciers. The sediment was left on the land after debris-laden glacier ice melted at the end of the last ice age. This material, which is an unsorted mixture of clay, silt, sand, and boulders, covers much of the northern United States. The soils that developed from this glacial material are extremely fertile. They are important for agriculture in many regions, such as the Midwest. How does this soil profile differ from the one described earlier? If you were to dig a deep hole, you would find bedrock as before, but it would not be the rock that the soil formed from. What material did this soil develop from?

Applying Math — Calculate Percentages

SOIL TEXTURE Some soil is coarse, and some soil is fine. This property of soil is called soil texture. The texture of soil often is determined by finding the percentages of sand, silt, and clay. What percentage of the soil represented by the circle graph is clay?

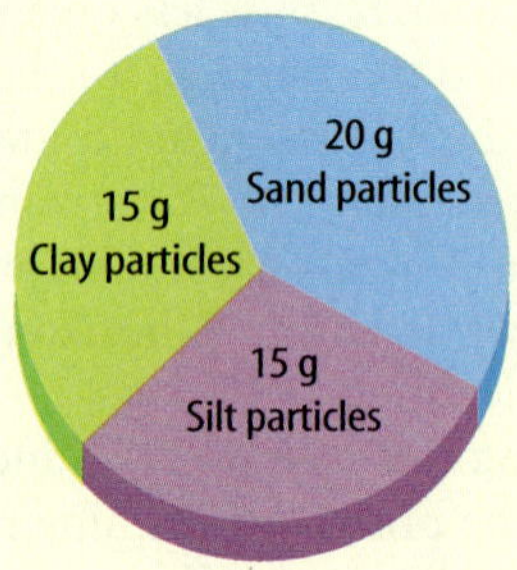

Solution

1. *This is what you know:*
   - sand weight: 20 g
   - clay weight: 15 g
   - silt weight: 15 g
2. *This is what you need to find:*
   - total weight of the sample
   - percentage of clay particles
3. *This is the procedure you need to use:*
   - Add the masses for each particle size to determine the total mass of the sample:
     20 g sand + 15 g silt + 15 g clay = 50 g sample
   - Divide the mass of clay by the mass of the sample and multiply by 100:
     15 g clay/50 g sample × 100 = 30% clay in the sample

Practice Problems

1. Calculate the percentage of sand in the sample.
2. Calculate the percentage of silt in the sample.
3. List all three percentages to describe the soil's texture. Check your work by adding the three values. Do you get 100%?

Science Online — For more practice, visit bookg.msscience.com/math_practice

LAB DEMONSTRATION

Purpose to make a soil profile model

Materials 250-mL beakers (4), coarse gravel, pea gravel, topsoil, mixture of light-colored sand and clay, large graduated cylinder

Preparation Fill each beaker with a different soil component.

Procedure Layer soil components into the cylinder to show the order of sediments from bottom to top in a normal soil profile: coarse gravel, pea gravel, sand and clay mixture, topsoil.

Expected Outcome Students will observe a soil profile.

Assessment

What does each layer represent? coarse gravel: bedrock; pea gravel: C horizon; sand and clay: B horizon; topsoil: A horizon

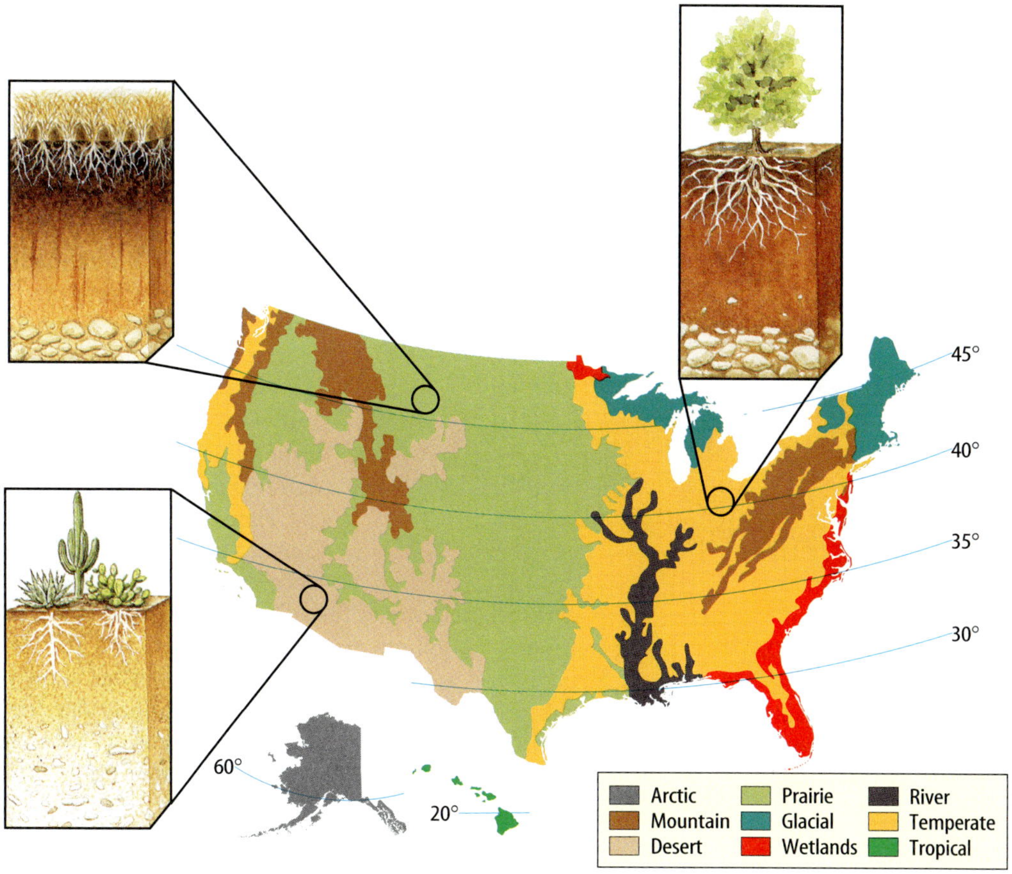

Figure 13 The United States has many different soil types. They vary in color, depth, texture, and fertility.
Identify *the soil type in your region.*

Soil Types

If you travel across the country and look at soils, you will notice that they are not all the same. Some soils are thick and red. Some are brown with hard rock nodules, and some have thick, black A horizons. Many different types of soils exist, as shown in **Figure 13.**

Soil Types Reflect Climate Different regions on Earth have different climates. Deserts are dry, prairies are semidry, and temperate forests are mild and moist. These places also have different types of soils. Soils in deserts contain little organic material and also are thinner than soils in wetter climates. Prairie soils have thick, dark A horizons because the grasses that grow there contribute lots of organic matter. Temperate forest soils have thinner A horizons than prairie soils do. Tree roots and leaves cannot deliver as much organic matter to deeper parts of the soil as grasses can deliver. Other regions such as tundra and tropical areas also have distinct soils.

Visual Learning

Figure 13 Have students compare the soil profiles shown in the figure. Desert soils are brown and often contain nodules of calcium carbonate. Prairie soils have thick, dark A horizons. Temperate soils are thick, but they have thinner A horizons than prairie soils do.

Caption Answer

Figure 13 Answers will vary.

Inquiry Lab

Soil Type and Plant Growth

Question How does soil type affect plant growth?

Possible Materials deep pots, various types of surface soil, plant seeds

Estimated Time several weeks

Teaching Strategies

- Students likely will plant the same type of seed in a variety of different soils.
- Make certain that students change only one variable.
- Encourage different student groups to use different types of seeds. Soil preferences often vary with type of plant.
- Students should wash their hands thoroughly after the activity. L2

For additional inquiry activities, see *Science Inquiry Labs*.

Differentiated Instruction

Visually Impaired Obtain several soil samples with a variety of textures and compositions. Ask visually impaired students to feel and describe each soil sample. Sighted students also will benefit from this activity. L1

IDENTIFYING Misconceptions

Soil and the Rock Cycle Many students think that soils are the same age as Earth. Explain that soil is forming and being eroded today. Soil is an important part of the rock cycle.

Caption Answer

Figure 14 Sediments move downhill and collect in valleys.

Text Question Answer

a sandy soil

3 Assess

DAILY INTERVENTION

Check for Understanding

Auditory-Musical Ask students to write and perform a song that includes the five factors that affect soil development. L2

Reteach

Soil Profile Take students to an area near the school where they can examine an exposed soil profile. Badly eroded gullies, road cuts, and steep slopes are good places to observe soil horizons. Ask students to identify as many soil horizons as possible. L2 LS **Visual-Spatial**

Assessment

Performance Have students make an events chain concept map to show how materials in the A horizon leach into the B horizon. Use **Performance Assessment in the Science Classroom,** p. 163. L2

Figure 14 The slope of the land affects soil development. Thin, poorly developed soils form on steep slopes, but valleys often have thick, well-developed soils. **Infer** *why this is so.*

Other Factors Parent material has an effect on the soils that develop from it. Clay soils often develop on rocks like basalt, because minerals in the rock weather to form clay. What type of soil do you think might develop on sandstone? Rock type also affects the type of vegetation that grows in a region, because different rocks provide different amounts of nutrients for plant growth. Type of vegetation then affects soil formation.

Time also affects soil development. If weathering has been occurring for only a short time, the parent rock determines the soil characteristics. As weathering continues, the soil resembles the parent rock less and less.

Slope also is a factor affecting soil profiles, as shown in **Figure 14.** On steep slopes, soils often are poorly developed, because material moves downhill before it can be weathered much. In bottomlands, sediment and water are plentiful. Bottomland soils are often thick, dark, and full of organic material.

section 2 review

Summary

Formation of Soil

- Soil is a mixture of rock and mineral fragments, decayed organic matter, water, and air.

Composition of Soil

- Organic matter in soil is gradually changed to humus.
- Soil moisture is important for plant growth.

Soil Profile

- The layers in a soil profile are called horizons.
- Most soils have an A, B, and C horizon.

Soil Types

- Many different types of soils occur in the United States.
- Climate and other factors determine the type of soil that develops.

Self Check

1. **List** the five factors that affect soil development.
2. **Explain** how soil forms.
3. **Explain** why A horizons often are darker than B horizons or C horizons.
4. **Describe** how leaching affects soil.
5. **Think Critically** Why is a soil profile in a tropical rain forest different from one in a desert? A prairie?

Applying Skills

6. **Use Statistics** A farmer collected five soil samples from a field and tested their acidity, or pH. His data were the following: 7.5, 8.2, 7.7, 8.1, and 8.0. Calculate the mean of these data. Also, determine the range and median.

Science Online bookg.msscience.com/self_check_quiz

section 2 review

1. climate, slope, type of rock, type of vegetation, length of time for weathering
2. Soil forms when bedrock or other parent material weathers. Plants and other organisms add organic matter to soil.
3. A horizons have more dark-colored humus than other soil horizons.
4. Leaching removes material from the soil. Some of the ions that are released by leaching might precipitate to form new minerals in the B horizon.
5. Deserts and rain forests have different types of vegetation and different climates, causing different amounts of humus to form and weathering to occur.
6. Mean = 7.9
   Range = 8.2 − 7.5 = 0.7
   Median = 8.0

LAB

Soil Texture

Soils have different amounts of different sizes of particles. When you determine how much sand, silt, and clay a soil contains, you describe the soil's texture.

Real-World Question

What is the texture of your soil?

Goals

- **Estimate** soil texture by making a ribbon.

Materials

soil sample (100 g)
water bottle

Safety Precautions

Procedure

1. Take some soil and make it into a ball. Work the soil with your fingers. Slowly add water to the soil until it is moist.
2. After your ball of soil is moist, try to form a thin ribbon of soil. Use the following descriptions to categorize your soil:
   - **a.** If you can form a long, thin ribbon, you have a clay soil.
   - **b.** If you formed a long ribbon but it breaks easily, you have a clay loam soil.
   - **c.** If you had difficulty forming a long ribbon, you have loam soil.
3. Now make your soil classification more detailed by selecting one of these descriptions:
   - **a.** If the soil feels smooth, add the word *silty* to your soil name.
   - **b.** If the soil feels slightly gritty, don't add any word to your soil name.
   - **c.** If the soil feels very gritty, add the word *sandy* before your soil name.

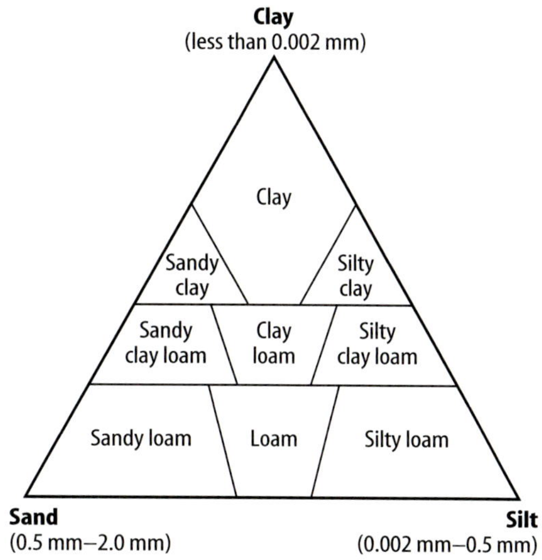

Conclude and Apply

1. **Classify** Which texture class name did you assign to your soil?
2. **Observe** Find your soil texture class name on the triangle above. Notice that the corners of the triangle are labeled *sand, silt,* and *clay.*
3. **Determine** Is your soil texture class close to one of the three corners or near the middle of the diagram? If your soil texture class is close to a corner, which one?
4. **Describe** Does your soil contain mostly sand, silt, or clay, or does it have nearly equal amounts of each? *Hint: If your soil name is close to a corner, it has mostly that size of sediment. If your soil name is in the middle of the triangle, it has nearly equal amounts of each sediment size.*

BENCH TESTED

Real-World Question

Purpose Students will estimate soil texture using the ribbon method.

Process Skills interpret scientific illustrations, collect data, interpret data, classify

Time Required 30 minutes

Safety Precautions Have students wash their hands after completing the lab.

Procedure

Teaching Strategy Demonstrate the ribbon-making technique before students conduct the activity.

Troubleshooting Students frequently add too much or too little water to the soil. This makes it difficult to form a good ribbon.

Conclude and Apply

1. Answers will vary depending on soil.
2. Students should locate the appropriate category in the ternary diagram.
3. Answers will vary.
4. Answers will vary.

Differentiated Instruction

Behaviorally Disordered Have behaviorally disordered students use soil samples that they collected themselves. This will increase their interest in the outcome of the activity.

Assessment

Oral Ask students to describe the texture of their soil sample and explain why the ribbon method can be used to estimate soil texture. Use **Performance Assessment in the Science Classroom,** p. 89 L2

section 3

Soil Erosion

1 Motivate

Bellringer

INTERACTIVE CHALKBOARD PowerPoint® Presentations

Section Focus Transparencies also are available on the Interactive Chalkboard CD-ROM. L2 ELL

Section 3 Section Focus Transparency **Blowing in the Wind**

During the 1930s, catastrophic "black blizzards" were experienced in many states of the southern Great Plains. One storm carried over 300 million tons of dirt all the way to the east coast.

1. What is occurring in this picture?
2. Why is the loss of topsoil harmful?
3. What steps have been taken to prevent such serious dust storms from happening today?

L2

Tie to Prior Knowledge

Human Use of Soil Have students list ways humans use soil. Possible answers: grow food and other plants, make structures with materials such as adobe Inform students that they will learn about the importance of soil and the problems of preserving soil in this section. L1

as you read

What **You'll Learn**

- **Explain** why soil is important.
- **Evaluate** ways that human activity has affected Earth's soil.
- **Describe** ways to reduce soil erosion.

Why **It's Important**

If topsoil is eroded, soil becomes less fertile.

Review Vocabulary
erosion: the picking up and moving of sediment or soil

New Vocabulary
- no-till farming
- contour farming
- terracing

Soil—An Important Resource

While picnicking at a local park, a flash of lightning and a clap of thunder tell you that a storm is upon you. Watching the pounding rain from the park shelter, you notice that the water flowing off of the ball diamond is muddy, not clear. The flowing water is carrying away some of the sediment that used to be on the field. This process is called soil erosion. Soil erosion is harmful because plants do not grow as well when topsoil has been removed.

Causes and Effects of Soil Erosion

Soil erodes when it is moved from the place where it formed. Erosion occurs as water flows over Earth's surface or when wind picks up and transports sediment. Generally, erosion is more severe on steep slopes than on gentle slopes. It's also more severe in areas where there is little vegetation. Under normal conditions, a balance between soil production and soil erosion often is maintained. This means that soil forms at about the same rate as it erodes. However, humans sometimes cause erosion to occur faster than new soil can form. One example is when people remove ground cover. Ground cover is vegetation that covers the soil and protects it from erosion. When vegetation is cleared, as shown in **Figure 15,** soil erosion often increases.

Figure 15 Removing vegetation can increase soil erosion.

Trees protect the soil from erosion in forested regions. When forest is removed, soil erodes rapidly.

Section 3 Resource Manager

Chapter *FAST FILE* Resources
- Transparency Activity, p. 42
- Directed Reading for Content Mastery, pp. 17, 18
- Enrichment, p. 28
- Reinforcement, p. 25
- Lab Worksheets, pp. 7–8

Lab Management and Safety, p. 65

Figure 16 Tropical rain forests often are cleared by burning. **Explain** *how this can increase soil erosion.*

Agricultural Cultivation Soil erosion is a serious problem for agriculture. Topsoil contains many nutrients, holds water well, and has a porous structure that is good for plant growth. If topsoil is eroded, the quality of the soil is reduced. For example, plants need nutrients to grow. Each year, nutrients are both added to the soil and removed from the soil. The difference between the amount of nutrients added and the amount of nutrients removed is called the nutrient balance. If topsoil erodes rapidly, the nutrient balance might be negative. Farmers might have to use more fertilizer to compensate for the nutrient loss. In addition, the remaining soil might not have the same open structure and water-holding ability that topsoil does.

Forest Harvesting When forests are removed, soil is exposed and erosion increases. This creates severe problems in many parts of the world, but tropical regions are especially at risk. Each year, thousands of square kilometers of tropical rain forest are cleared for lumber, farming, and grazing, as shown in **Figure 16.** Soils in tropical rain forests appear rich in nutrients but are almost infertile below the first few centimeters. The soil is useful to farmers for only a few years before the topsoil is gone. Farmers then clear new land, repeating the process and increasing the damage to the soil.

Overgrazing In most places, land can be grazed with little damage to soil. However, overgrazing can increase soil erosion. In some arid regions of the world, sheep and cattle raised for food are grazed on grasses until almost no ground cover remains to protect the soil. When natural vegetation is removed from land that receives little rain, plants are slow to grow back. Without protection, soil is carried away by wind, and the moisture in the soil evaporates.

INTEGRATE Career

Soil Scientist Elvia Niebla is a soil scientist at the U.S. Environmental Protection Agency (EPA). Soil scientists at the EPA work to reduce soil erosion and pollution. Niebla's research even helped keep hamburgers safe to eat. How? In a report for the EPA, she explained how meat can be contaminated when cattle graze on polluted soil.

Science Online

Topic: Land Use

Visit bookg.msscience.com for Web links to information about how land use affects Earth's soil and about measures taken to reduce the impact.

Activity Debate with classmates about the best ways to protect rich farmland. Consider advantages and disadvantages of each method.

2 Teach

Caption Answer

Figure 16 Burning removes the cover of vegetation, exposing the soil to erosion by wind and flowing water.

INTEGRATE Career

Soil Scientist Careers in soil science also can be found at universities, the U.S. Department of Agriculture, state and county agencies, and in the private sector.

Discussion

Nutrient Balance How are nutrients added to soil? decaying vegetation, fertilizer, manure How are nutrients removed from soil? crop harvesting, soil erosion, leaching

Activity

Erosion Control Take students to an area on the school grounds where soil is eroding. Have students write suggestions to the principal for solving the problem. With permission, organize students to carry out the plan. L2 LS **Logical-Mathematical**

Visual Learning

Figure 15 How do trees protect soil from erosion? Trees provide ground cover. Tree leaves and litter protect soil from rain, and roots bind the soil together.

Active Reading

Quickwrites This strategy, sometimes called freewrites, lets students use spontaneous writing to discover what they already know. Have students list ideas about a topic, then share with the class. Next, have students write their ideas freely in a paragraph, not worrying about punctuation, spelling, and grammar. Have students use a Quickwrite to share ideas after learning about soil erosion. L2

Caption Answer

Figure 17 It could cover stream-dwelling organisms and eggs, clog the stream, or reduce downstream water quality.

Teacher FYI

Erosion Rate Soil is eroding in the United States about 17 times faster than it can form. Erosion rates in Asia, Africa, and South America are estimated to be twice as high.

Make a Model

Erosion Control Ask students to make scale models that illustrate the erosion-control method of their choice. Students might model contour farming, terracing, or other methods. L3

Reading Check

Answer They can plant shelter belts of trees, plant cover crops, graze animals on vegetation instead of plowing it under, or use no-till farming.

Figure 17 Erosion from exposed land can cause streams to fill with excessive amounts of sediment. **Explain** *how this could damage streams.*

Excess Sediment If soil erosion is severe, sediment can damage the environment. Severe erosion sometimes occurs where land is exposed. Examples might include strip-mined areas or large construction sites. Eroded soil is moved to a new location where it is deposited. If the sediment is deposited in a stream, as shown in **Figure 17,** the stream channel might fill.

Preventing Soil Erosion

Each year more than 1.5 billion metric tons of soil are eroded in the United States. Soil is a natural resource that must be managed and protected. People can do several things to conserve soil.

Manage Crops All over the world, farmers work to slow soil erosion. They plant shelter belts of trees to break the force of the wind and plant crops to cover the ground after the main harvest. In dry areas, instead of plowing under crops, many farmers graze animals on the vegetation. Proper grazing management can maintain vegetation and reduce soil erosion.

In recent years, many farmers have begun to practice no-till farming. Normally, farmers till or plow their fields one or more times each year. Using **no-till farming,** seen in **Figure 18,** farmers leave plant stalks in the field over the winter months. At the next planting, they seed crops without destroying these stalks and without plowing the soil. Farm machinery makes a narrow slot in the soil, and the seed is planted in this slot. No-till farming provides cover for the soil year-round, which reduces water runoff and soil erosion. One study showed that no-till farming can leave as much as 80% of the soil covered by plant residue. The leftover stalks also keep weeds from growing in the fields.

Figure 18 No-till farming decreases soil erosion because fields are not plowed.

Reading Check *How can farmers reduce soil erosion?*

Cultural Diversity

Managing with Stones In Burkina Faso, farmers line fields with stones to keep soil from blowing away. They also build centimeter-high rock dams that direct water into the ground instead of across it. As a result of these practices, crop yields have increased as much as 90 percent. Have students use an atlas to locate Burkina Faso. Burkina Faso is a landlocked state in western Africa. L2

Differentiated Instruction

Challenge Have students research why it often is difficult to re-establish ground cover on land that has been strip-mined and what can be done to solve this problem. Students should give oral reports to the class. Strip-mining often exposes the mineral pyrite, which reacts with water to make soil acidic. Lime or crushed limestone is applied to neutralize acid. L3

52 ◆ G CHAPTER 2 Weathering and Soil

Reduce Erosion on Slopes On gentle slopes, planting along the natural contours of the land, called **contour farming**, reduces soil erosion. This practice, shown in **Figure 19,** slows the flow of water down the slope and helps prevent the formation of gullies.

Where slopes are steep, terracing often is used. **Terracing** (TER uh sing) is a method in which steep-sided, level topped areas are built onto the sides of steep hills and mountains so that crops can be grown. These terraces reduce runoff by creating flat areas and shorter sections of slope. In the Philippines, Japan, China, and Peru, terraces have been used for centuries.

Figure 19 This orchard was planted along the natural contours of the land.
Summarize *the benefits of using contour farming on slopes.*

Reduce Erosion of Exposed Soil A variety of methods are used to control erosion where soil is exposed. During the construction process water is sometimes sprayed onto bare soil to prevent erosion by wind. When construction is complete, topsoil is added in areas where it was removed and trees are planted. At strip mines, water flow can be controlled so that most of the eroded soil is kept from leaving the mine. After mining is complete, the land is reclaimed. This means that steep slopes are flattened and vegetation is planted.

section 3 review

Summary

Soil—An Important Resource

- Soil erosion is a serious problem because topsoil is removed from the land.

Causes and Effects of Soil Erosion

- Soil erosion occurs rapidly on steep slopes and areas that are not covered by vegetation.
- The quality of farmland is reduced when soil erosion occurs.

Preventing Soil Erosion

- Farmers reduce erosion by planting shelter belts, using no-till farming, and planting cover crops after harvesting.
- Contour farming and terracing are used to control erosion on slopes.

Self Check

1. **Explain** why soil is important.
2. **Explain** how soil erosion damages soil.
3. **Describe** no-till farming.
4. **Explain** how overgrazing increases soil erosion.
5. **Think Critically** How does contour farming help water soak into the ground?

Applying Skills

6. **Communicate** Do research to learn about the different methods that builders use to reduce soil erosion during construction. Write a newspaper article describing how soil erosion at large construction sites is being controlled in your area.

Science Online bookg.msscience.com/self_check_quiz

Caption Answer

Figure 19 Planting in this way slows the flow of water down the slope and reduces soil erosion.

3 Assess

DAILY INTERVENTION

Check for Understanding

Linguistic Ask students to list causes of soil erosion on one side of a sheet of paper. Then, ask them to list several effects for each cause. L2

Reteach

Soil Erosion Have students research and report on the problem of soil erosion in your state. How much is occurring? How does it affect the state's economy? What is being done to combat the problem? L3

Assessment

Process Have students make a spider concept map showing the ways people can prevent soil erosion. Use **Performance Assessment in the Science Classroom,** p. 161. L2 P

section 3 review

1. It is necessary for growing plants.
2. Soil erosion removes topsoil, which is highly fertile, has an open structure that is good for plant growth, and has high water-holding capacity.
3. Farmers do not plow fields. Stalks are left through the winter, and the next year's crop is seeded without removing them.
4. Overgrazing reduces ground cover. The soil is more vulnerable to erosion by water and wind.
5. Contour farming slows water flow, allowing the water time to soak into the ground.
6. Students should mimic the writing style and format of a newspaper column.

BENCH TESTED

Real-World Question

Purpose Students design and carry out an experiment to study variables that affect chemical weathering. L2 ELL

COOP LEARN LS **Interpersonal**

Process Skills communicate, form a hypothesis, design an experiment to test a hypothesis, identify and manipulate variables, observe and infer, relate cause and effect, interpret data, record, use numbers, make and use graphs

Time Required 45 minutes to plan, 45 minutes to do the experiment

Materials If materials are in short supply, place one or two sets of materials in a corner of the classroom. Have groups take turns completing the activity.

Safety Precautions Caution students to take care with the hot plate and heated vinegar. Tie back long hair and wear goggles.

Form a Hypothesis

Possible Hypotheses Students will probably expect chalk to dissolve faster if exposed to vinegar and heat. They will probably also expect smaller pieces (more surface area) to dissolve faster than larger pieces.

LAB Design Your Own

WEATHERING CHALK

Goals

- **Design** experiments to evaluate the effects of acidity, surface area, and temperature on the rate of chemical weathering of chalk.
- **Describe** factors that affect chemical weathering.
- **Explain** how the chemical weathering of chalk is similar to the chemical weathering of rocks.

Possible Materials

pieces of chalk (6)
small beakers (2)
metric ruler
water
white vinegar (100 mL)
hot plate
computer probe for temperature
*thermometer
Alternate materials

Safety Precautions

Wear safety goggles when pouring vinegar. Be careful when using a hot plate and heated solutions.

WARNING: *If mixing liquids, always add acid to water.*

Real-World Question

Chalk is a type of limestone made of the shells of microscopic organisms. The famous White Cliffs of Dover, England, are made up of chalk. This lab will help you understand how chalk can be chemically weathered. How can you simulate chemical weathering of chalk?

Form a Hypothesis

How do you think acidity, surface area, and temperature affect the rate of chemical weathering of chalk? What happens to chalk in water? What happens to chalk in acid (vinegar)? How will the size of the chalk pieces affect the rate of weathering? What will happen if you heat the acid? Make hypotheses to support your ideas.

Sample Data Table:

Test	Observations
<u>Acid Test</u> Chalk in water	no reaction
Chalk in vinegar	some reaction bubbles
<u>Surface Area Test</u> Whole chalk	some reaction
Broken chalk	more reaction than whole
<u>Temperature Test</u> Room temperature vinegar	some reaction
Warm vinegar	more reaction than cooler

Using Scientific Methods

Test Your Hypothesis

Make a Plan

1. **Develop** hypotheses about the effects of acidity, surface area, and temperature on the rate of chemical weathering.
2. **Decide** how to test your first hypothesis. List the steps needed to test the hypothesis.
3. Repeat step 2 for your other two hypotheses.
4. **Design** data tables in your Science Journal. Make one for acidity, one for surface area, and one for temperature.
5. **Identify** what remains constant in your experiment and what varies. Change only one variable in each procedure.
6. **Summarize** your data in a graph. Decide from reading the Science Skill Handbook which type of graph to use.

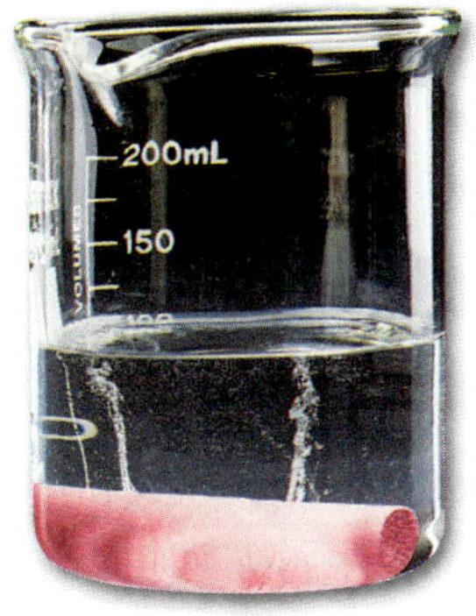

Follow Your Plan

1. Make sure your teacher approves your plan before you start.
2. Carry out the three experiments as planned.
3. While you are conducting the experiments, record your observations and complete the data tables in your Science Journal.
4. Graph your data to show how each variable affected the rate of weathering.

Analyze Your Data

1. **Analyze** your graph to find out which substance—water or acid—weathered the chalk more quickly. Was your hypothesis supported by your data?
2. **Infer** from your data whether the amount of surface area makes a difference in the rate of chemical weathering. Explain.

Conclude and Apply

1. **Explain** how the chalk was chemically weathered.
2. How does heat affect the rate of chemical weathering?
3. What does this imply about weathering in the tropics and in polar regions?

Communicating Your Data

Compare your results with those of your classmates. How were your data similar? How were they different? **For more help, refer to the Science Skill Handbook.**

Test Your Hypotheses

Possible Procedures Acidity test: Place equal-sized pieces of chalk into beakers of water and vinegar. Surface area test: Place a whole piece of chalk into a beaker and an equal-sized piece that is broken into pieces into a second beaker. Pour vinegar into both beakers. Temperature test: Place equal-sized pieces of chalk into room-temperature vinegar and heated vinegar.

Teaching Strategy Explain that the settling of chalk flakes is not evidence of chemical weathering.

Expected Outcome Water has no effect on chalk. Chalk bubbles and reacts with vinegar. Small chalk pieces react faster. Heated acid causes a faster reaction.

Analyze Your Data

Answers to Questions

1. Acid; answers will vary.
2. The greater the surface area, the faster the weathering because there is more contact with the acid.

Error Analysis Concentration of vinegar can affect results, as can the number of pieces into which chalk is broken.

Conclude and Apply

1. Material in the chalk reacted with vinegar.
2. Heat increases the rate.
3. Chemical weathering is faster in the tropics and slower in polar regions.

Assessment

Content Have students write new lyrics to a familiar song describing the variables that affect chemical weathering. Use **Performance Assessment in the Science Classroom,** p. 151. L2

Communicating Your Data

Exact data will vary, but all students should have similar results.

Understanding Literature

Repetition Silko is reminding the reader of the common link between rocks, clay, plants, and animals.

Respond to the Reading

1. dust
2. Answers will vary but might suggest that repeating the word *dust* reminds the reader of how all living things return to the earth.
3. **Linking Science and Writing** A goal for this activity is to give step-by-step instructions to someone for which the subject is new knowledge.

INTEGRATE Earth Science **Pedology** *Pedology* is the scientific discipline concerned with all aspects of soils, including their physical and chemical properties, the role of organisms in soil production, the description and mapping of soil units, and the origin and evolution of soils. Accordingly, pedology embraces several sub-disciplines, namely, soil chemistry, soil physics and soil microbiology.

Science and Language Arts

Landscape, History, and the Pueblo Imagination

by Leslie Marmon Silko

Leslie Marmon Silko, a woman of Pueblo, Hispanic, and American heritage, explains what ancient Pueblo people believed about the circle of life on Earth.

You see that after a thing is dead, it dries up. It might take weeks or years, but eventually if you touch the thing, it crumbles under your fingers. It goes back to dust. The soul of the thing has long since departed. With the plants and wild game the soul may have already been borne back into bones and blood or thick green stalk and leaves. Nothing is wasted. What cannot be eaten by people or in some way used must then be left where other living creatures may benefit. What domestic animals or wild scavengers can't eat will be fed to the plants. The plants feed on the dust of these few remains.

. . . Corn cobs and husks, the rinds and stalks and animal bones were not regarded by the ancient people as filth or garbage. The remains were merely resting at a mid-point in their journey back to dust. . . .

The dead become dust . . . . The ancient Pueblo people called the earth the Mother Creator of all things in this world. Her sister, the Corn mother, occasionally merges with her because all . . . green life rises out of the depths of the earth.

Rocks and clay . . . become what they once were. Dust.

A rock shares this fate with us and with animals and plants as well.

Understanding Literature

Repetition The recurrence of sounds, words, or phrases is called repetition. What is Silko's purpose of the repeated use of the word *dust?*

Respond to the Reading

1. What one word is repeated throughout this passage?
2. What effect does the repetition of this word have on the reader?
3. **Linking Science and Writing** Using repetition, write a one-page paper on how to practice a type of soil conservation.

INTEGRATE Earth Science This chapter discusses how weathered rocks and mineral fragments combine with organic matter to make soil. Silko's writing explains how the ancient Pueblo people understood that all living matter returns to the earth, or becomes dust. Lines such as "green life rises out of the depths of the earth," show that the Pueblo people understood that the earth, or rocks and mineral fragments, must combine with living matter in order to make soil and support plant life.

Resources for Teachers and Students

Ceremony, by Leslie Marmon Silko, Viking Penguin, 1996

Gardens in the Dunes, by Leslie Marmon Silko, Simon and Schuster, 2000

56 ◆ G CHAPTER 2 Weathering and Soil

Reviewing Main Ideas

Section 1 Weathering

1. Weathering helps to shape Earth's surface.
2. Mechanical weathering breaks apart rock without changing its chemical composition. Plant roots, animals, and ice wedging are agents of mechanical weathering.
3. Chemical weathering changes the chemical composition of rocks. Natural acids and oxygen in the air can cause chemical weathering.

Section 2 The Nature of Soil

1. Soil is a mixture of rock and mineral fragments, organic matter, air, and water.
2. A soil profile contains different layers that are called horizons.
3. Climate, parent rock, slope of the land, type of vegetation, and the time that rock has been weathering are factors that affect the development of soil.

Section 3 Soil Erosion

1. Soil is eroded when it is moved to a new location by wind or water.
2. Human activities can increase the rate of soil erosion.
3. Windbreaks, no-till farming, contour farming, and terracing reduce soil erosion on farm fields.

Visualizing Main Ideas

Copy and complete the following concept map about weathering.

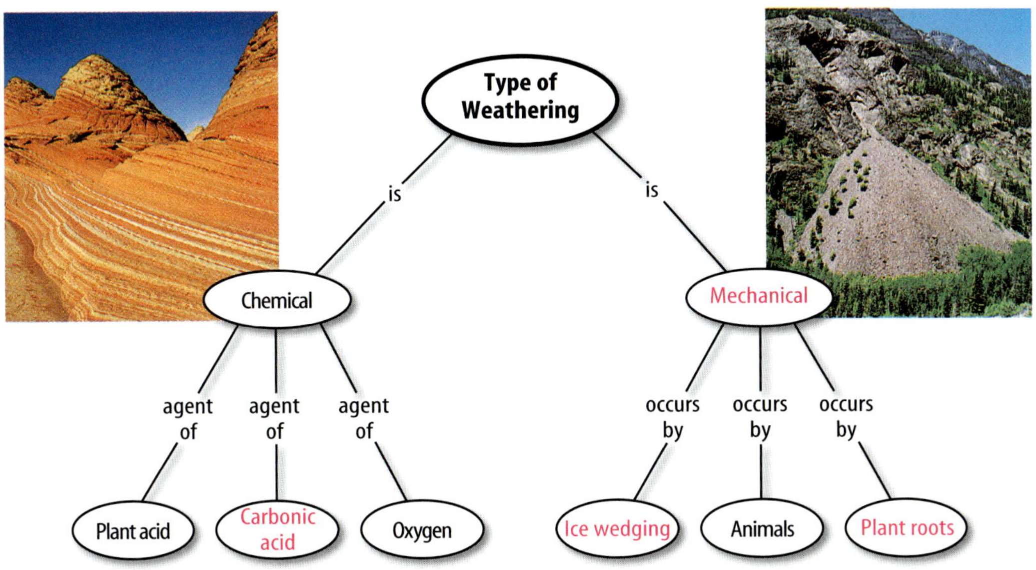

chapter 2 Study Guide

Reviewing Main Ideas

Summary statements can be used by students to review the major concepts of the chapter.

Visualizing Main Ideas

See student page.

Science Online

Visit bookg.msscience.com
/self_check_quiz
/interactive_tutor
/vocabulary_puzzlemaker
/chapter_review
/standardized_test

Assessment Transparency

For additional assessment questions, use the *Assessment Transparency* located in the transparency book.

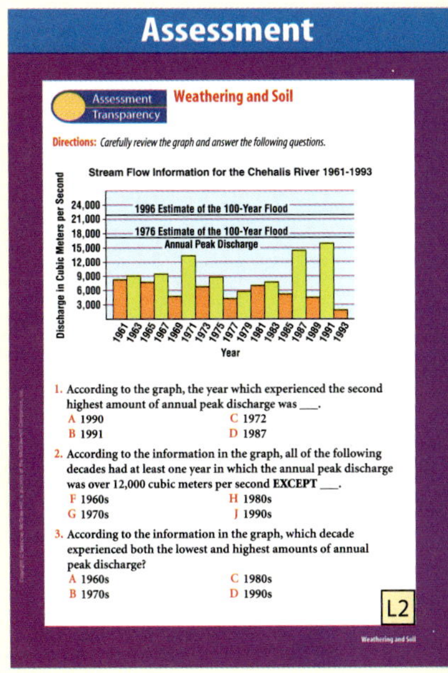

Assessment

Assessment Transparency **Weathering and Soil**

Directions: *Carefully review the graph and answer the following questions.*

1. According to the graph, the year which experienced the second highest amount of annual peak discharge was ___.
   A 1990 C 1972
   B 1991 D 1987
2. According to the information in the graph, all of the following decades had at least one year in which the annual peak discharge was over 12,000 cubic meters per second **EXCEPT** ___.
   F 1960s H 1980s
   G 1970s J 1990s
3. According to the information in the graph, which decade experienced both the lowest and highest amounts of annual peak discharge?
   A 1960s C 1980s
   B 1970s D 1990s

L2

Weathering and Soil

Identifying Misconceptions

Assess

Activity Provide students with a map of world climate zones. Ask them to characterize each climate by describing temperature and precipitation patterns. Students should observe that climate zones are related to latitude, proximity to oceans, and other factors.

After students are familiar with the various climate zones, have them make inferences about rates of mechanical and chemical weathering in each zone. Students can attach notes to the map that summarize their inferences. L2

CHAPTER STUDY GUIDE G ◆ 57

Using Vocabulary

1. Chemical weathering
2. Humus
3. soil profile
4. Leaching
5. Oxidation
6. Contour farming
7. Climate

Checking Concepts

8. D
9. C
10. A
11. C
12. A
13. B
14. C

chapter 2 Review

Using Vocabulary

chemical weathering p. 39
climate p. 40
contour farming p. 53
horizon p. 44
humus p. 44
ice wedging p. 38
leaching p. 45
litter p. 45
mechanical weathering p. 37
no-till farming p. 52
oxidation p. 40
soil p. 42
soil profile p. 44
terracing p. 53
weathering p. 36

Fill in the blanks with the correct vocabulary word or words.

1. _________ changes the composition of rock.
2. _________ forms from organic matter such as leaves and roots.
3. The horizons of a soil make up the _________.
4. _________ transports material to the B horizon.
5. _________ occurs when many materials containing iron are exposed to oxygen and water.
6. _________ means that crops are planted along the natural contours of the land.
7. _________ is the pattern of weather that occurs in a particular area for many years.

Checking Concepts

Choose the word or phrase that best answers the question.

8. Which of the following can be caused by acids produced by plant roots?
   A) soil erosion
   B) oxidation
   C) mechanical weathering
   D) chemical weathering

Use the graph below to answer question 9.

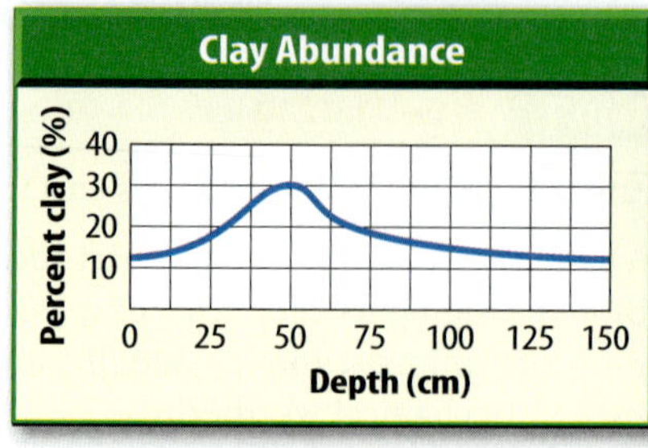

9. The above graph shows the percentage of clay in a soil profile at varying depths. Which depth has the highest amount of clay?
   A) 25 cm
   B) 150 cm
   C) 50 cm
   D) 100 cm
10. Which of the following is an agent of mechanical weathering?
    A) animal burrowing
    B) carbonic acid
    C) leaching
    D) oxidation
11. In which region is chemical weathering most rapid?
    A) cold, dry
    B) cold, moist
    C) warm, moist
    D) warm, dry
12. What is a mixture of rock and mineral fragments, organic matter, air, and water called?
    A) soil
    B) limestone
    C) horizon
    D) clay
13. What is organic matter in soil?
    A) leaching
    B) humus
    C) horizon
    D) profile
14. What is done to reduce soil erosion on steep slopes?
    A) no-till farming
    B) contour farming
    C) terracing
    D) grazing

Science Online bookg.msscience.com/vocabulary_puzzlemaker

Use the ExamView® Pro Testmaker CD-ROM to:

- create multiple versions of tests
- create modified tests with one mouse click for inclusion students
- edit existing questions and add your own questions
- build tests aligned with state standards using built-in State Curriculum Tags
- change English tests to Spanish with one mouse click and vice versa

Thinking Critically

15. Predict which type of weathering—mechanical or chemical—you would expect to have a greater effect in a polar region. Explain.

16. Recognize Cause and Effect How does soil erosion reduce the quality of soil?

17. Concept Map Copy and complete the concept map about layers in soil.

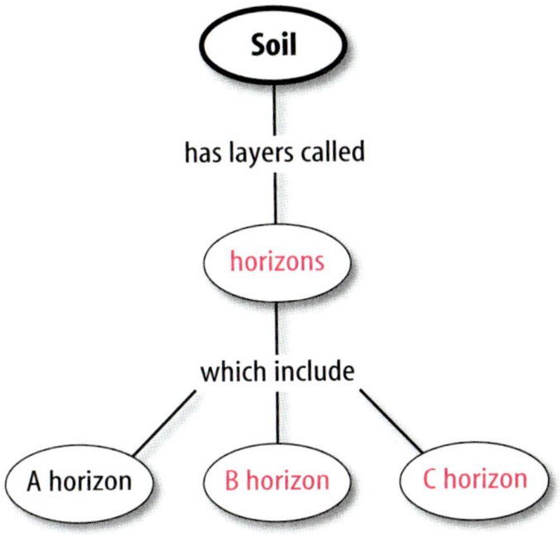

18. Recognize Cause and Effect Why do rows of trees along the edges of farm fields reduce wind erosion of soil?

19. Form a Hypothesis A pile of boulders lies at the base of a high-mountain cliff. Form a hypothesis explaining how the pile of rock might have formed.

20. Test a Hypothesis How would you test your hypothesis from question 19?

21. Identify a Question Many scientists are conducting research to learn more about how soil erosion occurs and how it can be reduced. Write a question about soil erosion that you would like to research. With your teacher's help, carry out an investigation to answer your question.

Performance Activities

22. Design a Landscape Find a slope in your area that might benefit from erosion maintenance. Develop a plan for reducing erosion on this slope. Make a map showing your plan.

23. Describing Peds Natural clumps of soil are called peds. Collect a large sample of topsoil. Describe the shape of the peds. Sketch the peds in your Science Journal.

Applying Math

Use the illustration below to answer questions 24–26.

24. Fertilizer Nutrients A bag of fertilizer is labeled to list the nutrients as three numbers. The numbers represent the percentages of nitrogen, phosphate, and potash in that order. What are the percentages of these nutrients for a fertilizer with the following information on the label: 5-10-10?

25. Fertilizer Ratio The fertilizer ratio tells you the proportions of the different nutrients in a fertilizer. To find the fertilizer ratio, divide each nutrient value by the lowest value. Calculate the fertilizer ratio for the fertilizer in question 24.

26. Relative Amounts of Nutrients Which nutrient is least abundant in the fertilizer? Which nutrients are most abundant? How many times more potash does the fertilizer contain than nitrogen?

Science Online bookg.msscience.com/chapter_review

Assessment Resources

Reproducible Masters

Chapter *Fast File* Resources
Chapter Review, pp. 33–34
Chapter Tests, pp. 35–38
Assessment Transparency Activity, p. 45

Glencoe Science Web site
Chapter Review Test
Standardized Test Practice

Glencoe Technology

Assessment Transparency
Exam*View*® Pro Testmaker
MindJogger Videoquiz
Interactive Chalkboard

chapter 2 Review

Thinking Critically

15. Mechanical; Cold temperatures slow most chemical weathering reactions.

16. Soil erosion removes topsoil. The soil below the topsoil generally is not well suited for plant growth.

17. See student page.

18. The trees reduce the speed of the wind, which lowers its ability to erode soil.

19. Mechanical weathering such as ice wedging might have loosened boulders higher on the cliff face. Gravity caused the boulders to fall and accumulate at the base of the cliff.

20. Possible answer: examine the rock exposure for evidence of mechanical weathering. Search records for reports of rockfall.

21. Student questions should be testable with available materials.

Performance Activities

22. Designs will vary. Students might consider planting ground cover, improving drainage, and lowering the angle of the slope.

23. The most common ped shape in topsoil is granular. These sometimes are called crumb peds. They generally range from a few millimeters to just less than a centimeter in diameter.

Applying Math

National Math Standards

1, 9

24. 5% nitrogen, 10% phosphate, 10% potash

25. 1-2-2

26. nitrogen; phosphate and potash; two times more

CHAPTER REVIEW G ◆ 59

chapter 2 Standardized Test Practice

FAST FILE

Answer Sheet A practice answer sheet can be found at bookg.msscience.com/answer_sheet.

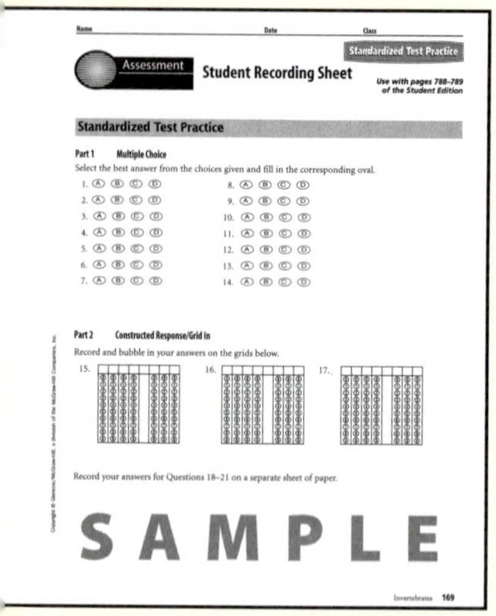

Name Date Class

Assessment — Student Recording Sheet — Standardized Test Practice

Use with pages 788–789 of the Student Edition

Standardized Test Practice

Part 1 Multiple Choice

Select the best answer from the choices given and fill in the corresponding oval.

1. Ⓐ Ⓑ Ⓒ Ⓓ 8. Ⓐ Ⓑ Ⓒ Ⓓ
2. Ⓐ Ⓑ Ⓒ Ⓓ 9. Ⓐ Ⓑ Ⓒ Ⓓ
3. Ⓐ Ⓑ Ⓒ Ⓓ 10. Ⓐ Ⓑ Ⓒ Ⓓ
4. Ⓐ Ⓑ Ⓒ Ⓓ 11. Ⓐ Ⓑ Ⓒ Ⓓ
5. Ⓐ Ⓑ Ⓒ Ⓓ 12. Ⓐ Ⓑ Ⓒ Ⓓ
6. Ⓐ Ⓑ Ⓒ Ⓓ 13. Ⓐ Ⓑ Ⓒ Ⓓ
7. Ⓐ Ⓑ Ⓒ Ⓓ 14. Ⓐ Ⓑ Ⓒ Ⓓ

Part 2 Constructed Response/Grid In

Record and bubble in your answers on the grids below.

15. 16. 17.

Record your answers for Questions 18–21 on a separate sheet of paper.

SAMPLE

Invertebrates 169

Part 1 Multiple Choice

1. C
2. B
3. D
4. C
5. D
6. A
7. C
8. B
9. C
10. B

chapter 2 Standardized Test Practice

Part 1 Multiple Choice

Record your answers on the answer sheet provided by your teacher or on a separate sheet of paper.

Use the photo below to answer question 1.

1. Which method for reducing soil erosion is shown on the hillsides above?
A. no-till farming **C.** contour farming
B. terracing **D.** shelter belts

2. Which of the following terms might describe a soil's texture?
A. red **C.** porous
B. coarse **D.** wet

3. Which soil horizon often has a dark color because of the presence of humus?
A. E horizon **C.** B horizon
B. C horizon **D.** A horizon

4. Which of the following is an agent of chemical weathering?
A. ice wedging
B. burrowing animals
C. carbonic acid
D. growing tree roots

Test-Taking Tip

Come Back To It Never skip a question. If you are unsure of an answer, mark your best guess on another sheet of paper and mark the question in your test booklet to remind you to come back to it at the end of the test.

5. Which of the following might damage a soil's structure?
A. a gentle rain **C.** earthworms
B. organic matter **D.** compaction

6. In which of the following types of rock are caves most likely to form?
A. limestone **C.** granite
B. sandstone **D.** basalt

7. Which of the following is most likely to cause erosion of farmland during a severe drought?
A. water runoff
B. soil creeping downhill
C. wind
D. ice

Use the table below to answer questions 8–10.

Texture Data for a Soil Profile

Horizon	Percent		
	Sand	Silt	Clay
A	16.2	54.4	29.4
B	10.5	50.2	39.3
C	31.4	48.4	20.2
R (bedrock)	31.7	50.1	18.2

8. According to the table, which horizon in this soil has the lowest percentage of sand?
A. A horizon **C.** C horizon
B. B horizon **D.** R horizon

9. Which of the following is the R horizon?
A. topsoil **C.** bedrock
B. humus **D.** gravel

10. Which of the following is the best description of the soil represented by the table?
A. sandy **C.** clayey
B. silty **D.** organic

Part 2 Short Response/Grid In

11. Student answers will vary. The ped names are descriptive of their shapes. For example, prismatic peds are shaped like prisms.

12. Caves form below the water table as slightly acidic groundwater dissolves limestone or dolomite bedrock. Carbonic acid forms when water reacts with carbon dioxide in soil air. It makes the groundwater acidic.

13. 19.8 kg

14. 16.8 kg

15. 3.0 kg

Part 3 Open Ended

16. agriculture, clearing forest, and development; by practicing erosion control and by making wise decisions about land use

17. Through time, weathering lowers Earth's surface.

18. Because stalks remain in the field all year, runoff is reduced.

Part 2 Short Response/Grid In

Record your answers on the answer sheet provided by your teacher or on a sheet of paper.

Use the illustration below to answer question 11.

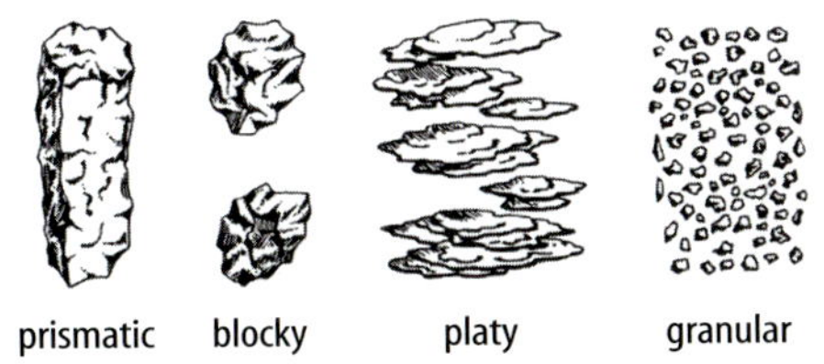

11. Natural clumps of soil are called peds. Compare and contrast the different types of peds in the sketch. Explain how the names of the peds describe their shape.

12. Explain how caves form. What role does carbonic acid have?

Use the diagram below to answer questions 13–15.

2003 Phosphorus Budget per Acre
Jones Family Farm

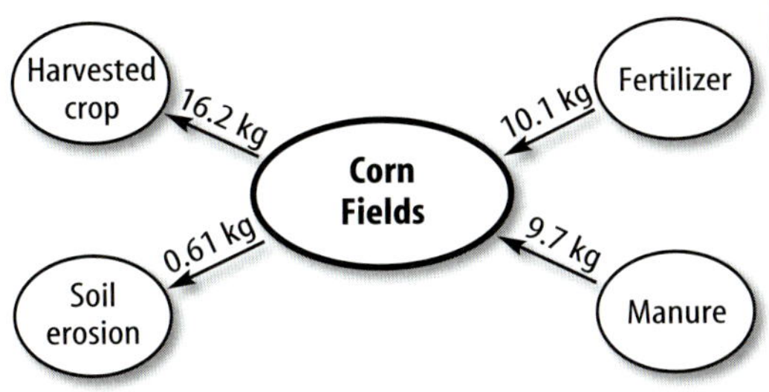

13. According to the diagram above, what was the total amount of phosphorus added to each acre?

14. What is the total amount of phorphorus lost from each acre?

15. What is the difference between the amount of phosphorus added and the amount of phosphorus lost?

Part 3 Open Ended

Record your answers on a sheet of paper.

16. Describe ways that humans affect Earth's soil. How can damage to soil be reduced?

17. How does weathering change Earth's surface?

18. How does no-till farming reduce soil erosion?

19. How does time affect soil development?

20. How does humus form? What does it form from?

Use the graph below to answer questions 21–23.

Variation of a Soil's Temperature with Time of Year

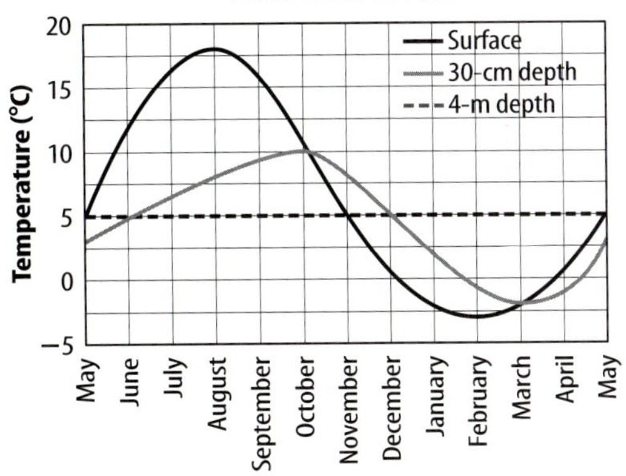

21. During which month was the surface soil warmest? During which month was it coldest? Explain.

22. During which month was the soil at 30 cm warmest? During which month was it coldest? How is this different than surface soil?

23. Why didn't soil temperature vary at a depth of 4 m?

19. The longer the time, the more highly developed the soil.

20. Humus forms as microorganisms alter organic matter, such as plant roots and leaves.

21. The surface soil was warmest in August and coldest in February. August is near the end of summer, and February is near the end of winter.

22. The soil at 30-cm depth was warmest in October. It was coolest in March. The soil temperature at 30 cm lags behind the surface soil temperature.

23. Near the surface, soil temperature varies in response to annual changes in atmospheric temperature. At some depth, which varies from place to place, soil temperature is not affected by short-term changes in atmospheric temperature. An equilibrium exists between average atmospheric temperature and Earth's heat flow.

Rubrics

The following rubrics are sample scoring devices for short response and open-ended questions.

Short Response

Points	Description
2	The student demonstrates a thorough understanding of the science of the task. The response may contain minor flaws that do not detract from the demonstration of a thorough understanding.
1	The student has provided a response that is only partially correct.
0	The student has provided a completely incorrect solution or no response at all.

Open Ended

Points	Description
4	The student demonstrates a thorough understanding of the science of the task. The response may contain minor flaws that do not detract from the demonstration of a thorough understanding.
3	The student demonstrates an understanding of the science of the task. The response is essentially correct and demonstrates an essential but less than thorough understanding of the science.
2	The student demonstrates only a partial understanding of the science of the task. Although the student may have used the correct approach to a solution or may have provided a correct solution, the work lacks an essential understanding of the underlying science concepts.
1	The student demonstrates a very limited understanding of the science of the task. The response is incomplete and exhibits many flaws.
0	The student provides a completely incorrect solution or no response at all.

chapter 3 Organizer

Section/Objectives	Standards		Labs/Features
	National	**State/Local**	
Chapter Opener	See p. 9T–10T for a Key to Standards.		**Launch Lab:** Demonstrate Sediment Movement? p. 63 **Foldables,** p. 63
Section 1 Erosion by Gravity 2 sessions 1 block 1. **Explain** the differences between erosion and deposition. 2. **Compare** and contrast slumps, creep, rockfalls, rock slides, and mudflows. 3. **Explain** why building on steep slopes might not be wise.	National Content Standards: UCP.1, UCP.2, UCP.3, UCP.5, A.1, A.2		**MiniLAB:** Modeling Slump, p. 65 **Integrate Physics,** p. 67
Section 2 Glaciers 2 sessions 1 block 4. **Explain** how glaciers move. 5. **Describe** evidence of glacial erosion and deposition. 6. **Compare** and contrast till and outwash.	National Content Standards: UCP.1, UCP.2, UCP.3, UCP.5, A.1, A.2		**Science Online,** p. 74 **Lab:** Glacial Grooving, p. 75
Section 3 Wind 2 sessions 1 block 7. **Explain** how wind causes deflation and abrasion. 8. **Recognize** how loess and dunes form.	National Content Standards: UCP.1, UCP.2, UCP.3, UCP.5, A.1, A.2, B.2, F.3, F.4		**Integrate History,** p. 77 **Applying Science:** What factors affect wind erosion?, p. 77 **MiniLAB:** Observing How Soil Is Held in Place, p. 78 **Science Online,** p. 78 **Visualizing How Dunes Form and Migrate,** p. 80 **Lab:** Blowing in the Wind, p. 82 **Science Stats:** Losing Against Erosion, p. 84

Glencoe Exclusive! **TeacherWorks**™ All-In-One Planner and Resource Center

Lab Materials	Reproducible Resources	Section Assessment	Technology
Launch Lab: sand, gravel, shoe-box lid	**Chapter *FAST FILE* Resources** Foldables Worksheet, p. 13 Directed Reading Overview, p. 15 Note-taking Worksheets, pp. 29–31	Glencoe's Assessment Advantage	TeacherWorks includes: • Interactive Teacher Edition • Lesson Planner with calendar • Access to all program blacklines • Correlations to standards • Web links
MiniLAB: baking pan; bricks (2); sink; gelatin powder; aquarium gravel; large, flat rock; watering can, water *Need materials?* Contact Science Kit at 1-800-828-7777 or www.sciencekit.com on the Internet.	**Chapter *FAST FILE* Resources** Transparency Activity, p. 40 MiniLAB, p. 3 Enrichment, p. 26 Reinforcement, p. 23 Directed Reading, p. 16 Lab Activity, pp. 9–10 **Science Inquiry Labs,** pp. 23–24	**Portfolio** Differentiated Instruction, p. 67 **Performance** MiniLAB, p. 65 Applying Skills, p. 68 **Content** Section Review, p. 68	Section Focus Transparency Guided Reading Audio Program Interactive Chalkboard CD-ROM
Lab: sand, large plastic or metal tray or stream table, ice block, books (2 or 3) or wood block, metric ruler, overhead light source with reflector	**Chapter *FAST FILE* Resources** Transparency Activity, p. 41 Enrichment, p. 27 Reinforcement, p. 24 Directed Reading, p. 16 Lab Worksheets, pp. 5–6 Lab Activity, pp. 11–12 Transparency Activity, pp. 43–44	**Portfolio** Curriculum Connection, p. 71 Differentiated Instruction, p. 72 **Performance** Applying Skills, p. 74 **Content** Section Review, p. 74	Section Focus Transparency Teaching Transparency Virtual Labs CD-ROM Guided Reading Audio Program Interactive Chalkboard CD-ROM
MiniLAB: sod, trowel, magnifying lens, old newspaper, bucket of water, soap **Lab:** flat pans (4), fine sand (400 mL), gravel (400 mL), three-speed hair dryer, sprinkling can, water, 28-cm × 35-cm cardboard sheets (4), tape, mixing bowl, metric ruler, wind speed indicator	**Chapter *FAST FILE* Resources** Transparency Activity, p. 42 MiniLAB, p. 4 Enrichment, p. 28 Reinforcement, p. 25 Directed Reading, pp. 17, 18 Lab Worksheets, pp. 7–8 **Mathematics Skill Activities,** p. 9 **Lab Management and Safety,** p. 65	**Portfolio** Cultural Diversity, p. 79 **Performance** Applying Science, p. 77 MiniLAB, p. 78 Applying Math, p. 81 **Content** Section Review, p. 81	Section Focus Transparency Guided Reading Audio Program Interactive Chalkboard CD-ROM Video Lab

End of Chapter Assessment

Glencoe's Assessment Advantage

Blackline Masters	Technology	Professional Series
Chapter *FAST FILE* Resources Chapter Review, pp. 33–34 Chapter Tests, pp. 35–38 **Standardized Test Practice** pp. 15–18	MindJogger Videoquiz Virtual Labs CD-ROM Exam*View*® Pro Testmaker TeacherWorks CD-ROM Interactive Chalkboard CD-ROM	**Performance Assessment in the Science Classroom (PASC)**

chapter 3 Erosional Forces

Transparencies

Section Focus

SECTION 1 Section Focus Transparency **Avalanche**

Avalanches occur when snow on a mountain slope breaks loose and slides downhill. Because of avalanche danger, people in snowy, mountainous areas put a lot of effort into avalanche safety and control.

1. Describe what is happening in the photo. How is the land changed as snow and ice move over it?
2. What is the key force in pulling an avalanche downhill?
3. How do you think the steepness of the slope affects avalanches?

L2

SECTION 2 Section Focus Transparency **I'll Be Back in 5,000 Years**

Meet Ötzi, a human from the Stone Age who lived about 5,000 years ago. He was somehow trapped in a rock depression and subsequently covered by snow and ice. Ötzi was eventually discovered by hikers on a ridge that divides Austria from Italy.

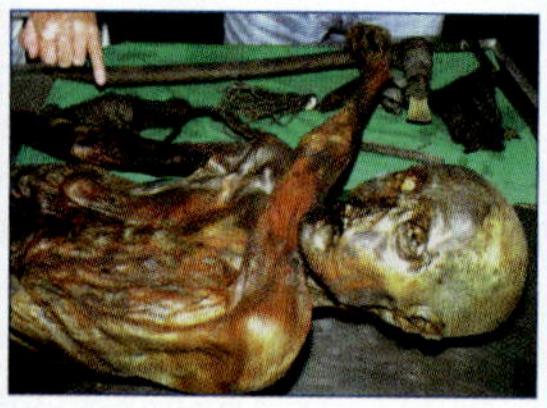

1. How was Ötzi preserved? What do you think modern scientists have learned from studying Ötzi?
2. What happened that resulted in Ötzi's eventual discovery?
3. In what areas do you most commonly see a lot of snow and ice?

L2

SECTION 3 Section Focus Transparency **Rock On!**

You may have noticed animal tracks in the sand, but have you ever seen boulder tracks? If you go to the Racetrack in Death Valley, this is exactly what you'll see. Rocks at the Racetrack seem to move without any help at all!

1. No one has ever seen a rock move at the Racetrack. What evidence do you have that these rocks are moving without human interference?
2. Pretend you are scientists studying Death Valley. Develop a theory that explains the mysterious movements at the Racetrack.

L2

This is a representation of key blackline masters available in the Teacher Classroom Resources. See Resource Manager boxes within the chapter for additional information.

Key to Teaching Strategies

The following designations will help you decide which activities are appropriate for your students.

L1 Level 1 activities should be appropriate for students with learning difficulties.

L2 Level 2 activities should be within the ability range of all students.

L3 Level 3 activities are designed for above-average students.

ELL ELL activities should be within the ability range of English Language Learners.

COOP LEARN Cooperative Learning activities are designed for small group work.

LS Multiple Learning Styles logos, as described on page 6T, are used throughout to indicate strategies that address different learning styles.

P These strategies represent student products that can be placed into a best-work portfolio.

PBL Problem-Based Learning activities apply real-world situations to learning.

Assessment

Assessment Transparency **Erosional Forces**

Directions: *Carefully review the table and answer the following questions.*

Wind Erosion Experiment Data (Using Fans) Height of Soil Mound (cm)

Soil Mound	After 1 minute	After 2 minutes	After 3 minutes	After 4 minutes	After 5 minutes
A	29.5	28.5	27.5	26.5	25.5
B	28.0	26.5	24.0	22.0	20.0
C	30.0	29.5	29.0	28.5	28.0
D	30.5	30.5	30.5	30.5	30.5

1. According to this information, which mound of soil lost over 5 cm in height due to wind erosion?
   A Mound A C Mound C
   B Mound B D Mound D
2. If everything remains the same, what would be the height of Mound B after 10 minutes?
   F 12.0 cm H 5.0 cm
   G 8.0 cm J 3.0 cm
3. According to the table, Mound D could probably be composed of ___.
   A dry temperate soil
   B loosely packed semi-moist soil
   C fine grain sand
   D moist clay-like soil

L2

Teaching

SECTION 2 Teaching Transparency **Glacial Margin**

L2

Hands-on Activities

Student Text Lab Worksheet

Name Date Class

Activity **Glacial Grooving**

Hands-On Activities

Lab Preview

Directions: *Answer these questions before you begin the Activity.*

1. Why is the safety symbol for thermal safety shown for this activity?
2. Would you expect the glacier channel to have a different shape than the meltwater channel? why or why not?

Throughout the world's mountainous regions, 200,000 valley glaciers are moving in response to gravity.

What You'll Investigate

How is the land affected when a valley glacier moves downslope?

Materials

sand
large plastic or metal tray
**stream table*
ice block
books (2 or 3)
**wood block*
metric ruler
overhead light source with reflector
**Alternate materials*

Goals

- **Compare** stream and glacial valleys.

Safety Precautions

CAUTION: *Do not pour sand down the drain. Make sure source is plugged into a GFI electrical outlet. Do not touch light source—it may be hot.*

Procedure

1. Set up the large tray as shown. Place books under one end of the tray to make it slope.
2. Cut a narrow channel, like a river, through the sand. Measure and record its width and depth. Draw a sketch that includes these measurements on the next page.
3. Position the overhead light source to shine on the channel as shown.
4. Force the ice block into the channel at the upper end of the tray.
5. Gently push the ice along the channel until it's halfway between the top and bottom of the tray, and directly under the light.
6. Turn on the light and allow the ice to melt. Record what happens.
7. **Record** the width and depth of the ice channel. Make a scale drawing

Figure 1

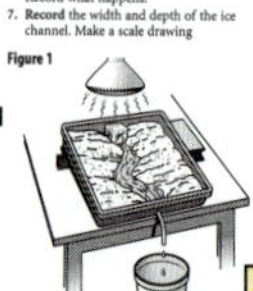

L2

Laboratory Activities

Name Date Class

LAB 1 Laboratory Activity **Mass Movements**

Hands-On Activities

The force of gravity causes loose material to move down slope. Sometimes water helps to move the material. Water makes the material heavier and more slippery.

Down slope movements of earth materials may be sudden or slow. Landslides and mudflows are sudden movements. Rocky slopes tend to move as landslides; clay and sand materials may become mudflows.

Creep is an example of slow earth movement. Even when a slope is covered by vegetation, the soil may creep to a lower level.

Strategy

You will cause mass movements.
You will classify the mass movements.

Materials

stream table with hose
4 wood blocks
plastic bucket
protractor
1 L clay
1 L sand
sprinkling can
water
1 L gravel
meterstick

Figure 1

Procedure

1. Set up the stream table as shown in Figure 1.
2. Place the protractor with its flat edge down on the table that is supporting the stream table. Position the protractor next to the lower end of the stream table. Use the protractor to measure the slope angle of the stream table. Record the angle in Table 1.
3. Place the clay in the upper end of the stream table.
4. Pour the sand over the clay. Wet the sand and clay thoroughly until it moves.
5. Observe and record in Table 1 how the mass moves.
6. Add two more blocks under the stream table. Measure and record the new slope angle of the stream table.
7. Move the sand and clay back to the upper end of the stream table.
8. Pour water over the sand and clay until it moves. Record how the mass moves.
9. Remove the sand and clay. Spread a thin layer of clay in the upper end of the table.
10. Spread gravel over the clay. Pour water over the clay and gravel and observe the motion. Record your observations.

L2

Meeting Different Ability Levels

Content Outline

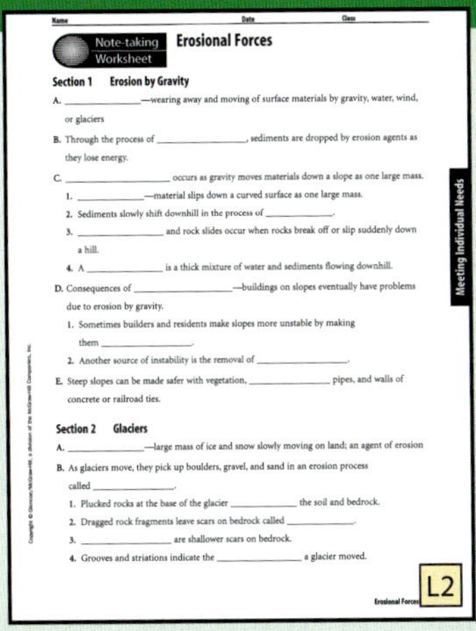

Reinforcement

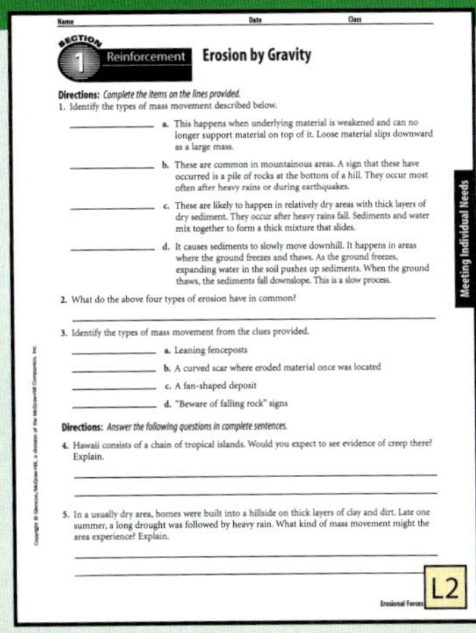

Enrichment

Directed Reading (English/Spanish)

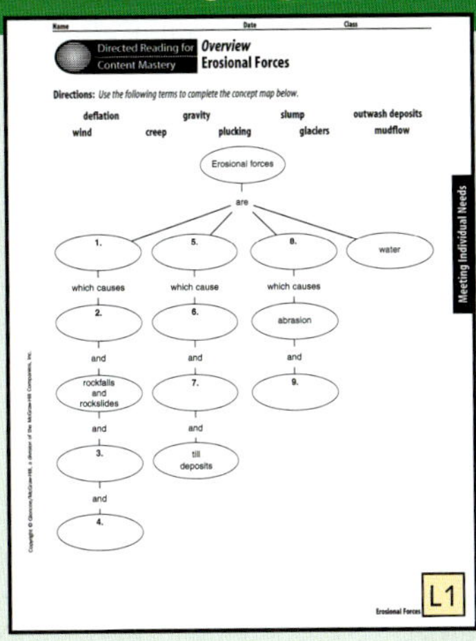

Study Guide

Reading Essentials

Assessment

Test Practice Workbook

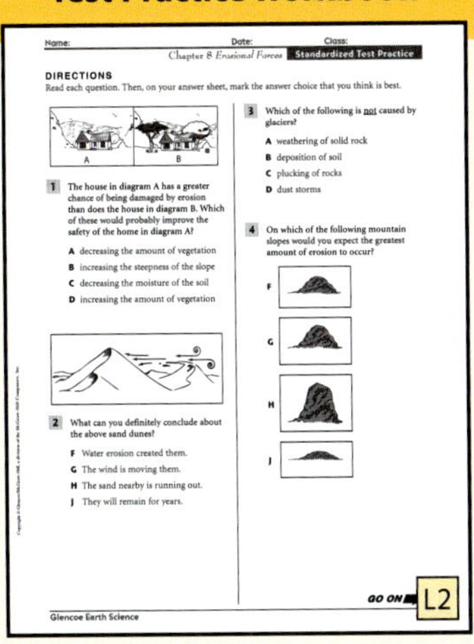

Chapter Review

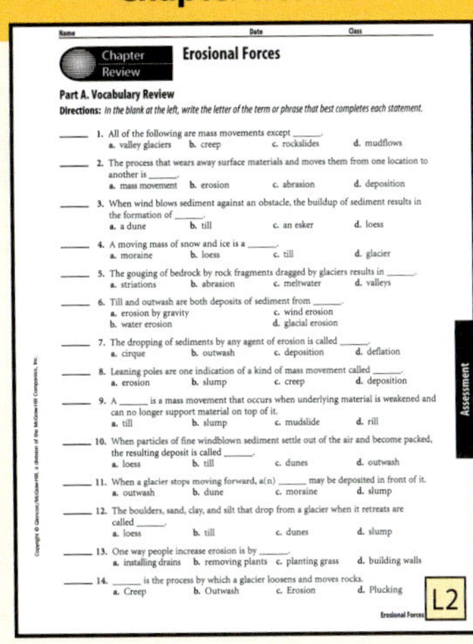

Chapter Tests

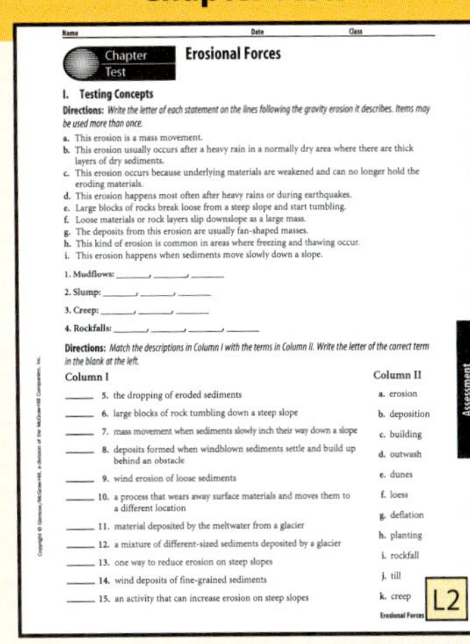

Science Content Background

section 1 Erosion by Gravity

Erosion and Deposition

All erosion requires energy, which comes from wind, water, glaciers, or gravity. When energy of motion is reduced, deposition occurs. Energy is reduced when wind, glaciers, or water speed slows or when less steep terrain is encountered. Mass movement occurs under all geological conditions. For example, it occurs on steep slopes of mountains, along rivers, on gently rolling plains, on cliffs by the sea, on slopes under the ocean, and even on other planets and the Moon.

section 2 Glaciers

Glacial Movement

Glaciers flow as gravity causes the ice to deform internally and to slide along the glacier bed. These gravitational forces increase as the ice thickness and surface slope increase. Internal deformation takes place as slippage occurs along internal planes within the ice crystals. Gravity causes glacial ice to flow from high to low ice-surface elevation, just as pancake batter spreads out on a griddle. Glaciers in which basal ice is near its melting point can also move by sliding over the substrate.

Glacial Erosion

Continental glaciers formed many lakes. As ice scoured the landscape, depressions were left in bedrock. As the ice melted, these holes filled with water. Mountain peaks and sharp ridges were formed by alpine glaciers. Meltwater froze around protruding rocks. The rocks then were plucked away as the ice advanced.

section 3 Wind

Abrasion

Abrasion by wind depends on the availability of suspended particles that act as scouring tools. Gentle winds can move dust-sized particles. If the wind is strong enough, sand-sized particles can be bounced along the ground in a series of intermittent leaps of varying lengths. These particles scour the ground over which they move. During the 1930s Dust Bowl in the United States, a dust cloud near Wichita, Kansas, extended 3,650 m above the ground and was estimated to contain 50,000 metric tons of dust per cubic kilometer.

Wind Erosion and Sand Dunes

Wind erosion can be reduced by placing obstacles to divert the wind. As sand is blown into or around an obstacle, its energy of motion decreases, sand is dropped, and a dune forms. Once this process begins, the dune itself acts as a barrier to the wind.

Dune migration occurs when wind blows sand up the windward side of a dune. When the sand piles too high, a sheetlike mass of sand avalanches down the leeward side of the dune, and the dune moves forward.

Tom Bean/DRK Photo

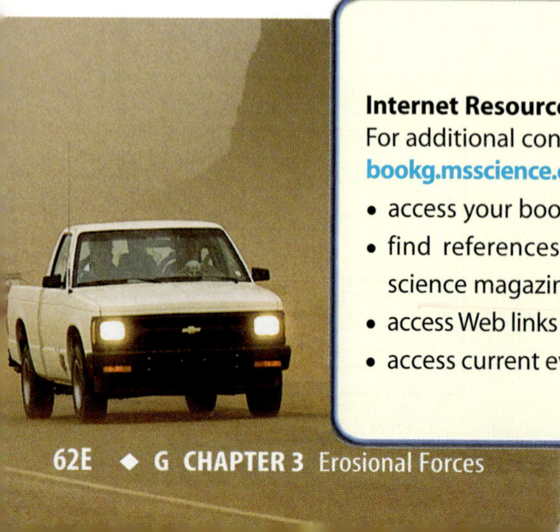

chapter content resources

Internet Resources

For additional content background, visit bookg.msscience.com to:

- access your book online
- find references to related articles in popular science magazines
- access Web links with related content background
- access current events with science journal topics

Print Resources

A Sand County Almanac, by Aldo Leopold Oxford University Press, 1949

American Exodus: The Dust Bowl Migration and Okie Culture in California, by James N. Gregory, Oxford University Press, 1991

Rich Land, Poor Land: A Study of Waste in the Natural Resources of America, by Stuart Chase, Whittlesey Publisher, 1936

IDENTIFYING Misconceptions

Find Out What Students Think

Students may think that . . .

Erosion is caused solely by the wind.

Students develop this misconception as they experience the wind blowing sand and soil across Earth's surface. Because the movement of these particles is readily observed, they may think that only the wind is responsible for the erosion of Earth's surface.

Discussion

Ask students to describe what happens before, during, and after a thunderstorm. Encourage them to describe the darkening of the sky, the increase of the wind, the pattern of rainfall, and the subsequent passing of the storm. List events on the board, placing them in sequential order, as students mention them.

When the list is complete, review the events with the class. Have students now describe how objects in the path of the storm move during each event. If necessary, prompt students with questions such as, "What happens to the trees as the wind speed increases?" "Are materials on the ground moved by the wind?" "How does the falling water affect materials on the ground?" Keep a list of all ideas offered by students.

Promote Understanding

Discussion

Show students a series of pictures of a town experiencing a flood. Include pictures of the river before, during, and immediately after the flood. Have students discuss how the river changes as the flood progresses.

- Point out that the river becomes more muddy and that larger materials are transported by the water as the flood grows stronger.
- As the flood wanes, the river carries smaller items only, and eventually runs clear again.

Activity

Organize students into small groups. Have each group use a stream table, a sprinkling can, and a hair dryer to compare wind and water erosion.

- Direct students to compare the amount of erosion that occurs from wind and water when the sand is dry and when the sand is wet.
- Have each group present their observations of erosion by wind and water in dry and wet sand. Most should find that dry sand is eroded strongly by wind, and somewhat by water. Wet sand is eroded much less by wind but is still eroded by water.

Tom Bean/DRK Photo

As an extension, have students debate whether wind erosion is more effective during a drought or after a series of heavy rains.

Assess

After completing the chapter, see *Identifying Misconceptions* in the Study Guide at the end of the chapter.

chapter 3

Chapter Vocabulary

erosion, p. 64
deposition, p. 67
mass movement, p. 65
slump, p. 65
creep, p. 66
glacier, p. 69
plucking, p. 70
till, p. 70
moraine, p. 71
outwash, p. 71
deflation, p. 76
abrasion, p. 76
loess, p. 79
dune, p. 79

Science Journal The Grand Canyon, Channeled Scablands, and Waterpocket Fold in Utah were all created by the erosional force of water.

INTERACTIVE CHALKBOARD with Image Bank

PowerPoint® Presentations

This CD-ROM is an editable Microsoft® PowerPoint® presentation that includes:

- a pre-made presentation for every chapter
- interactive graphics
- animations
- audio clips
- image bank
- all new section and chapter questions
- Standardized Test Practice
- transparencies
- pre-lab questions for all labs
- Foldables directions
- links to bookg.msscience.com

chapter 3

Erosional Forces

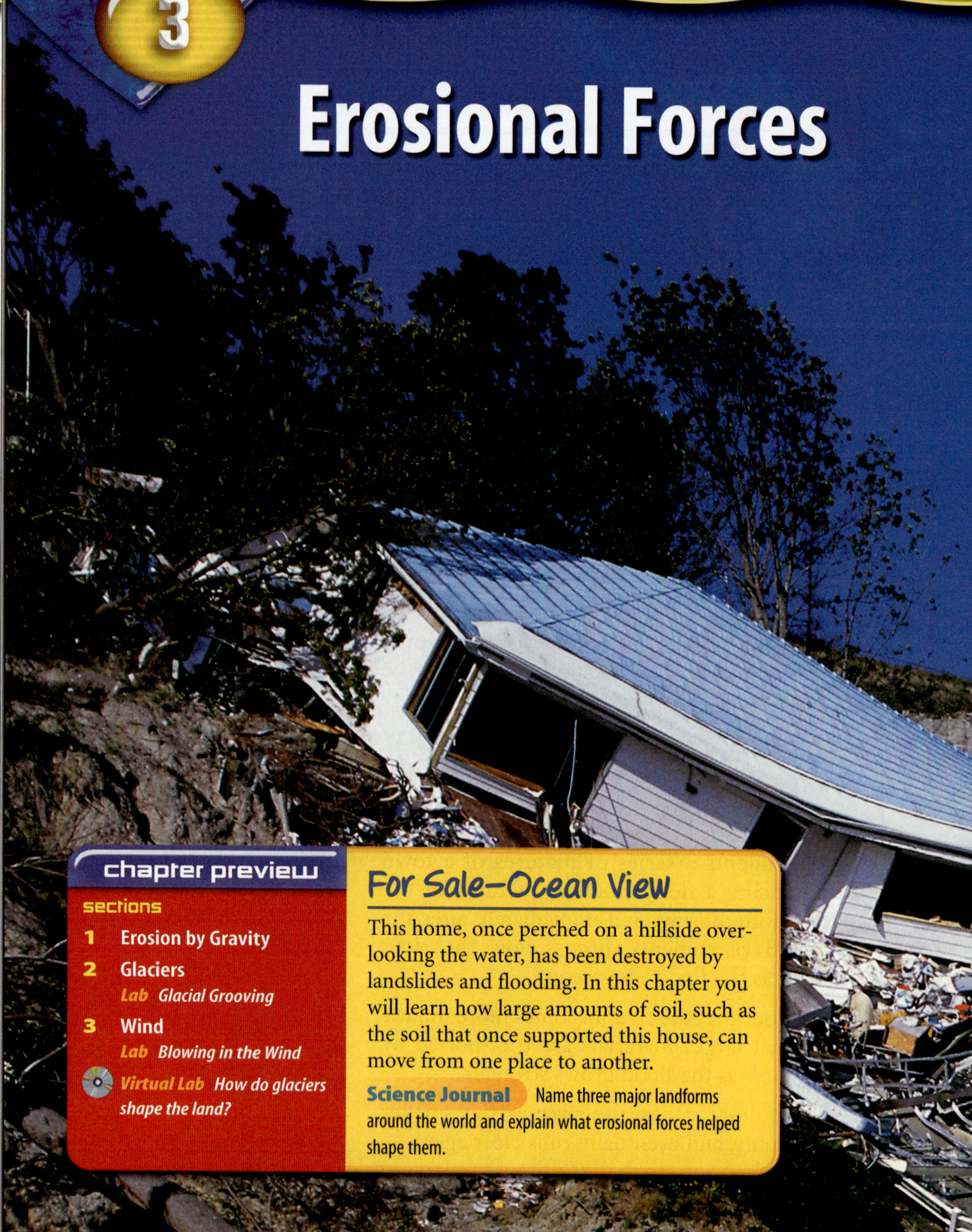

chapter preview

sections

1 Erosion by Gravity
2 Glaciers
  Lab *Glacial Grooving*
3 Wind
  Lab *Blowing in the Wind*
Virtual Lab *How do glaciers shape the land?*

For Sale—Ocean View

This home, once perched on a hillside overlooking the water, has been destroyed by landslides and flooding. In this chapter you will learn how large amounts of soil, such as the soil that once supported this house, can move from one place to another.

Science Journal Name three major landforms around the world and explain what erosional forces helped shape them.

Theme Connection

Stability and Change Students learn that although landforms on Earth's surface are generally stable in the short term, in the long term (thousands or even millions of years) erosional forces make great changes in the landforms on Earth's surface.

About the Photo

Erosional Impact Wind, water, and ice can cause soil and rocks to move, impacting the lives of humans.

Start-Up Activities

Launch LAB

Demonstrate Sediment Movement

Can you think of ways to move something without touching it? In nature, sediment is moved from one location to another by a variety of forces. What are some of these forces? In this lab, you will investigate to find out the answers to these questions.

WARNING: *Do not pour sand or gravel down the drain.*

1. Place a small pile of a sand-and-gravel mixture into a large shoe-box lid.
2. Move the sediment pile to the other end of the lid without touching the particles with your hands. You can touch and manipulate the box lid.
3. Try to move the mixture in a number of different ways.
4. **Think Critically** In your Science Journal, describe the methods you used to move the sediment. Which method was most effective? Explain how your methods compare with forces of nature that move sediment.

FOLDABLES™ Study Organizer

Erosion and Deposition Make the following Foldable to help you identify the examples of erosion and deposition.

STEP 1 **Fold** one piece of paper widthwise into thirds.

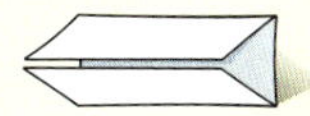

STEP 2 **Fold** the paper lengthwise into fourths.

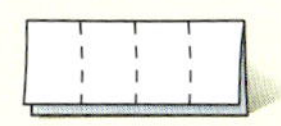

STEP 3 **Unfold,** lay the paper lengthwise, and draw lines along the folds.

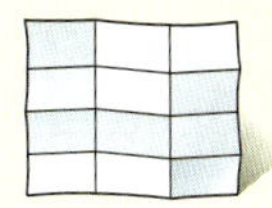

STEP 4 **Label** your table as shown.

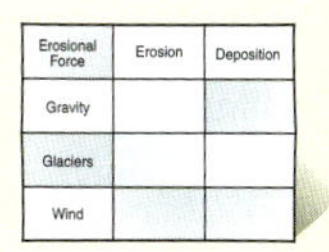

Make a Table As you read the chapter, complete the table, listing specific examples of erosion and deposition for each erosional force.

Science Online — Preview this chapter's content and activities at bookg.msscience.com

Launch LAB

Purpose Use the Launch Lab to introduce students to the ways that sediments can be moved from place to place. L2 ELL

COOP LEARN LS **Kinesthetic**

Preparation Make sure sand and gravel are dry.

Materials sand (250 g), gravel (250 g), shoe-box lid, safety goggles

Teaching Strategy If students have trouble coming up with ideas for moving sediments, suggest the following: blowing, tilting the box lid, or pushing with a piece of paper.

Safety Precautions Have all students wear safety goggles.

Think Critically

Possible answers: blowing resembles wind; tilting the box lid creates a steep slope that allows gravity to pull sediments downhill; pushing resembles the effects of glaciers.

Assessment

Process Have students draw and label diagrams that explain the methods they used to move sediments in the lab. Use **Performance Assessment in the Science Classroom,** p. 165.

FOLDABLES™ Study Organizer — Dinah Zike Study Fold

Student preparation materials for this Foldable are available in the Chapter *FAST FILE* Resources.

1 Motivate

Bellringer

Section Focus Transparencies also are available on the Interactive Chalkboard CD-ROM. L2 ELL

SECTION 1 Section Focus Transparency **Avalanche**

Avalanches occur when snow on a mountain slope breaks loose and slides downhill. Because of avalanche danger, people in snowy, mountainous areas put a lot of effort into avalanche safety and control.

1. Describe what is happening in the photo. How is the land changed as snow and ice move over it?
2. What is the key force in pulling an avalanche downhill?
3. How do you think the steepness of the slope affects avalanches?

L2

Erosional Forces

Tie to Prior Knowledge

Tumbling Sediments Ask students to recall seeing sediment tumble down a hillside. Invite them to discuss whether these actions could have a great effect on changing the landscape. Explain that they will discover the answer as they read this section.

Caption Answers

Figure 1 gravity

Figure 2 Possible answer: The base was eroded away.

section 1

Erosion by Gravity

as you read

What You'll Learn

- **Explain** the differences between erosion and deposition.
- **Compare** and contrast slumps, creep, rockfalls, rock slides, and mudflows.
- **Explain** why building on steep slopes might not be wise.

Why It's Important

Many natural features throughout the world were shaped by erosion.

Review Vocabulary

sediment: loose materials, such as mineral grains and rock fragments, that have been moved by erosional forces

New Vocabulary

- erosion
- deposition
- mass movement
- slump
- creep

Erosion and Deposition

Do you live in an area where landslides occur? As **Figure 1** shows, large piles of sediment and rock can move downhill with devastating results. Such events often are triggered by heavy rainfall. The muddy debris at the lower end of the slide comes from material that once was further up the hillside. The displaced soil and rock debris is a product of erosion (ih ROH zhun). **Erosion** is a process that wears away surface materials and moves them from one place to another.

What wears away sediments? How were you able to move the pile of sediments in the Launch Lab? If you happened to tilt the pan, you took advantage of an important erosional force—gravity. Gravity is the force of attraction that pulls all objects toward Earth's center. Other causes of erosion, also called agents of erosion, are water, wind, and glaciers.

Water and wind erode materials only when they have enough energy of motion to do work. For example, air can't move much sediment on a calm day, but a strong wind can move dust and even larger particles. Glacial erosion works differently by slowly moving sediment that is trapped in solid ice. As the ice melts, sediment is deposited, or dropped. Sometimes sediment is carried farther by moving meltwater.

Figure 1 The jumbled sediment at the base of a landslide is material that once was located farther uphill.
Define *the force that moves materials toward the center of Earth.*

64 ◆ G

Section 1 Resource Manager

Chapter *FAST FILE* Resources

- Transparency Activity, p. 40
- Directed Reading for Content Mastery, pp. 15,16
- MiniLAB, p. 3
- Reinforcement, p. 23
- Enrichment, p. 26
- Note-taking Worksheets, pp. 29–31
- Lab Activity, pp. 9–10

Science Inquiry Labs, pp. 23–24

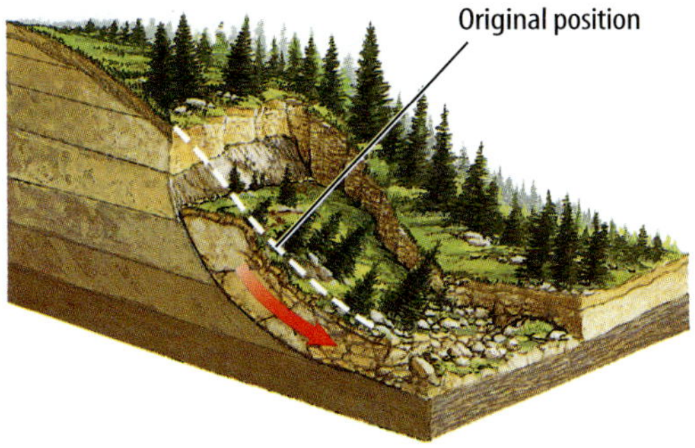

Figure 2 Slump occurs when material slips downslope as one large mass.
Infer *What might have caused this slump to happen?*

Dropping Sediments Agents of erosion drop the sediments they are carrying as they lose energy. This is called **deposition.** When sediments are eroded, they are not lost from Earth—they are just relocated.

Mass Movement

The greater an object's mass is, the greater its gravitational force is. Earth has such a great mass that gravity is a major force of erosion and deposition. Rocks and other materials, especially on steep slopes, are pulled toward the center of Earth by gravity.

A **mass movement** is any type of erosion that happens as gravity moves materials downslope. Some mass movements are so slow that you hardly notice they're happening. Others happen quickly—possibly causing catastrophes. Common types of mass movement include slump, creep, rockfalls, rock slides, and mudflows. Landslides are mass movements that can be one of these types or a combination of these types of mass movement.

Reading Check *What is a mass movement?*

Slump When a mass of material slips down along a curved surface, the mass movement is called **slump.** Often, when a slope becomes too steep, the base material no longer can support the rock and sediment above it. The soil and rock slip downslope as one large mass or break into several sections.

Sometimes a slump happens when water moves to the base of a slipping mass of sediment. This water weakens the slipping mass and can cause movement of material downhill. Or, if a strong rock layer lies on top of a weaker layer—commonly clay—the clay can weaken further under the weight of the rock. The clay no longer can support the strong rock on the hillside. As shown in **Figure 2,** a curved scar is left where the slumped materials originally rested.

Mini LAB

Modeling Slump

Procedure

WARNING: *Do not pour lab materials down the drain.*

1. Place one end of a **baking pan** on **two bricks** and position the other end over a sink with a sealed drain.
2. Fill the bottom half of the pan with **gelatin powder** and the top half of the pan with **aquarium gravel.** Place a large, **flat rock** on the gravel.
3. Using a **watering can,** sprinkle water on the materials in the pan for several minutes. Record your observations in your **Science Journal.**

Analysis

1. What happened to the different sediments in the pan?
2. Explain how your experiment models slump.

2 Teach

Reading Check

Answer any type of erosion that occurs as gravity moves materials downslope

Mini LAB

Purpose to model slump L2
ELL COOP LEARN LS
Visual-Spatial

Materials baking pan; two bricks; gelatin powder; aquarium gravel; large, flat rock; watering can; water

Teaching Strategy If there is no slump evident in the model, suggest students gradually add a little more water.

Analysis

1. The gelatin powder washed away, causing the gravel and pebbles to slip down the pan.
2. The inclined pan models a steep slope and the water from the watering can models rain. Erosion of base sediments combined with gravity causes slump.

Assessment

Performance Have students take before-and-after photos of their models to show the effects of slump. Use **Performance Assessment in the Science Classroom,** p. 149.

Teacher FYI

Steep Slopes Slopes commonly become too steep when people remove materials near their bases. When this happens, a slope becomes unstable and gravity pulls the mass downward. Many natural agents, such as rivers and waves, also can undercut slopes.

Identifying Misconceptions

Causes of Erosion Students may think that erosion is caused solely by the wind. Refer to page F at the beginning of this chapter for teaching strategies that address this misconception.

SECTION 1 Erosion by Gravity G ◆ 65

Use Science Words

Word Meaning Have students use dictionaries to define *slump*. Ask volunteers to read aloud the definitions of slump that describe a person's physical appearance and the condition of economic activity. Then ask students what all meanings of the word have in common. *Slump* describes a decline or collapse of something. LS **Linguistic**

Discussion

Creep and Slopes Does creep occur in our area? Answers will depend on local climate and topography. How does building on slopes increase erosion? When people build on slopes, they often make the slopes steeper. Sometimes, they remove the vegetation that provides stability to the slope.

Inquiry Lab

Splash Erosion

Purpose To observe and explore the force of water, have students make a target to measure how far soil will move when splashed with water.

Possible Materials target with 4 or 5 rings, meterstick, dropper, soil, water, compass

Estimated Time 1 class session

Teaching Strategies

- Students can create their own target using the compass to draw a circle, a bigger one around that, and so on until there are 4 to 5 rings.
- Students place soil in the middle of the target.
- Students can hold the dropper at various measurements above the target and measure how far soil moves from the middle.
- Students could use other containers that could let smaller or bigger drops hit the soil.

For additional inquiry activities, see *Science Inquiry Labs*.

Figure 3 Over time, creep has caused these tree trunks to lean downhill. The trees then curved back toward the Sun.

Creep The next time you travel, look along the roadway or trail for slopes where trees and fence posts lean downhill. Leaning trees and human-built structures show another mass movement called creep. **Creep** occurs when sediments slowly shift their positions downhill, as **Figure 3** illustrates. Creep is common in areas of frequent freezing and thawing.

Rockfalls and Rock Slides Signs along mountainous roadways warn of another type of mass movement called rockfalls. Rockfalls happen when blocks of rock break loose from a steep slope and tumble through the air. As they fall, these rocks crash into other rocks and knock them loose. More and more rocks break loose and tumble to the bottom. The fall of a single, large rock down a steep slope can cause serious damage to structures at the bottom. During the winter, when ice freezes in the cracks of rocks, the cracks expand and extend. In the spring, the pieces of rock break loose and fall down the mountainside, as shown in **Figure 4.**

Rock slides occur when layers of rock—usually steep layers—slip downslope suddenly. Rock slides, like rockfalls, are fast and can be destructive in populated areas. They commonly occur in mountainous areas or in areas with steep cliffs, also as shown in **Figure 5.** Rock slides happen most often after heavy rains or during earthquakes, but they can happen on any rocky slope at any time without warning.

Figure 4 Rockfalls, such as this one, occur as material free falls through the air.

Figure 5 Rock slides are common in regions where layers of rock are steep.

Science Journal

Mass Movement Impact on Slope Have students use **Figure 4** to explain in their journals how mass movement affects slopes. Have them infer what the slope of the cliff looked like before weathering and erosion. LS **Visual-Spatial**

Visual Learning

Figure 4 Have students use a dictionary to define the term *talus.* Then have them use the figure to explain what talus slopes are and how they form. Talus slopes are piles of loose material that often form at the bases of steep cliffs. They are composed of the materials that weathered and eroded from the cliff.

Mudflows What would happen if you took a long trip and forgot to turn off the sprinkler in your hillside garden before you left? If the soil is usually dry, the sprinkler water could change your yard into a muddy mass of material much like chocolate pudding. Part of your garden might slide downhill. You would have made a mudflow, a thick mixture of sediments and water flowing down a slope. The mudflow in **Figure 6** caused a lot of destruction.

Mudflows usually occur in areas that have thick layers of loose sediments. They often happen after vegetation has been removed by fire. When heavy rains fall on these areas, water mixes with sediment, causing it to become thick and pasty. Gravity causes this mass to flow downhill. When a mudflow finally reaches the bottom of a slope, it loses its energy of motion and deposits all the sediment and everything else it has been carrying. These deposits often form a mass that spreads out in a fan shape. Why might mudflows cause more damage than floodwaters?

Figure 6 Mudflows, such as these in the town of Sarno, Italy, have enough energy to move almost anything in their paths.
Explain *how mudflows differ from slumps, creep, and rock slides.*

Reading Check ***What conditions are favorable for triggering mudflows?***

Mudflows, rock slides, rockfalls, creep, and slump are similar in some ways. They all are most likely to occur on steep slopes, and they all depend on gravity to make them happen. Also, all types of mass movement occur more often after a heavy rain. The water adds mass and creates fluid pressure between grains and layers of sediment. This makes the sediment expand—possibly weakening it.

INTEGRATE Physics

Driving Force The force that drives most types of erosion is gravity. Water at an elevation has potential, or stored energy. When water drops in elevation this energy changes to kinetic energy, or energy of motion. Water may then become a powerful agent of erosion. Find out how water has shaped the region in which you live.

Consequences of Erosion

People like to have a great view and live in scenic areas away from noise and traffic. To live this way, they might build or move into houses and apartments on the sides of hills and mountains. When you consider gravity as an agent of erosion, do you think steep slopes are safe places to live?

Building on Steep Slopes When people build homes on steep slopes, they constantly must battle naturally occurring erosion. Sometimes builders or residents make a slope steeper or remove vegetation. This speeds up the erosion process and creates additional problems. Some steep slopes are prone to slumps because of weak sediment layers underneath.

Text Question Answer

Because they carry large amounts of sediment, mudflows are denser and more destructive than floodwaters.

Reading Check

Answer A normally dry area, with thick layers of dry sediments and vegetation removed by fire or other means, suddenly experiences a heavy rainfall.

Caption Answer

Figure 6 During a mudflow, wet sediment flows down a steep slope. Rock slides are usually dry. Creep acts more slowly, and slumps move en masse.

INTEGRATE Physics

The student may answer that water has shaped the region they live in by large rivers or streams.

Activity

Erosion Prevention Have students design and carry out an experiment to demonstrate how plant roots help hold soil and prevent erosion on slopes. Accept all reasonable designs. LS **Kinesthetic**

Quick Demo

Force of Gravity

Materials foil baking pans (2) with a hole on one end, potting soil, water, measuring cups (2)

Estimated Time 15 minutes

Procedure Place equal amounts of potting soil in the pans. Place one pan flat and elevate the other pan so that the end without the hole is raised 5–10 cm. Place measuring cups under the holes. Pour water over the soil. Measure how much soil comes out.

Differentiated Instruction

Challenge Have students use the new science words presented in this chapter to develop a crossword puzzle. They can use definitions from the chapter for clues. L3 P

Fun Fact

A raindrop, while falling through the air, is in the shape of a hamburger bun.

Reading Check

Answer Place drainage pipes or tiles in slopes to drain water; put walls or railroad ties on hillsides or plant vegetation to hold soil in place.

3 Assess

DAILY INTERVENTION

Check for Understanding

Linguistic Have students research friction and describe how it affects sediments on a slope.

Reteach

Classify Movement Have students examine each photograph in this section, identify the mass movement shown, and classify each type of movement as slow, fast, or very fast.

Assessment

Process Instruct students to draw a concept map that compares and contrasts the types of mass movement covered in this section. Use **Performance Assessment in the Science Classroom,** p. 161.

Figure 7 Some slopes are stabilized by building walls made from concrete or stone.

Making Steep Slopes Safe Plants can be beautiful or weedlike—but they all have root structures that hold soil in place. One of the best ways to reduce erosion is to plant vegetation. Deep tree roots and fibrous grass roots bind soil together, reducing the risk of mass movement. Plants also absorb large amounts of water. Drainage pipes or tiles inserted into slopes can prevent water from building up, too. These materials help increase the stability of a slope by allowing excess water to flow out of a hillside more easily.

Walls made of concrete or boulders also can reduce erosion by holding soil in place, as shown in **Figure 7.** However, preventing mass movements on a slope is difficult because rain or earthquakes can weaken all types of Earth materials, eventually causing them to move downhill.

Reading Check *What can be done to slow erosion on steep slopes?*

People who live in areas with erosion problems spend a lot of time and money trying to preserve their land. Sometimes they're successful in slowing down erosion, but they never can eliminate erosion and the danger of mass movement. Eventually, gravity wins. Sediment moves from place to place, constantly reducing elevation and changing the shape of the land.

section 1 review

Summary

Erosion and Deposition

- Gravity is the force that pulls all objects toward Earth's center.
- Water and wind erode materials only when they have enough energy of motion to do work.
- Agents of erosion drop sediment as they lose energy.

Mass Movement

- The greater an object's mass is, the greater its gravitational force is.
- Gravity is a major force of erosion and deposition.
- Common types of mass movement include slump, creep, rockfalls, rock slides, and mudflows.

Self Check

1. **Define** the term *erosion* and name the forces that cause it.
2. **Explain** how deposition changes the surface of Earth.
3. **Describe** the characteristics that all types of mass movements have in common.
4. **Describe** ways to help slow erosion on steep slopes.
5. **Think Critically** When people build houses and roads, they often pile up dirt or cut into the sides of hills. Predict how this might affect sediment on a slope. Explain how to control the effects of such activities.

Applying Skills

6. **Compare and Contrast** What are the similarities and differences between rock falls and rock slides?

Science Online bookg.msscience.com/self_check_quiz

section 1 review

1. Erosion: a process that wears away and moves surface materials; agents include gravity, glaciers, wind, and water.
2. Possible answer: Old surfaces and depressions are covered and filled in by sediments.
3. occur on slopes, depend on gravity, often occur after heavy rain
4. plant vegetation; control runoff; build walls
5. These activities produce steep, unstable slopes, which are subject to mass movements.
6. Both involve rock moving quickly downhill; rockfall: rocks break free from a steep slope and tumble through the air; rockslide: layers of rock slip downslope suddenly.

section 2

Glaciers

How Glaciers Form and Move

If you've ever gone sledding, snowboarding, or skiing, you might have noticed that after awhile, the snow starts to pack down under your weight. A snowy hillside can become icy if it is well traveled. In much the same way, glaciers form in regions where snow accumulates. Some areas of the world, as shown in **Figure 8,** are so cold that snow remains on the ground year-round. When snow doesn't melt, it piles up. As it accumulates slowly, the increasing weight of the snow becomes great enough to compress the lower layers into ice. Eventually, there can be enough pressure on the ice so that it becomes plasticlike. The mass slowly begins to flow in a thick, plasticlike lower layer, and ice slowly moves away from its source. A large mass of ice and snow moving on land under its own weight is a **glacier.**

Ice Eroding Rock

Glaciers are agents of erosion. As glaciers pass over land, they erode it, changing features on the surface. Glaciers then carry eroded material along and deposit it somewhere else. Glacial erosion and deposition change large areas of Earth's surface. How is it possible that something as fragile as snow or ice can push aside trees, drag rocks along, and slowly change the surface of Earth?

as you read

What You'll Learn

- **Explain** how glaciers move.
- **Describe** evidence of glacial erosion and deposition.
- **Compare and contrast** till and outwash.

Why It's Important

Glacial erosion and deposition create many landforms on Earth.

Review Vocabulary

plasticlike: not completely solid or liquid; capable of being molded or changing form

New Vocabulary

- glacier
- plucking
- till
- moraine
- outwash

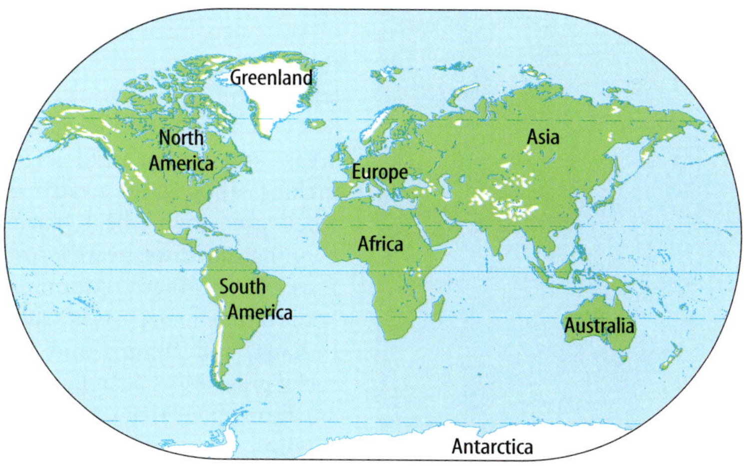

Figure 8 The white regions on this map show areas that are glaciated today.

Section 2 Resource Manager

Chapter FAST FILE Resources

Transparency Activity, pp. 41, 43–44
Directed Reading for Content Mastery, p. 16
Enrichment, p. 27
Reinforcement, p. 24
Lab Worksheets, pp. 5–6
Lab Activity, pp. 11–12

1 Motivate

INTERACTIVE CHALKBOARD PowerPoint® Presentations

Bellringer

Section Focus Transparencies also are available on the Interactive Chalkboard CD-ROM. L2 ELL

Section 2 Section Focus Transparency **I'll Be Back in 5,000 Years**

Meet Ötzi, a human from the Stone Age who lived about 5,000 years ago. He was somehow trapped in a rock depression and subsequently covered by snow and ice. Ötzi was eventually discovered by hikers on a ridge that divides Austria from Italy.

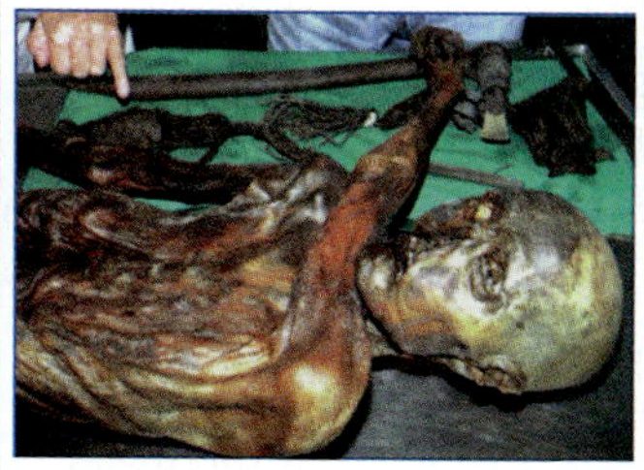

1. How was Ötzi preserved? What do you think modern scientists have learned from studying Ötzi?
2. What happened that resulted in Ötzi's eventual discovery?
3. In what areas do you most commonly see a lot of snow and ice?

L2

Erosional Forces

Tie to Prior Knowledge

Glaciers Show pictures of glaciers and ask students to tell what they know about them. In this section, students will learn how glaciers can change Earth's surface.

2 Teach

Quick Demo

Solid or Not

Materials cornstarch, 2 plastic cups, water

Estimated Time 15 minutes

Procedure Mix cornstarch with water in a clear, plastic cup to make a thick paste. Have students observe the mixture. Then have a volunteer pick up a handful of the mixture and tilt his or her palm to deposit the mixture into a second plastic cup. How is the cornstarch mixture like the ice at the base of a glacier? They both appear to be solids, but they flow like liquids.

Use an Analogy

Sandpaper Erosion Sandpaper is sand glued to paper. When sandpaper is pushed along a piece of wood, its sand particles leave small grooves in the wood. This is similar to rocks frozen in the bottom of glaciers. These rocks also leave grooves as they pass. Grooves from sandpaper and glaciers indicate the direction of movement.

Reading Check

Answer removal of rock pieces by ice

Fun Fact

It takes approximately 500 years to form 2.5 centimeters of topsoil.

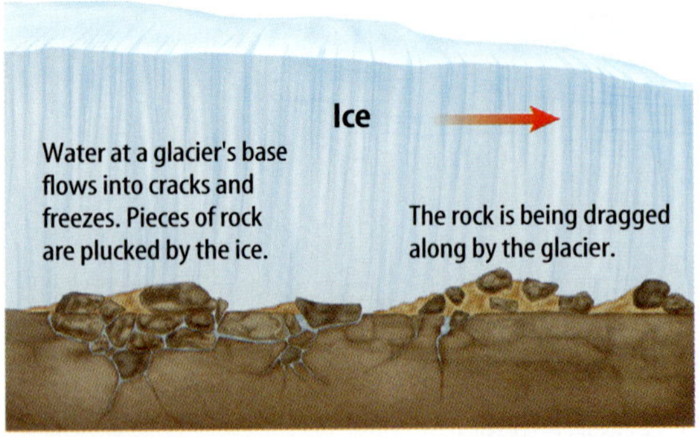

Figure 9 Plucking is a process that occurs when a moving glacier picks up loosened rock particles.

Plucking Glaciers weather and erode solid rock. When glacial ice melts, water flows into cracks in rocks. Later, the water refreezes in these cracks, expands, and fractures the rock. Pieces of rock then are lifted out by the ice, as shown in **Figure 9.** This process, called **plucking,** results in boulders, gravel, and sand being added to the bottom and sides of a glacier.

Reading Check *What is plucking?*

Transporting and Scouring As it moves forward over land, a glacier can transport huge volumes of sediment and rock. Plucked rock fragments and sand at its base scour and scrape the soil and bedrock like sandpaper against wood, eroding the ground below even more. When bedrock is gouged deeply by rock fragments being dragged along, marks such as those in **Figure 10** are left behind. These marks, called grooves, are deep, long, parallel scars on rocks. Shallower marks are called striations (stri AY shunz). Grooves and striations indicate the direction in which the glacier moved.

Ice Depositing Sediment

When glaciers begin to melt, they are unable to carry much sediment. The sediment drops, or is deposited, on the land. When a glacier melts and begins to shrink back, it is said to retreat. As it retreats, a jumble of boulders, sand, clay, and silt is left behind. This mixture of different-sized sediments is called **till.** Till deposits can cover huge areas of land. Thousands of years ago, huge ice sheets in the northern United States left enough till behind to fill valleys completely and make these areas appear flat. Till areas include the wide swath of what are now wheat farms running northwestward from Iowa to northern Montana. Some farmland in parts of Ohio, Indiana, and Illinois and the rocky pastures of New England are also regions that contain till deposits.

Figure 10 When glaciers melt, striations or grooves can be found on the rocks beneath. These glacial grooves on Kelley's Island, Ohio, give evidence of past glacial erosion and movement.

LAB DEMONSTRATION

Purpose to show the deposition of outwash sediments

Materials sand and gravel, rectangular loaf pan, large rectangular cake pan, hair dryer, book

Preparation Put a thin layer of sand and gravel in the loaf pan. Fill the pan with water and freeze it overnight.

Procedure Remove the ice-and-gravel-block and place it sediment-side down in the cake pan. Use a book to elevate the end of the pan under the ice. Use a hair dryer to slowly melt the ice.

Expected Outcome Students will see outwash deposits form at the bottom of the slope.

Assessment

How are the outwash materials deposited? Outwash is deposited from meltwater in layers—the heaviest sediments are closest to the "glacier," lighter sediments are carried farther away.

Moraine Deposits Till also is deposited at the end of a glacier when it is not moving forward. Unlike the till that is left behind as a sheet of sediment over the land, this type of deposit doesn't cover such a wide area. Rocks and soil are moved to the end of the glacier, much like items on a grocery store conveyor belt. Because of this, a big ridge of material piles up that looks as though it has been pushed along by a bulldozer. Such a ridge is called a **moraine.** Moraines also are deposited along the sides of a glacier, as shown in **Figure 11.**

Figure 11 Moraines are forming along the sides of this glacier. Unlike the moraines that form at the ends of glaciers, these moraines form as rock and sediment fall from nearby slopes.

Outwash Deposits When glacial ice starts to melt, the meltwater can deposit sediment that is different from till. Material deposited by the meltwater from a glacier, most often beyond the end of the glacier, is called **outwash.** Meltwater carries sediments and deposits them in layers. Heavier sediments drop first, so bigger pieces of rock are deposited closer to the glacier. The outwash from a glacier also can form into a fan-shaped deposit when the stream of meltwater deposits sand and gravel in front of the glacier.

Reading Check *What is outwash?*

Eskers Another type of outwash deposit looks like a long, winding ridge. This deposit forms in a melting glacier when meltwater forms a river within the ice, as shown in the diagram in **Figure 12.** This river carries sand and gravel and deposits them within its channel. When the glacier melts, a winding ridge of sand and gravel, called an esker (ES kur), is left behind. An esker is shown in the photograph in **Figure 12.**

Figure 12 Eskers are glacial deposits formed by meltwater.

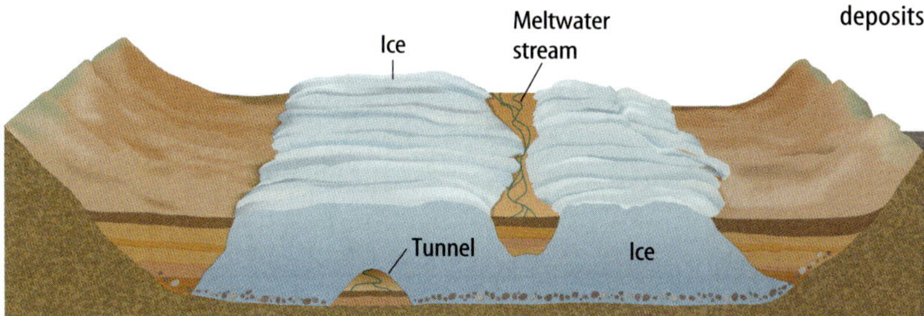

Eskers form when sediment deposited in ice tunnels or by streams on top of the ice is left behind on Earth's surface.

The snake-like shape of this esker in North Dakota is characteristic of this type of glacial deposit.

IDENTIFYING Misconceptions

Missing Glaciers Students may think that Earth's polar regions have always been covered with ice. In fact, for most of Earth's history, there have been no glaciers.

Reading Check

Answer material deposited by the meltwater from a glacier

Discussion

Confirming an Esker How might a geologist confirm that a ridge is an esker and not a ridge formed by some other means? Students should infer that geologists would have to dig into the ridge to discover its composition. If it is found to be made of deposits of sand and gravel, this is one clue that the ridge could be an esker.

Activity

Esker Creation Place a small steel rod in the bottom of a shallow pan. Fill the pan with water and put it into the freezer. Carefully remove the ice block from the pan. Allow the ice to thaw just until you can remove the steel rod from the ice. Fill the trench left by the rod with sand. Allow the ice to melt and watch an "esker" form.

Curriculum Connection

Music Have students work in small groups to write lyrics that describe glacial erosion and deposition. Encourage interested students to perform their songs in class. COOP LEARN LS **Auditory-Musical** P

Reading Check

Answer mostly near the poles

Discussion

Ice Age Air Temperature Infer what happened to the average air temperature on Earth at the end of the last ice age. Average air temperature rose.

Caption Answer

Figure 14 If your area shows evidence of glaciation, have students look for landforms such as grooves in rocks, moraines, and outwash deposits such as fans and eskers.

Visual Learning

Figure 14 Would you expect to find landforms in Mexico that formed by glaciation during the last ice age? Explain. No; glaciers didn't extend that far south.

Figure 13 Continental glaciers and valley glaciers are agents of erosion and deposition. This continental glacier covers a large area in Antarctica.

Continental Glaciers

The two types of glaciers are continental glaciers and valley glaciers. Today, continental glaciers, like the one in **Figure 13** cover ten percent of Earth, mostly near the poles in Antarctica and Greenland. These continental glaciers are huge masses of ice and snow. Continental glaciers are thicker than some mountain ranges. Glaciers make it impossible to see most of the land features in Antarctica and Greenland.

Reading Check *In what regions on Earth would you expect to find continental glaciers?*

Climate Changes In the past, continental glaciers covered as much as 28 percent of Earth. **Figure 14** shows how much of North America was covered by glaciers during the most recent ice advance. These periods of widespread glaciation are known as ice ages. Extensive glaciers have covered large portions of Earth many times over the last 2 million to 3 million years. During this time, glaciers advanced and retreated many times over much of North America. The average air temperature on Earth was about 5°C lower during these ice ages than it is today. The last major advance of ice reached its maximum extent about 18,000 years ago. After this last advance of glaciers, the ends of the ice sheets began to recede, or move back, by melting.

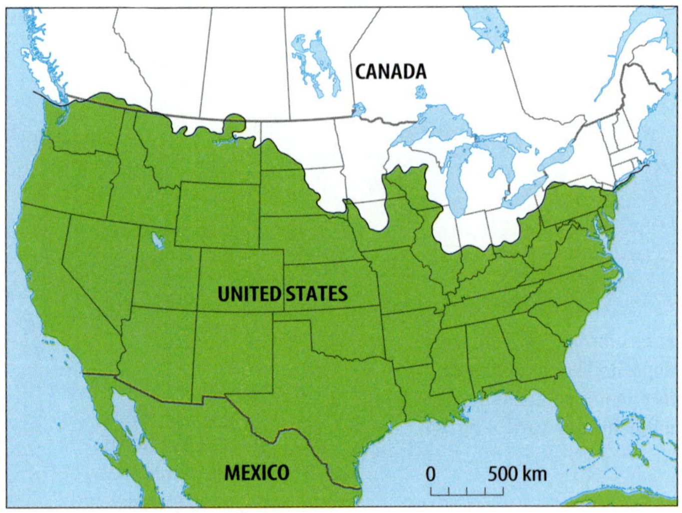

Figure 14 This map shows how much of North America was covered by continental glaciers about 18,000 years ago.
Observe *Was your location covered? If so, what evidence of glaciers does your area show?*

Differentiated Instruction

Learning Disabled Have students pinpoint on a map where they live and report whether the area exhibits any evidence of glaciation.

Challenge Have students research sea levels during the last ice age. Have them draw an outline map of North America and sea levels during the last ice age. Place it alongside an outline map of North America and current sea levels. L3 P

Curriculum Connection

History During the last ice advance, people traveled from Asia to North America across a strip of land that connected the two continents. This Bering land bridge was located where the Bering Strait is today. Have students locate the Bering Strait on a map and explain why there was a land bridge during the last ice advance. Much of Earth's water was locked up in glaciers, so sea level was lower.

Valley Glaciers

Valley glaciers occur even in today's warmer global climate. In the high mountains where the average temperature is low enough to prevent snow from melting during the summer, valley glaciers grow and creep along. **Figure 15** shows valley glaciers in Africa.

Evidence of Valley Glaciers If you visit the mountains, you can tell whether valley glaciers ever existed there. You might look for striations, then search for evidence of plucking. Glacial plucking often occurs near the top of a mountain where a glacier is mainly in contact with solid rock. Valley glaciers erode bowl-shaped basins, called cirques (SURKS), into the sides of mountains. If two valley glaciers side by side erode a mountain, a long ridge called an arête (ah RAYT) forms between them. If valley glaciers erode a mountain from several directions, a sharpened peak called a horn might form. **Figure 16** shows some features formed by valley glaciers.

Valley glaciers flow down mountain slopes and along valleys, eroding as they go. Valleys that have been eroded by glaciers have a different shape from those eroded by streams. Stream-eroded valleys are normally V-shaped. Glacially eroded valleys are U-shaped because a glacier plucks and scrapes soil and rock from the sides as well as from the bottom. A large U-shaped valley and smaller hanging valleys are illustrated in **Figure 16.**

Figure 15 Valley glaciers, like these on Mount Kilimanjaro in north Tanzania, Africa, form between mountain peaks that lie above the snow line, where snow lasts all year.

Figure 16 Valley glaciers transform the mountains over which they pass.

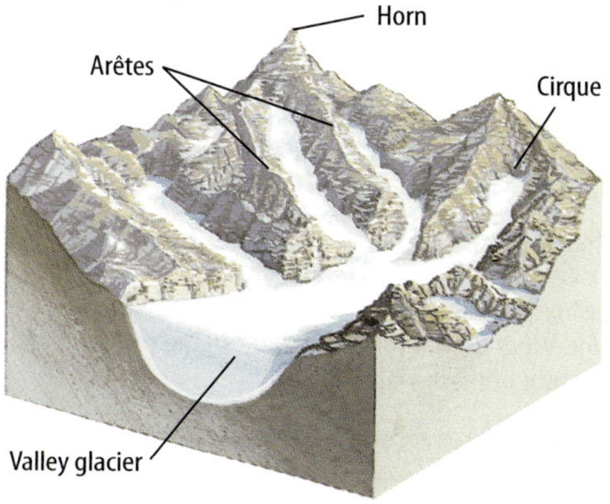

Bowl-shaped basins called cirques form by erosion at the start of a valley glacier. Arêtes form where two adjacent valley glaciers meet and erode a long, sharp ridge. Horns are sharpened peaks formed by glacial action in three or more cirques.

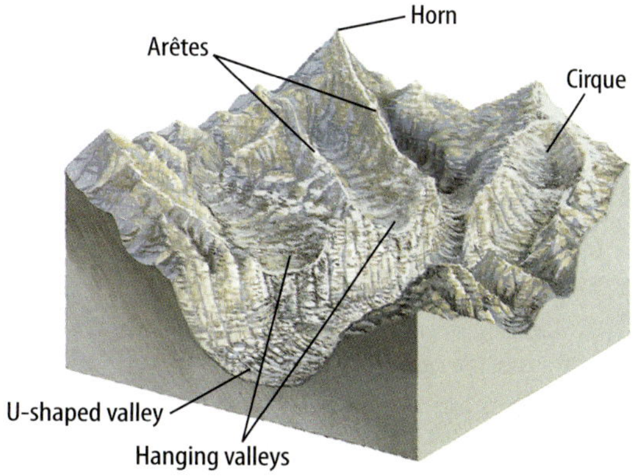

U-shaped valleys result when valley glaciers move through regions once occupied by streams. A tributary glacial valley whose mouth is high above the floor of the main valley is called a hanging valley. The discordance between the different valley floors is due to the greater erosive power of the trunk glacier.

Teacher FYI

Continental Glaciers Continental glaciation in Antarctica is almost 4,200 m thick in places and covers more than 13.9 million km^2. The continental glaciers in Greenland and Antarctica cover about 10 percent of Earth's land surface.

Discussion

Valley Glaciers How can a valley glacier form in an area that is tropical or subtropical? The glacier can form if it is at a high enough elevation so that the average temperature is low. LS **Logical-Mathematical**

Curriculum Connection

Geography Inform students that the Alps, Sierra Nevada, Rockies, and Himalaya all were greatly eroded by glaciers during the last ice advance. Have students locate these mountain ranges on a map and describe some of the glacial features they are likely to find in each location. L2

3 Assess

DAILY INTERVENTION

Check for Understanding

Linguistic Have students write a story describing changes if a new ice age were to begin.

Reteach

Glacial Landforms What landforms might you see in an area once covered by glaciers? grooves gouged in rocks, moraines, eskers, cirques, arêtes, horns, U-shaped valleys

Assessment

Content Have students compare and contrast the appearance of outwash and moraine deposits. Use **Performance Assessment in the Science Classroom**, p. 89.

Virtual Labs

Glacial Erosion *How do glaciers shape the land?*

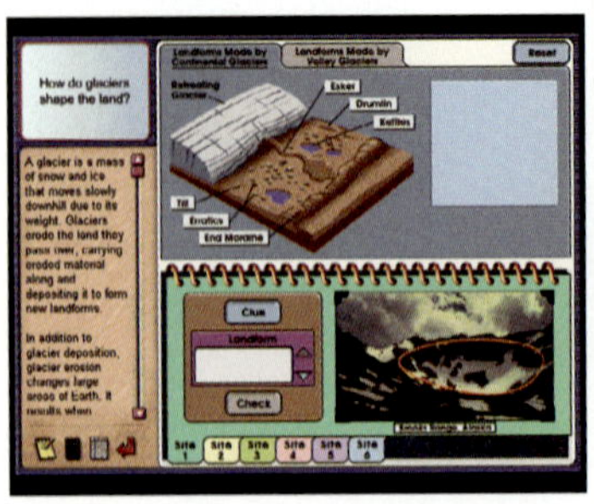

Figure 17 Sand and gravel deposits left by glaciers are important starting materials for the construction of roadways and buildings.

Science Online

Topic: Glacial Deposits

Visit bookg.msscience.com for Web links to information about uses of glacial deposits.

Activity List various uses of glacial deposits and name the methods of removing this material.

Importance of Glaciers

Glaciers have had a profound effect on Earth's surface. They have eroded mountaintops and transformed valleys. Vast areas of the continents have sediments that were deposited by great ice sheets. Today, glaciers in polar regions and in mountains continue to change the surface features of Earth.

In addition to changing the appearance of Earth's surface, glaciers leave behind sediments that are economically important, as illustrated in **Figure 17.** The sand and gravel deposits from glacial outwash and eskers are important resources. These deposits are excellent starting materials for the construction of roads and buildings.

section 2 review

Summary

How Glaciers Move and Form

- Glaciers form in regions where snow accumulates and remains year round.
- The weight of snow compresses the lower layers into ice and causes the ice to become plasticlike.

Ice Eroding Rock

- Glaciers are agents of erosion.
- Glacial erosion and deposition change large areas of the Earth's surface.

Ice Depositing Sediment

- Glaciers melt and retreat, leaving behind sediment.
- Forms of glacial deposits include till, moraines, outwash deposits and eskers.

Continental and Valley Glaciers

- Continental and valley glaciers are the two types of glaciers.

Self Check

1. **Describe** how glaciers move.
2. **Identify** two common ways in which a glacier can cause erosion.
3. **Determine** Till and outwash are glacial deposits. Explain how till and outwash are different.
4. **Discuss** How do moraines form? What are moraines made of?
5. **Think Critically** Many rivers and lakes that receive water from glacial meltwater often appear milky blue in color. What do you think might cause the milk appearance of these waters?

Applying Skills

6. **Recognize Cause and Effect** Since 1900, the Alps have lost 50 percent of their ice caps, and New Zealand's glaciers have shrunk by 26 percent. Describe what you think some causes and effects of this glacial melting have been.

Science Online bookg.msscience.com/self_check_quiz

section 2 review

1. The weight of the glacier causes ice to flow outward like pancake batter on a griddle. Under some conditions, glaciers also can slide over the ground.
2. by plucking and scouring
3. Till is a mixture of different-sized sediments dropped directly from the front, sides, and base of a glacier. Outwash consists of sediments deposited by meltwater.
4. Ice laden with sediment flows to the margins of a glacier, melts, and deposits the sediment. Moraines are made of till.
5. sediments in the glacial meltwater
6. Possible answer: The melting of the glaciers could have adversely affected water supplies and caused a rise in sea level.

74 ◆ G CHAPTER 3 Erosional Forces

LAB

GLACIAL GROOVING

Throughout the world's mountainous regions, 200,000 valley glaciers are moving in response to gravity.

Real-World Question

How is the land affected when a valley glacier moves downslope?

Goals

- **Compare** stream and glacial valleys.

Materials

sand
large plastic or metal tray
**stream table*
ice block
books (2 or 3)
**wood block*
metric ruler
overhead light source with reflector

**Alternate materials*

Safety Precautions

WARNINGS: *Do not pour sand down the drain. Make sure source is plugged into a GFI electrical outlet. Do not touch light source—it may be hot.*

Procedure

1. Set up the large tray of sand as shown above. Place books under one end of the tray to make a slope.
2. Cut a narrow riverlike channel through the sand. Measure and record its width and depth in a table similar to the one shown. Draw a sketch that includes these measurements.
3. Position the overhead light source to shine on the channel as shown.
4. Force the ice block into the channel at the upper end of the tray.
5. Gently push the ice along the channel until it's halfway between the top and bottom of the tray, and directly under the light.

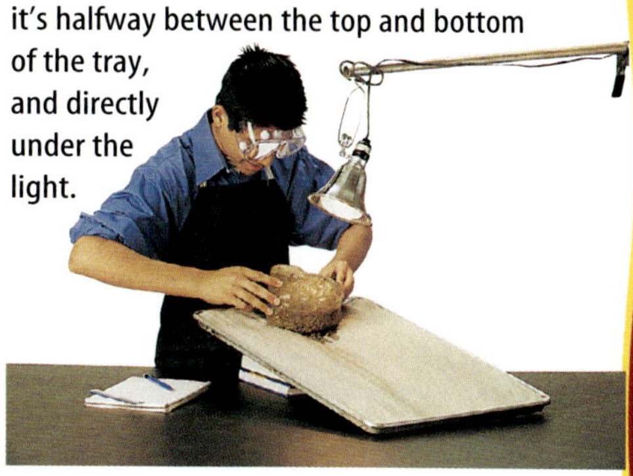

6. Turn on the light and allow the ice to melt. Record what happens.
7. Record the width and depth of the ice channel in the table. Make a scale drawing.

Conclude and Apply

1. **Explain** how you can determine the direction that a glacier traveled from the location of deposits.
2. **Explain** how you can determine the direction of glacial movement from sediments deposited by meltwater.
3. **Describe** how valley glaciers affect the surface over which they move.

Glacier Data

Sample Data	Width (cm)	Depth (cm)	Observations
Original channel	1–2	3	Stream channel looked V-shaped
Glacier channel	8	4	U-shaped channel
Meltwater channel	1–2	3.5	V-shaped channel

BENCH TESTED

Real-World Question

Purpose Students observe how valley glaciers affect the land. L2 ELL COOP LEARN

LS **Visual-Spatial** P

Process Skills observe, record and interpret data, make models, compare and contrast, relate cause and effect, measure in SI

Time Required 50 to 60 minutes

Procedure

Safety Precaution Make sure light source is plugged into a GFI electrical outlet. Remind students not to touch the light source.

Teaching Strategy Make blocks of ice a day before students do this lab.

Troubleshooting Have students adjust the overhead light to provide enough heat to melt the ice.

Conclude and Apply

1. Small hills of till mark the end position of a glacier in its valley.
2. Large outwash sediments are deposited closest to the glacial front. Smaller sediments are carried farther from the front.
3. Valley glaciers produce U-shaped valleys with steep sides and relatively flat bottoms.

Assessment

Performance Have students de-sign an experiment to test the effects of one of the following variables: slope, pressure applied to the ice block, mass of the ice block, amount of sediments in the ice block, or the width of the original channel. Use **PASC**, p. 95.

1 Motivate

INTERACTIVE CHALKBOARD
PowerPoint® Presentations

Bellringer

Section Focus Transparencies also are available on the Interactive Chalkboard CD-ROM. L2 ELL

Section 3 Section Focus Transparency **Rock On!**

You may have noticed animal tracks in the sand, but have you ever seen boulder tracks? If you go to the Racetrack in Death Valley, this is exactly what you'll see. Rocks at the Racetrack seem to move without any help at all!

1. No one has ever seen a rock move at the Racetrack. What evidence do you have that these rocks are moving without human interference?
2. Pretend you are scientists studying Death Valley. Develop a theory that explains the mysterious movements at the Racet[rack.] L2

Tie to Prior Knowledge

Dust in the Wind Have students recall seeing dust carried by the wind. Ask students how wind speed affects the amount and type of material that the wind can transport. Stronger winds carry more and heavier materials. Tell students they will find out how wind can alter Earth's surface in this section.

section 3

Wind

as you read

***What* You'll Learn**

- **Explain** how wind causes deflation and abrasion.
- **Recognize** how loess and dunes form.

***Why* It's Important**

Wind erosion and deposition change landscapes, especially in dry climates.

Review Vocabulary
friction: force that opposes the motion of an object when the object is in contact with another object or surface

New Vocabulary
- deflation
- abrasion
- loess
- dune

Wind Erosion

When air moves, it picks up loose material and transports it to other places. Air differs from other erosional forces because it usually cannot pick up heavy sediments. Unlike rivers that move in confined places like channels and valleys, wind carries and deposits sediments over large areas. For example, wind is capable of picking up and carrying dust particles from fields or volcanic ash high into the atmosphere and depositing them thousands of kilometers away.

Deflation Wind erodes Earth's surface by deflation (dih FLAY shun) and abrasion (uh BRAY zhun). When wind erodes by **deflation,** it blows across loose sediment, removing small particles such as silt and sand. The heavier, coarser material is left behind.

Abrasion When windblown sediment strikes rock, the surface of the rock gets scraped and worn away by a process called **abrasion.** Abrasion, shown in **Figure 18,** is similar to sandblasting. Workers use machines that spray a mixture of sand and water under high pressure against a building. The friction wears away dirt from stone, concrete, or brick walls. It also polishes the walls of buildings by breaking away small pieces and leaving an even, smooth finish. Wind acts like a sandblasting machine, bouncing and blowing sand grains along. These sand grains strike against rocks and break off small fragments. The rocks become pitted and are worn down gradually.

Reading Check *How is wind abrasion similar to sandblasting?*

Deflation and abrasion happen to all land surfaces but occur mostly in deserts, beaches, and plowed fields. These areas have fewer plants to hold the sediments in place. When winds blow over them, they can be eroded rapidly. Grassland or pasture land have many plants that hold the soil in place, therefore there is little soil erosion caused by the wind.

Figure 18 The odd shape of this boulder was produced by wind abrasion.

Section 3 Resource Manager

Chapter *FAST FILE* Resources
- Transparency Activity, p. 42
- Directed Reading for Content Mastery, pp. 17, 18
- MiniLAB, p. 4
- Enrichment, p. 28
- Reinforcement, p. 25
- Lab Worksheets, pp. 7–8

Mathematics Skill Activities, p. 9
Home and Community Involvement, p. 41
Performance Assessment in the Science Classroom, p. 44
Lab Management and Safety, p. 65

Sandstorms Even when the wind blows strongly, it seldom carries sand grains higher than 0.5 m from the ground. However, sandstorms do occur. When the wind blows forcefully in the sandy parts of deserts, sand grains bounce along and hit other sand grains, causing more and more grains to rise into the air. These windblown sand grains form a low cloud just above the ground. Most sandstorms occur in deserts, but they can occur in other arid regions.

Dust Storms When soil is moist, it stays packed on the ground, but when it dries out, it can be eroded by wind. Soil is composed largely of silt- and clay-sized particles. Because these small particles weigh less than sand-sized particles of the same material, wind can move them high into the air.

Silt and clay particles are small and stick together. A faster wind is needed to lift these fine particles of soil than is needed to lift grains of sand. However, after they are airborne, the wind can carry them long distances. Where the land is dry, dust storms can cover hundreds of kilometers. These storms blow topsoil from open fields, overgrazed areas, and places where vegetation has disappeared. In the 1930s, silt and dust that was picked up in Kansas fell in New England and in the North Atlantic Ocean. Dust blown from the Sahara has been traced as far away as the West Indies—a distance of at least 6,000 km.

INTEGRATE History

Dust Bowl Poor agricultural practices and a long period of sustained drought caused the Dust Bowl of the 1930s. Research how this affected the livelihood of the people of the southern plains.

Applying Science

What factors affect wind erosion?

Many factors compound the effects of wind erosion. But can anything be done to minimize erosion?

Identifying the Problem

Wind velocity and duration, the size of sediment particles, the size of the area subjected to the wind, and the amount of vegetation present all affect how much soil is eroded by wind. The table shows different combinations of these factors. It also includes an erosion rating that depends upon what factors pertain to an area.

Factors That Affect Wind Erosion

Factor	Descriptions				
Wind velocity	high	high	low	low	low
Duration of wind	long	long	short	long	long
Particle size	coarse	medium	coarse	coarse	medium
Surface area	large	large	small	small	large
Amount of vegetation	high	low	high	high	high
Erosion rating	some	a lot	a little	some	?

Solving the Problem

1. Looking at the table, can you figure out which factors increase and which factors decrease the amount of erosion?
2. From what you've discovered, can you estimate the missing erosion rating?

2 Teach

Reading Check

Answer The impact of the sand being blown against the rocks wears the rocks down.

INTEGRATE History

Dust Bowl Poor agricultural practices and a long period of sustained drought caused the Dust Bowl of the 1930s. Students will find that small midwestern farmers lost their livelihood as their farmlands dried up and the topsoil blew away.

Career Have students research various careers that teach and help people with soil conservation.

Discussion

Erosion Vulnerability How does over-farming and over-grazing make some lands vulnerable to wind erosion? Is the same land as easily harmed by rain as it is by wind? Over-farming and over-grazing exposes the soil. During dry periods, the soil can blow away. Likewise, heavy rain can wash exposed soil away.

Applying Science

Answers

1. High wind velocity, long duration, and large surface area all increase wind erosion. Coarse particle size and a high amount of vegetation decrease erosion.
2. Students may say "a little" or "some." Low winds blowing over a large surface area for a long period of time would cause some erosion. However, the presence of a large amount of vegetation would reduce these effects.

Visual Learning

Figure 19 Ask students how this sandstorm is affecting the local area. Possible answers: Visibility is reduced; sand is removed from some places and deposited in others; structures may be sandblasted.

Activity

Dry Areas On a U.S. map, have students locate dry areas that are most likely to be affected by sandstorms or dust storms. L3

Mini LAB

Purpose Students will observe how roots help hold soil in place and reduce erosion. L2 ELL

COOP LEARN LS **Kinesthetic**

Materials sod, trowel, magnifying lens, old newspaper, bucket of water, soap

Teaching Strategy Have students hold the sod in a bucket of water and carefully massage the roots to remove the soil. Cover work areas with newspaper to hasten cleanup.

Analysis

1. Students' drawings should clearly show a branching, fibrous root system.
2. Grass roots branch out under the shaft of the grass, forming a root network that holds soil particles.

Assessment

Performance Have students pour water over soil with and without grass and compare erosional results. Use **Performance Assessment in the Science Classroom**, p. 97.

Try at Home

Fun Fact

A plant's root system may equal or exceed the size of the plant seen above the ground. For example, a tree that is 50 m tall might have a root system that covers an area the size of a football field.

Mini LAB

Observing How Soil Is Held in Place

Procedure

1. Obtain a piece of **sod** (a chunk of soil about 5 cm thick with grass growing from it).
2. Carefully remove soil from the sod roots by hand. Examine the roots with a **magnifying lens.**
3. Wash hands thoroughly with soap and water.

Analysis

1. Draw several of these roots in your **Science Journal.**
2. What characteristics of grass roots help hold soil in place and thus reduce erosion?

Try at Home

Science Online

Topic: Conservation Practices

Visit bookg.msscience.com for Web links to collect data on various methods of protecting soil from wind erosion.

Activity List methods that conserve soil. What is the most commonly used method by farmers?

Reducing Wind Erosion

INTEGRATE Life Science As you've learned, wind erosion is most common where there are no plants to protect the soil. Therefore, one of the best ways to slow or stop wind erosion is to plant vegetation. This practice helps conserve soil and protect valuable farmland.

Windbreaks People in many countries plant vegetation to reduce wind erosion. For centuries, farmers have planted trees along their fields to act as windbreaks that prevent soil erosion. As the wind hits the trees, its energy of motion is reduced. It no longer is able to lift particles.

In one study, a thin belt of cottonwood trees reduced the effect of a 25-km/h wind to about 66 percent of its normal speed, or to about 16.5 km/h. Tree belts also trap snow and hold it on land. This increases the moisture level of the soil, which helps prevent further erosion.

Roots Along many seacoasts and deserts, vegetation is planted to reduce erosion. Plants with fibrous root systems, such as grasses, work best at stopping wind erosion. Grass roots are shallow and slender with many fibers. They twist and turn between particles in the soil and hold it in place.

Planting vegetation is a good way to reduce the effects of deflation and abrasion. Even so, if the wind is strong and the soil is dry, nothing can stop erosion completely. **Figure 19** shows a project designed to decrease wind erosion.

Figure 19 Rows of grasses and rocks were installed on these dunes in Qinghai, China, to reduce wind erosion.

Active Reading

Jigsaw In this collaborative learning technique, individuals become experts on a portion of a text and share their expertise with a small group, called their home group. Everyone shares responsibility for learning the assigned reading. Assign each person in each home group an expert number (1 through 5, for example). Have students gather into the expert groups that correspond to their assigned number. Have them read, discuss, and master chapter concepts and determine how best to teach them to their home groups. Then have students return to their home groups and share the content they learned in their expert groups. Tell students to use the Jigsaw strategy to learn about ways to reduce wind erosion.

Deposition by Wind

Sediments blown away by wind eventually are deposited. Over time, these windblown deposits develop into landforms, such as dunes and accumulations of loess.

Loess Some examples of large deposits of windblown sediments are found near the Mississippi and Missouri Rivers. These wind deposits of fine-grained sediments known as **loess** (LES) are shown in **Figure 20.** Strong winds that blew across glacial outwash areas carried the sediments and deposited them. The sediments settled on hilltops and in valleys. Once there, the particles packed together, creating a thick, unlayered, yellowish-brown-colored deposit. Loess is as fine as talcum powder. Many farmlands of the midwestern United States have fertile soils that developed from loess deposits.

Figure 20 This sediment deposit is composed partially of windblown loess.

Dunes Do you notice what happens when wind blows sediments against an obstacle such as a rock or a clump of vegetation? The wind sweeps around or over the obstacle. Like a river, air drops sediment when its energy decreases. Sediment starts to build up behind the obstacle. The sediment itself then becomes an obstacle, trapping even more material. If the wind blows long enough, the mound will become a dune, as shown in **Figure 21.** A **dune** (DOON) is a mound of sediments drifted by the wind.

Reading Check *What is a dune?*

Dunes are common in desert regions. You also can see sand dunes along the shores of oceans, seas, or lakes. If dry sediments exist in an area where prevailing winds or sea breezes blow daily, dunes build up. Sand or other sediment will continue to build up and form a dune until the sand runs out or the obstruction is removed. Some desert sand dunes can grow to 100 m high, but most are much shorter.

Figure 21 Loose sediment of any type can form a dune if enough of it is present and an obstacle lies in the path of the wind.

Moving Dunes A sand dune has two sides. The side facing the wind has a gentler slope. The side away from the wind is steeper. Examining the shape of a dune tells you the direction from which the wind usually blows.

Unless sand dunes are planted with grasses, most dunes move, or migrate away from the direction of the wind. This process is shown in **Figure 22.** Some dunes are known as traveling dunes because they move rapidly across desert areas. As they lose sand on one side, they build it up on the other.

Discussion

Christmas Dunes Explain that along some beaches, people lay old Christmas trees in the sand to keep the sand from eroding. Why do large dunes develop in these areas? The trees act as barriers to the wind and trap the sand grains, which eventually form dunes.

Reading Check

Answer a mound of sediments drifted by the wind

Make a Model

Creating Dunes Have students wear goggles and use a large pan, cardboard, sand, and a hair dryer to create different sand dune shapes. Have students hold the cardboard against the far side of the pan to prevent sand from blowing out of the pan. By varying the amount of sand, the wind's velocity, and the direction of the wind, dune shapes will change. Have students match the dunes they make to the dunes shown in **Figures 22** and **23.**

Teacher FYI

Soil That We Use Of Earth's total surface, we use only 1/32 of it to grow the things we need.

Cultural Diversity

Wind Erosion In many areas of the world, deserts are spreading because of drought and the removal of native vegetation. To prevent wind erosion, people in Kenya are planting trees around cropland. Invite students to use the Internet to find out more about this project. Have them write short reports to share with the class. P

Differentiated Instruction

Challenge Have students research various ways people try to stop the migration of dunes into populated areas and have them give an oral report of their findings. L3

NATIONAL GEOGRAPHIC

Visualizing How Dunes Form and Migrate

Have students examine the pictures and read the captions. Then ask the following questions.

What do you think dunes on Hawaii's volcanic islands would be made from? eroded materials from lava flows or calcite shell fragments

Explain to students that the gently sloping side of the dune is the windward side, and the steep face is called the leeward side of the dune. Why do the sand grains stop blowing on the leeward side of a dune? Because the dune itself shelters them from the wind.

Activity

Dune Varieties Have small groups of students research the variety of shapes dunes form. Have them use sand to make models of transverse, barchan, longitudinal, star, linear, and parabolic dunes. Models should be made on sturdy cardboard.

NATIONAL GEOGRAPHIC

VISUALIZING HOW DUNES FORM AND MIGRATE

Figure 22

Sand blown loose from dry desert soil often builds up into dunes. A dune may begin to form when windblown sand is deposited in the sheltered area behind an obstacle, such as a rock outcrop. The sand pile grows as more grains accumulate. As shown in the diagram at right, dunes are mobile, gradually moved along by wind.

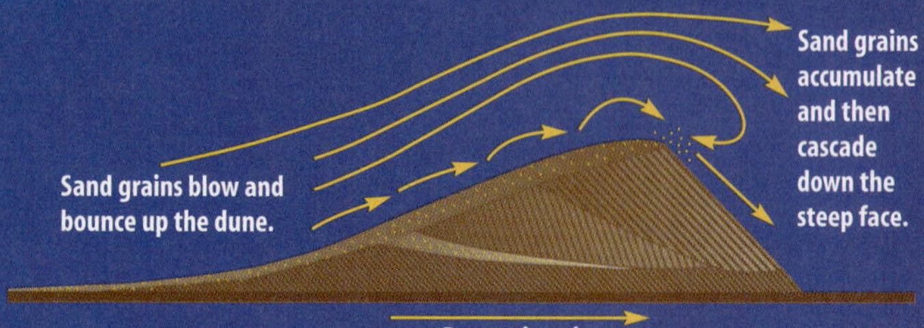

▲ A dune migrates as sand blows up its sloping side and then cascades down the steeper side. Gradually, a dune moves forward—in the same direction that the wind is blowing—as sand, lost from one side, piles up on the other side.

▲ Dunes are made of sediments eroded from local materials. Although many dunes are composed of quartz and feldspar, the brilliant white dunes in White Sands National Park, New Mexico, are made of gypsum.

▲ Deserts may expand when humans move into the transition zone between habitable land and desert. Here, villagers in Mauritania in northwestern Africa shovel the sand that encroaches on their schoolhouse daily.

◀ The dunes at left are coastal dunes from the Laguna Madre region of South Texas on the Gulf of Mexico. Note the vegetation in the photo, which has served as an obstacle to trap sand.

Differentiated Instruction

Learning Disabled Challenge students to make a poster of ways people restrict soil movement by planting vegetation, building fences, and using plant residue.

Dune Shape The shape of a dune depends on the amount of sand or other sediment available, the wind speed and direction, and the amount of vegetation present. One common dune shape is a crescent-shaped dune known as a barchan (BAR kun) dune. The open side of a barchan dune faces the direction that the wind is blowing. When viewed from above, the points of the crescent are directed downwind. This type of dune forms on hard surfaces where the sand supply is limited.

Another common type of dune, called a transverse dune, forms where sand is abundant. Transverse dunes are so named because the long directions of these dunes are perpendicular to the general wind direction. In regions where the wind direction changes, star dunes, shown in **Figure 23,** form pointed structures. Other dune forms also exist, some of which show a combination of features.

Figure 23 Star dunes form in areas where the wind blows from several different directions.

Shifting Sediments When dunes and loess form, the landscape changes. Wind, like gravity, running water, and glaciers, shapes the land. New landforms created by these agents of erosion are themselves being eroded. Erosion and deposition are part of a cycle of change that constantly shapes and reshapes the land.

section 3 review

Summary

Wind Erosion

- Air movement picks up loose material and transports it to other places.
- Deflation and abrasion happen mainly in deserts, beaches, and plowed fields.

Reducing Wind Erosion

- Planting vegetation can reduce wind erosion.
- Farmers use windbreaks to protect their crop fields from wind erosion.

Deposition by Wind

- Windblown deposits develop into landforms, such as dunes and accumulation of loess.
- Many farmlands of the midwestern United States have fertile soils that developed from loess deposits.

Self Check

1. **Compare and contrast** abrasion and deflation. Describe how they affect the surface of Earth.
2. **Explain** the differences between dust storms and sandstorms. Describe how energy of motion affects the deposition of sand and dust by these storms.
3. **Think Critically** You notice that sand is piling up behind a fence outside your apartment building. Explain why this occurs.

Applying Math

4. **Solve One-Step Equations** Between 1972 and 1992, the Sahara increased by nearly 700 km^2 in Mali and the Sudan. Calculate the average number of square kilometers the desert increased each year between 1972 and 1992.

Science Online bookg.msscience.com/self_check_quiz

3 Assess

DAILY INTERVENTION

Check for Understanding

Visual-Spatial Have students look at pictures of dunes in this section and explain the different shapes.

Reteach

Sand Dunes Have students work in groups to determine the effect of clumps of grass on the development of sand dunes. Each team will need the following items: safety goggles, a flat pan filled with sand, a hair dryer, cardboard, and clumps of grass. L2 ELL COOP LEARN

Quick Demo

Reducing Erosion

Materials small baking pan, clay, potting soil, cut up straw used for ground cover, water, small watering can

Estimated Time 15 minutes

Procedure Make 2 walls perpendicular in small baking pan with the clay. It should look like a small "T." In the 2 large sections, place potting soil, representing 2 crop fields. On one field, place straw over soil. In the 2 walls going across, make finger-width depressions so water can go through. Using a watering can, rain on both crop fields. The side with straw should have less erosion than the side without the straw.

Assessment

Content Assess students' abilities to sequence by having them list the steps involved in the formation of loess deposits. Use **Performance Assessment in the Science Classroom,** p. 163.

section 3 review

1. Deflation: wind picks up and transports small particles, leaving behind larger particles; abrasion: blowing sand grains strike rock and wear it away.
2. In sandstorms, sand remains close to the ground and doesn't blow very far. Dust storms pick up dust and carry it great distances. When energy of motion decreases, deposition occurs.
3. Wind striking the fence loses some of its energy of motion and deposits the sand.
4. $700\ km^2 \div 20 = 35\ km^2$ per year

SECTION 3 Wind G ◆ 81

BENCH TESTED

Real-World Question

Purpose Students observe how soil moisture and wind velocity affect wind erosion. L2 LS **Kinesthetic**

Process Skills design an experiment to test a hypothesis, identify and manipulate variables, make models, observe and infer, interpret data, relate cause and effect

Time Required 45 minutes for planning the experiment, 45 minutes for doing the experiment

Materials Three-speed hair dryers and wind-speed indicators are needed for this activity. Deep metal baking pans or paint trays and cardboard sheets will help contain the sediment.

Safety Precautions Have students wear safety goggles when sediments are being blown. Remind them to handle hot objects with care.

Form a Hypothesis

Possible Hypotheses Students might hypothesize that the greater the speed of the wind, the larger the sediments it can transport. Students might also suspect that moisture will add weight to the sediment and cause it to stick together. This should reduce erosion.

LAB Design Your Own

Blowing in the Wind

Goals

- **Observe** the effects of soil moisture and wind speed on wind erosion.
- **Design** and carry out experiments that test the effects of soil moisture and wind speed on wind erosion.

Possible Materials

flat pans (4)
fine sand (400 mL)
gravel (400 mL)
hair dryer
sprinkling can
water
28-cm × 35-cm cardboard sheets (4)
tape
mixing bowl
metric ruler
wind speed indicator

Safety Precautions

Wear your safety goggles at all times when using the hair dryer on sediments. Make sure the dryer is plugged into a GFI electrical outlet.

Real-World Question

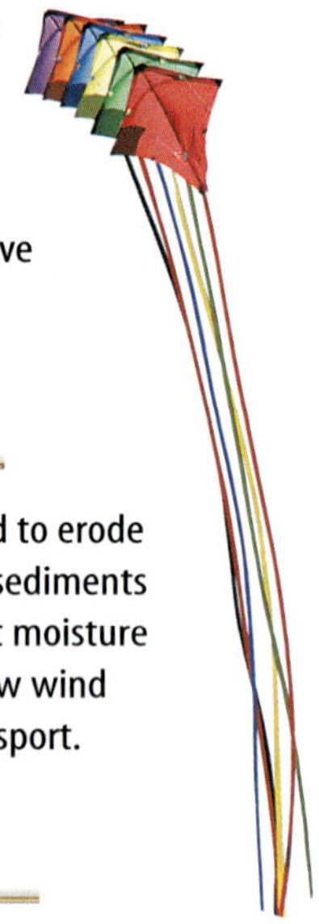

Have you ever played a sport outside and suddenly had the wind blow dust into your eyes? What did you do? Turn your back? Cover your eyes? How does wind pick up sediment? Why does wind pick up some sediments and leave others on the ground? What factors affect wind erosion?

Form a Hypothesis

How does moisture in sediment affect the ability of wind to erode sediments? Does the speed of the wind limit the size of sediments it can transport? Form a hypothesis about how sediment moisture affects wind erosion. Form another hypothesis about how wind speed affects the size of the sediment the wind can transport.

Test Your Hypothesis

Make a Plan

1. As a group, agree upon and write your hypothesis statements.
2. **List** the steps needed to test your first hypothesis. Plan specific steps and vary only one factor at a time. Then, list the steps needed to test your second hypothesis. Test only one factor at a time.

Test Your Hypothesis

Possible Procedures

- Mix equal amounts of sediments. Place half of the mixture in each of two pans. Tape a sheet of folded cardboard to one end of each pan. Sprinkle water onto the soil in one pan. Set the hair dryer at medium and blow the sediments toward the cardboard in each pan from the same angle and distance for two minutes.

Using Scientific Methods

3. Mix the sediments in the pans. Plan how you will fold cardboard sheets and attach them to the pans to keep sediments contained.
4. **Design** data tables in your Science Journal. Use them as your group collects data.
5. **Identify** all constants, variables, and controls of the experiment. One example of a control is a pan of sediment not subjected to any wind.

Sediment Movement

Sediment	Wind Speed	Sediment Moved
Fine sand (dry)	low	
	high	
Fine sand (wet)	low	
	high	
Gravel (dry)	low	
	high	Answers will vary.
Gravel (wet)	low	
	high	
Fine sand and gravel (dry)	low	
	high	
Fine sand and gravel (wet)	low	
	high	

Follow Your Plan

1. Make sure your teacher approves your plan before you start.
2. Carry out the experiments as planned.
3. While doing the experiments, write any observations that you or other members of your group make. Summarize your data in the data tables you designed in your Science Journal.

Analyze Your Data

1. **Compare** your results with those of other groups. Explain what might have caused any differences among the groups.
2. **Explain** the relationship that exists between the speed of the wind and the size of the sediments it transports.

Conclude and Apply

1. How does energy of motion of the wind influence sediment transport? What is the general relationship between wind speed and erosion?
2. **Explain** the relationship between the sediment moisture and the amount of sediment moved by the wind.

Communicating Your Data

Design a table that summarizes the results of your experiment, and use it to explain your interpretations to others in the class.

- To test the effect of wind speed, begin as previously described. Sediments in both pans should be dry. Blow air over one pan for two minutes on low. Repeat with the other pan, with the dryer on high.

Teaching Strategy Emphasize that the angle at which the dryers are held and the distance of the dryers from the pans must be the same each time.

Expected Outcome Soil moisture slows wind erosion. The greater the wind speed, the greater the size and amount of sediment transported.

Analyze Your Data

1. Results could vary slightly with variations in experimental design and procedure.
2. The greater the wind speed, the larger the sediments moved.

Error Analysis If higher wind speeds did not result in the movement of larger particles, wind speeds or particle sizes may have been too similar.

Conclude and Apply

1. When energy of motion increases, erosion increases. The greater the wind speed, the greater amount of erosion.
2. Moist sediments are not easily eroded.

Alternative Inquiry Lab

Textures and Wind How would soil with different textures (amount of sand, silt, clay) be affected by wind? Encourage students to design an experiment to explore this question or other questions that arise from the lab.

Communicating Your Data

Suggest that students use electronic spreadsheets for designing their tables.

Assessment

Performance To further assess students' understanding of wind erosion, have them design an experiment to collect data in order to graph the effects of different amounts of moisture on the amount of sediment that erodes. Use **PASC,** p. 95.

Science Stats

Content Background

About 10 percent of Earth's land is covered with glaciers. Today, continental glaciers cover most of Antarctica and Greenland. Glaciers can be up to 4,200 m deep and sometimes cover entire mountains. Several of Iceland's glaciers cover active volcanoes. When they erupt, floods of water and ice erode the land.

Erosion also occurs beneath Earth's surface, where groundwater causes limestone to dissolve. Water then carries away the dissolved materials, sometimes forming caves.

If it were not for the shifting of Earth's plates, which builds mountains and raises the land's surface in other ways, weathering and erosion might eventually wear the entire surface of Earth into a smooth, low plain.

Activity

Sculptures Have small groups of students research an area of the United States that has been sculpted by wind erosion or erosion by glaciers and prepare a travel brochure that would inform visitors to the area about how erosion affected the land.

Applying Math

Answer 2.5 m

Find Out About It

Mudflow A landslide is called a mudflow when heavy rain soaks into loose soil on slopes, lubricating the soil and causing it to flow like a thick liquid.

Science Stats

Losing Against Erosion

Did you know...

...Glaciers, one of nature's most powerful erosional forces, can move more than 30 m per day. In one week, a fast-moving glacier can travel the length of almost two football fields. Glaciers such as these are unusual—most move less than 10 cm per day.

...Some sand dunes migrate as much as 30 m per year. In a coastal region of France, traveling dunes have buried farms and villages. The dunes were halted by anti-erosion practices, such as planting grass in the sand and growing a barrier of trees between the dunes and farmland.

Applying Math If a sand dune is traveling at 30 m per year, how many meters does it travel in one month?

...In 1959, an earthquake triggered a mass movement in Madison River Canyon, Montana. About 21 million km^3 of rock and soil slid down the canyon at an estimated 160 km/h. This type of mass movement of earth is called a rock slide.

Find Out About It

Visit bookg.msscience.com/science_stats to learn about landslides. When is a landslide called a mudflow? In which U.S. states are mudflows most likely to occur?

84 ◆ G CHAPTER 3 Erosional Forces

Visual Learning

Homes in Danger from Shoreline Erosion by the Year 2060 Have students locate each coastline location on a U.S. map and ask them to compare the number of barrier islands along each coastline. Have them use an atlas or almanac to find the populations of states along each coast. Then ask why they think more homes along the Atlantic coastline are predicted to be in danger of shoreline erosion. Possible answer: More barrier islands on which homes are built occur along the Atlantic coastline. Hurricanes affect this coastline more often. The states along this coast line also have large populations.

Reviewing Main Ideas

Section 1 Erosion by Gravity

1. Erosion is the process that picks up and transports sediment.
2. Deposition occurs when an agent of erosion loses its energy and can no longer carry its load of sediment.
3. Slump, creep, rock slides, and mudflows are all mass movements caused by gravity.

Section 2 Glaciers

1. Glaciers are powerful agents of erosion. As water freezes and thaws in cracks, it breaks off pieces of surrounding rock. These pieces then are incorporated into glacial ice by plucking.
2. As sediment embedded in the base of a glacier moves across the land, grooves and striations form. Glaciers deposit two kinds of material—till and outwash.

Section 3 Wind

1. Deflation occurs when wind erodes only fine-grained sediments, leaving coarse sediments behind.
2. The pitting and polishing of rocks and grains of sediment by windblown sediment is called abrasion.
3. Wind deposits include loess and dunes. Loess consists of fine-grained particles such as silt and clay. Dunes form when windblown sediments accumulate behind an obstacle.

Visualizing Main Ideas

Copy and complete the following concept map on erosional forces. Use the following terms and phrases: striations, leaning trees and structures, curved scar on slope, deflation, *and* mudflows.

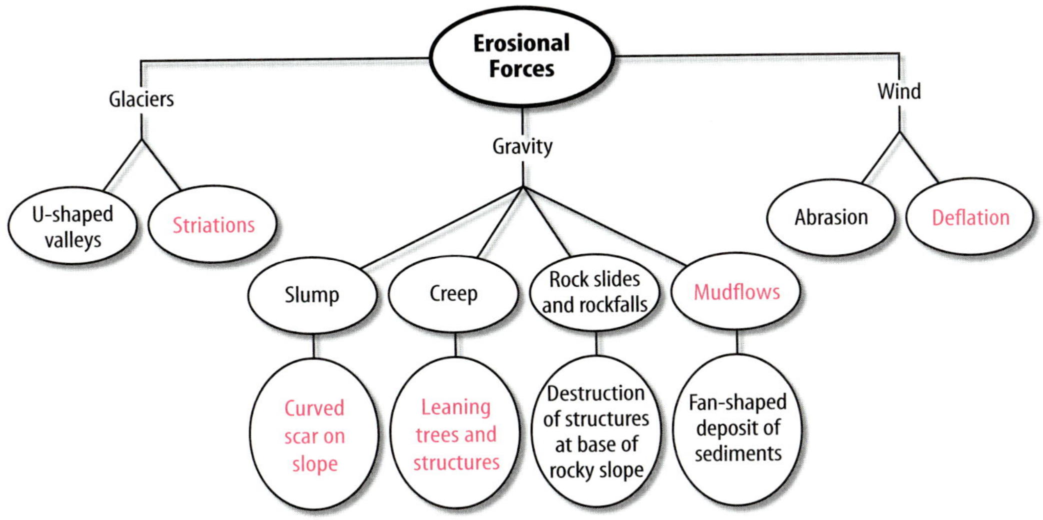

chapter 3 Study Guide

Reviewing Main Ideas

Summary statements can be used by students to review the major concepts of the chapter.

Visualizing Main Ideas

See student page.

Science Online

Visit bookg.msscience.com
/self_check_quiz
/interactive_tutor
/vocabulary_puzzlemaker
/chapter_review
/standardized_test

Identifying Misconceptions

Assess

Use the assessment as follow-up to page F at the beginning of this chapter.

Materials stream table, sand, ice cubes

Procedure Organize students into groups. Have them develop a procedure for investigating how an ice cube (a glacier) causes erosion. Ask each group to design a way to increase glacial erosion.

Expected Outcome Students should realize glaciers are agents of erosion. They should discover a direct correlation between ice thickness and how much erosion occurs. Help students understand that the slope of the valley (stream table) does not affect the amount of erosion significantly. Other factors that affect glacial scouring include the hardness of the substrate and the hardness of the rocks being carried by the glacier.

Using Vocabulary

1. deposition
2. creep
3. plucking
4. abrasion
5. mass movement
6. outwash
7. till

Checking Concepts

8. D
9. C
10. D
11. B
12. C
13. B
14. A
15. D
16. C

Assessment Transparency

For additional assessment questions, use the *Assessment Transparency* located in the transparency book.

Assessment

Assessment Transparency **Erosional Forces**

Directions: *Carefully review the table and answer the following questions.*

Wind Erosion Experiment Data (Using Fans) Height of Soil Mound (cm)					
Soil Mound	After 1 minute	After 2 minutes	After 3 minutes	After 4 minutes	After 5 minutes
A	29.5	28.5	27.5	26.5	25.5
B	28.0	26.5	24.0	22.0	20.0
C	30.0	29.5	29.0	28.5	28.0
D	30.5	30.5	30.5	30.5	30.5

1. According to this information, which mound of soil lost over 5 cm in height due to wind erosion?
   A Mound A C Mound C
   B Mound B D Mound D
2. If everything remains the same, what would be the height of Mound B after 10 minutes?
   F 12.0 cm H 5.0 cm
   G 8.0 cm J 3.0 cm
3. According to the table, Mound D could probably be composed of ___.
   A dry temperate soil
   B loosely packed semi-moist soil
   C fine grain sand
   D moist clay-like soil

L2

Erosional Forces

chapter 3 Review

Using Vocabulary

abrasion p. 76	loess p. 79
creep p. 66	mass movement p. 65
deflation p. 76	moraine p. 71
deposition p. 65	outwash p. 71
dune p. 79	plucking p. 70
erosion p. 64	slump p. 65
glacier p. 69	till p. 70

Each phrase below describes a vocabulary word from the list. In your Science Journal, write the term that matches each description.

1. loess, dunes, and moraines are examples
2. slowest mass movement
3. ice picking up pieces of rock
4. much like sandblasting
5. gravity transport of material downslope
6. sand and gravel deposited by meltwater
7. glacial deposit composed of sediment with many sizes and shapes

Checking Concepts

Choose the word or phrase that best answers the question.

8. Which term is an example of a feature created by deposition?
   A) cirque **C)** striation
   B) abrasion **D)** dune
9. The best plants for reducing wind erosion have what type of root system?
   A) taproot **C)** fibrous
   B) striated **D)** sheet
10. What does a valley glacier create at the point where it starts?
    A) esker **C)** till
    B) moraine **D)** cirque

Use the photo below to answer question 11.

11. Which of the following is suggested by leaning trees curving upright on a hillside?
    A) abrasion **C)** slump
    B) creep **D)** mudflow
12. What shape do glacier-created valleys have?
    A) V-shape **C)** U-shape
    B) L-shape **D)** S-shape
13. Which is formed by glacial erosion?
    A) eskers **C)** moraines
    B) arêtes **D)** warmer climate
14. What type of wind erosion leaves pebbles and boulders behind?
    A) deflation **C)** abrasion
    B) loess **D)** sandblasting
15. What is a ridge formed by deposition of till called?
    A) striation **C)** cirque
    B) esker **D)** moraine
16. What is the material called that is deposited by meltwater beyond the end of a glacier?
    A) esker
    B) cirque
    C) outwash
    D) moraine

Science Online bookg.msscience.com/vocabulary_puzzlemaker

Use the ExamView® Pro Testmaker CD-ROM to:

- create multiple versions of tests
- create modified tests with one mouse click for inclusion students
- edit existing questions and add your own questions
- build tests aligned with state standards using built-in State Curriculum Tags
- change English tests to Spanish with one mouse click and vice versa

86 ◆ G CHAPTER REVIEW

Thinking Critically

17. **Explain** how striations can give information about the direction that a glacier moved.

18. **Describe** how effective a retaining wall made of fine wire mesh would be against erosion.

19. **Determine** what can be done to prevent the migration of beach dunes.

20. **Recognize Cause and Effect** A researcher finds evidence of movement of ice within a glacier. Explain how this movement could occur.

21. **Think Critically** The end of a valley glacier is at a lower elevation than its point of origin is. How does this help explain melting at its end while snow and ice still are accumulating where it originated?

22. **Make and Use Tables** Make a table to contrast continental and valley glaciers.

23. **Concept Map** Copy and complete the events-chain concept map below to show how a sand dune forms. Use the terms and phrases: *sand accumulates, dune, dry sand,* and *obstruction traps.*

Dry sand
↓
Wind blows
↓
Obstruction traps
↓
Sand accumulates
↓
Dune

24. **Form a Hypothesis** Hypothesize why silt in loess deposits is transported farther than sand in dune deposits.

25. **Test a Hypothesis** Explain how to test the effect of glacial thickness on a glacier's ability to erode.

26. **Classify** the following as erosional or depositional features: loess, cirque, U-shaped valley, sand dune, abraded rock, striation, and moraine.

Performance Activities

27. **Poster** Make a poster with magazine photos showing glacial features in North America. Add a map to locate each feature.

28. **Design an experiment** to see how the amount of moisture added to sediments affects mass movement. Keep all variables constant except the amount of moisture in the sediment. Try your experiment.

Applying Math

29. **Slope** Gravity is a very powerful erosional force. This means the steeper a slope is, the more soil will move. A person can calculate how steep a slope is by using the height (rise) divided by the length (run). This answer is then multiplied by 100 to get percent slope. If you had a slope 15 m high and 50 m long, what would be the percent slope?

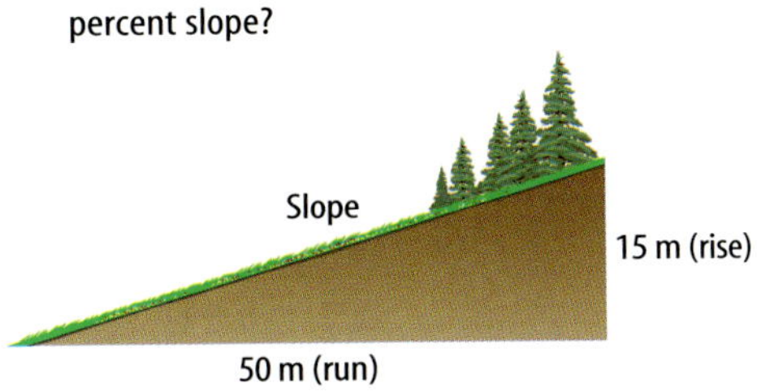

30. **Traveling Sand** A sand dune can travel up to 30 m per year. How far does the sand dune move per day?

Science Online bookg.msscience.com/chapter_review

Thinking Critically

17. Striations are generally parallel to glacial movement.
18. It only would prevent the movement of large particles.
19. Planting vegetation or erecting fences can reduce or prevent dune migration.
20. The mass of overlying snow and ice compresses underlying layers. Pressure causes the ice to become plastic and to flow.
21. Warm weather at lower elevations causes the front end to melt. Cold temperatures at higher elevations allow snow and ice to accumulate.
22. The table should contain two rows: *Continental Glacier* and *Valley Glacier*. Possible elements in the table include: *Continental:* large, Antarctica and Greenland; *Valley:* smaller, mountainous regions
23. See student page.
24. Loess is made of smaller sediments that wind can carry further.
25. Two ice blocks, one thicker than the other, could be guided over a pan containing sand.
26. erosional: cirque, U-shaped valley, abraded rock, striation; depositional: loess, sand dune, moraine

Performance Activities

27. Posters might show features such as eskers, moraines, U-shaped valleys, cirques, arêtes, horns, and striations in rock. Use **PASC,** p. 145.
28. Student designs will vary. Remind students to vary only the amount of moisture, with all other conditions remaining constant. Use **PASC,** p. 95.

Applying Math

National Math Standards

1, 6

29. 30%
30. .082 m/day

Assessment Resources

Reproducible Masters

Chapter *Fast File* Resources
- Chapter Review, pp. 39–40
- Chapter Tests, pp. 41–44
- Assessment Transparency Activity, p. 51

Glencoe Science Web site
- Chapter Review Test
- Standardized Test Practice

Glencoe Technology

- Assessment Transparency
- Exam*View*® Pro TestMaker
- MindJogger Videoquiz
- Interactive Chalkboard

FAST FILE

Answer Sheet A practice answer sheet can be found at bookg.msscience.com/answer_sheet.

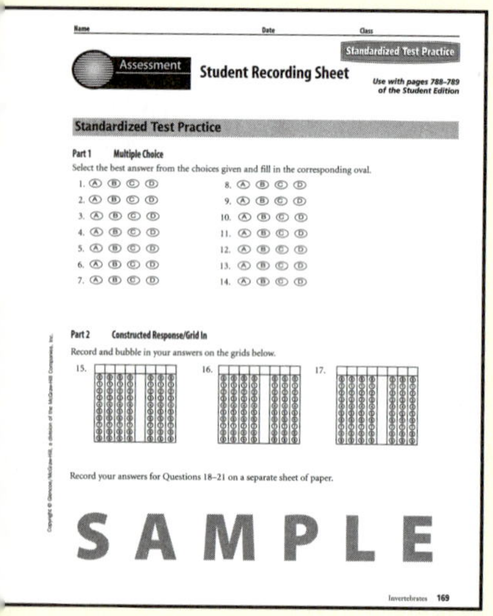
Assessment Student Recording Sheet

Standardized Test Practice

Use with pages 788–789 of the Student Edition

Standardized Test Practice

Part 1 Multiple Choice

Select the best answer from the choices given and fill in the corresponding oval.

Part 2 Constructed Response/Grid In

Record and bubble in your answers on the grids below.

Record your answers for Questions 18–21 on a separate sheet of paper.

SAMPLE

Part 1 Multiple Choice

1. C
2. B
3. C
4. A
5. D
6. C
7. A
8. B
9. A
10. D
11. D

Part 2 Short Response

12. Erosion is the picking up and carrying away of rock and/or sediment. Erosion lowers and sculpts Earth's surface.
13. Deposition is the process by which sediment that is being moved is dropped onto the land or into a body of water. till, moraine, outwash, eskers, rock slide, rockfalls, sand dunes
14. During a rockfall, boulders tumble through the air to the ground below. During a rock slide, sloping layers of rock break free and slide and tumble down a slope. Check student diagrams for accuracy.

chapter 3 Standardized Test Practice

Part 1 Multiple Choice

Record your answers on the answer sheet provided by your teacher or on a sheet of paper.

Use the illustration below to answer question 1.

1. Which type of mass movement is shown above?
   A. slump **C.** rock slide
   B. creep **D.** mudflow

2. Which term refers to sediment that is deposited by glacier ice?
   A. outwash **C.** loess
   B. till **D.** esker

3. During which process does wind pick up fine sediment?
   A. abrasion
   B. mass movement
   C. deflation
   D. deposition

4. On which of the following continents do continental glaciers exist today?
   A. Antarctica **C.** Australia
   B. Africa **D.** Europe

Test-Taking Tip

Take Your Time Stay focused during the test and don't rush, even if you notice that other students are finishing the test early.

5. Which forms when a rock in glacier ice slides over Earth's surface?
   A. moraine **C.** horn
   B. esker **D.** groove

6. What causes sediment and rock to move to lower elevations through time?
   A. sunlight **C.** gravity
   B. plant roots **D.** dust storms

7. Which can reduce wind erosion?
   A. windbreaks **C.** eskers
   B. dunes **D.** horns

8. Which consists of fine-grained, wind-blown sediment?
   A. moraine **C.** till
   B. loess **D.** rock fall

Use the diagram below to answer questions 9–11.

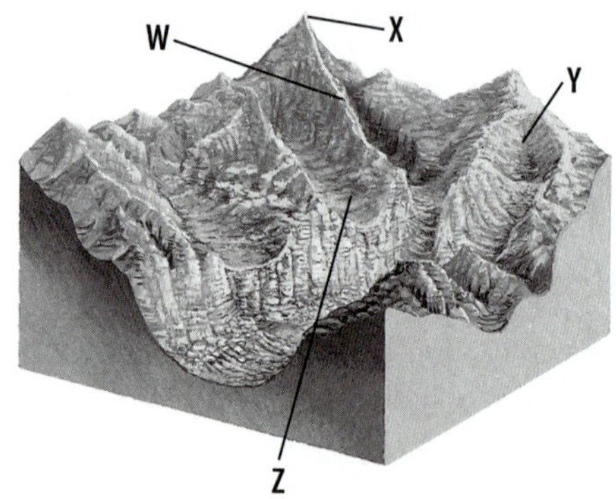

9. Which term describes point X?
   A. horn **C.** cirque
   B. arête **D.** hanging valley

10. Which term describes point Z?
    A. horn **C.** cirque
    B. arête **D.** hanging valley

11. Which agent of erosion created the landscape in the diagram?
    A. wind **C.** gravity
    B. water **D.** ice

Part 2 Short Response/Grid In

15. Glaciers pluck material from the substrate below. They also scratch and groove the rock beneath their base. till—sediment dropped from base of glacier, moraine—large ridge of sediment, outwash—deposited by meltwater, eskers—deposited from a glacial river
16. 7 m^3/s; 15 m^3/s
17. 5 m^3/s; 10 m^3/s
18. Stream flow is highest after the afternoon melting. It is lowest before daily temperatures begin to increase. Some days would be warmer than others, causing more melting.

Part 2 Short Response/Grid In

Record your answers on the answer sheet provided by your teacher or on a separate sheet of paper.

12. Give three examples of erosion? How does it affect Earth's surface?

13. What is deposition? Give three examples of how it changes Earth's surface.

14. How is a rock fall different from a rock slide? Use a labeled diagram to support your answer.

15. Explain how a glacier can erode the land, and then describe three forms of glacial deposition.

The graph below shows data about how much water flows through a stream. The stream is fed by glacial meltwater. Use the graph to answer questions 16–18.

Glacial Stream Flow

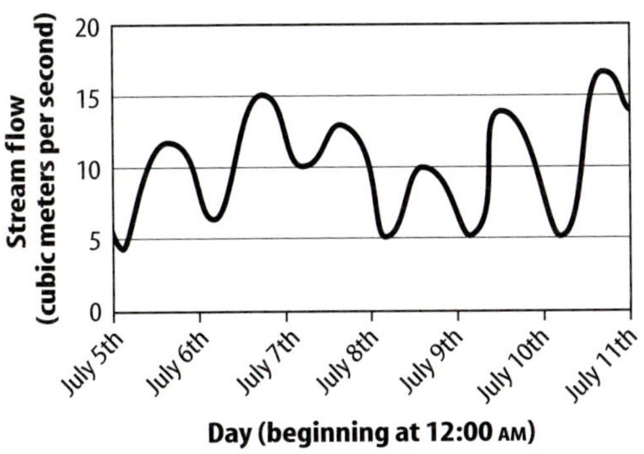

16. What were the lowest and highest amounts of stream flow on July 6th?

17. What were the lowest and highest amounts of stream flow on July 8th?

18. Notice that these data were obtained in July. Explain why the amount of stream flow from a glacier would vary each day. Give three examples to support your reasoning.

Part 3 Open Ended

Record your answers on a sheet of paper.

Use the diagram below to answer questions 19–22.

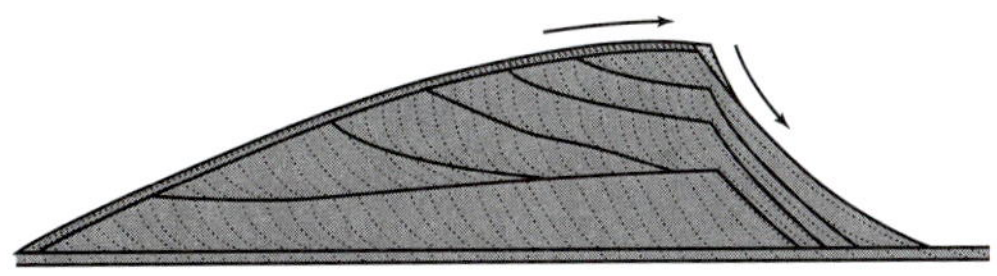

19. Describe the process that causes sand to move up the less steep side of the dune. Use a labeled diagram to support your answer.

20. Why does sand move down the steeper side of the dune? Use a labeled diagram to support your answer.

21. Design three time-lapse illustrations to show how sand dunes move across land.

22. Describe three ways to slow down the movement of sand dunes.

23. What types of damage are caused by landslides? How are people affected by landslides physically and economically?

24. Give three ways that damage from landslides can be reduced.

25. What is a dust storm? Where would you expect dust storms to occur? Give two safety suggestions for people caught in a dust storm.

26. Describe two ways that glacier ice can move across the surface.

27. How is creep different from most other types of mass movement? Explain the forces that cause creep, as well as the effect of creep.

28. Create a chart to show how continental glaciers are different from valley glaciers. Include their causes, physical features, and geological effects on the land.

Part 3 Open Ended

19. The force of the wind moving up and over the sand dune causes sand particles to roll and slide up the less-steep side of the dune.

20. Illustrations should show that after sand accumulates at the dune crest, it avalanches down the steep side under the influence of gravity.

21. Illustrations should show that dunes advance as sand is transferred from the less-steep side to the more steep side.

22. Answers may vary, but may include: sand grasses, planting trees, barriers in the sand–fences, breakfronts.

23. People can be killed or injured. Landslides destroy homes and other buildings, block roads and rail systems, and can trigger catastrophic flooding by destroying dams or levees. These effects may be economically devastating for victims, but in the long run, provide jobs for construction workers.

24. Students might indicate that the risk can be reduced by controlling drainage, maintaining appropriate vegetation cover, and building retaining and protective walls.

25. A dust storm is a cloud of silt and clay. They most often occur in dry areas, such as the southwestern U.S. Face masks, eye and ear protection.

26. Glacier ice can move in two ways: ice slowly deforms under the influence of gravity, or it can slide over its substrate.

27. Creep occurs much more slowly than other types of mass movement. Levels of water saturation and gravity cause soil to move downhill.

28. Continental glaciers–larger, thicker, often on flat land at low elevations, sliding over substrate. Valley glaciers–smaller, in alpine valleys, move by sliding and gravity. Both formed when lower air temperatures prevent ice from melting.

Rubrics

For more help evaluating open-ended assessment questions, see the rubric on p. 10T.

chapter 4 Organizer

Section/Objectives	Standards		Labs/Features
	National	State/Local	
Chapter Opener	See pp. 9T–10T for a Key to Standards.		**Launch Lab:** Model how Erosion Works, p. 91 **Foldables,** p. 91 A data-collection lab using Probeware technology can be found in the **Probeware Lab Manual,** pp. 25–28
Section 1 Surface Water 2 sessions 1 block 1. **Identify** the causes of runoff. 2. **Compare** rill, gully, sheet, and stream erosion. 3. **Identify** three different stages of stream development. 4. **Explain** how alluvial fans and deltas form.	National Content Standards: UCP.1–UCP.3, UCP.5, A.1, A.2, D.1, F.3		**Integrate Career,** p. 93 **Science Online,** p. 96 **Visualizing Stream Development,** p. 98 **Science Online,** p. 100 **MiniLAB:** Observing Runoff Collection, p. 101
Section 2 Groundwater 2 sessions 1 block 5. **Recognize** the importance of groundwater. 6. **Describe** the effect that soil and rock permeability have on groundwater movement. 7. **Explain** how groundwater dissolves and deposits minerals.	National Content Standards: UCP.1–UCP.3, UCP.5, A.1, A.2, D.1		**MiniLAB:** Measuring Pore Space, p. 104 **Applying Math:** Calculating the Rate of Groundwater Flow, p. 105 **Integrate Chemistry,** p. 107
Section 3 Ocean Shoreline 3 sessions 1.5 blocks 8. **Identify** the different causes of shoreline erosion. 9. **Compare** and contrast different types of shorelines. 10. **Describe** some origins of sand.	National Content Standards: UCP.1–UCP.3, UCP.5, A.1, A.2, D.1, F.3		**Lab:** Classifying Types of Sand, p. 113 **Lab:** Water Speed and Erosion, p. 114–115 **Science and Society:** Sands in Time, p. 116

Glencoe Exclusive! TeacherWorks™ All-In-One Planner and Resource Center

Lab Materials	Reproducible Resources	Section Assessment	Technology
Launch Lab: aluminum pie pan, dry soil, dropper, water	**Chapter *FAST FILE* Resources** Foldables Worksheet, p. 13 Directed Reading Overview, p. 15 Note-taking Worksheets, pp. 29–31	Glencoe's Assessment Advantage	TeacherWorks includes: • Interactive Teacher Edition • Lesson Planner with calendar • Access to all program blacklines • Correlations to standards • Web links
MiniLAB: plastic rain gauge, drinking glass, sink, sprinkling can, water, funnel	**Chapter *FAST FILE* Resources** Transparency Activity, p. 40 MiniLAB, p. 3 Enrichment, p. 26 Reinforcement, p. 23 Directed Reading, p. 16 Transparency Activity, pp. 43–44 **Cultural Diversity,** p. 33	**Portfolio** Science Journal, p. 95 **Performance** MiniLAB, p. 101 Applying Skills, p. 102 **Content** Section Review, p. 102	Section Focus Transparency Teaching Transparency Virtual Labs CD-ROM Guided Reading Audio Program Interactive Chalkboard CD-ROM
MiniLAB: 2 identical clear plastic containers, metric ruler, sand, gravel, measuring cup, water *Need materials?* Contact Science Kit at 1-800-828-7777 or www.sciencekit.com on the Internet.	**Chapter *FAST FILE* Resources** Transparency Activity, p. 41 MiniLAB, p. 4 Enrichment, p. 27 Reinforcement, p. 24 Directed Reading, p. 17 Lab Activities, pp. 9–12 **Mathematics Skill Activities,** p. 11 **Physical Science Critical Thinking/Problem Solving,** p. 12	**Portfolio** Differentiated Instruction, p. 104 **Performance** MiniLAB, p. 104 Applying Math, p. 105 Applying Skills, p. 108 **Content** Section Review, p. 108	Section Focus Transparency Guided Reading Audio Program Interactive Chalkboard CD-ROM
Lab: samples of different sands (3), magnifying lens, magnet **Lab:** paint roller pan, sand, 1-L beaker, rubber tubing (20 cm), metric ruler, water, stopwatch, fine-mesh screen, wood block	**Chapter *FAST FILE* Resources** Transparency Activity, p. 42 Enrichment, p. 28 Reinforcement, p. 25 Directed Reading, pp. 17, 18 Lab Worksheets, pp. 5–6, 7–8 **Lab Management and Safety,** p. 74	**Portfolio** Differentiated Instruction, p. 110 **Performance** Applying Math, p. 112 **Content** Section Review, p. 112	Section Focus Transparency Guided Reading Audio Program Interactive Chalkboard CD-ROM Video Lab Probeware Lab

End of Chapter Assessment

Glencoe's Assessment Advantage

Blackline Masters	Technology	Professional Series
Chapter *FAST FILE* Resources Chapter Review, pp. 33–34 Chapter Tests, pp. 35–38 **Standardized Test Practice,** pp. 19–22	MindJogger Videoquiz Virtual Labs CD-ROM Exam*View*® Pro Testmaker TeacherWorks CD-ROM Interactive Chalkboard CD-ROM	**Performance Assessment in the Science Classroom (PASC)**

Transparencies

Section Focus

Section 1 Section Focus Transparency **Take the Long Way Home**

Trail switchbacks are the parts of trails that zigzag up and down hills. People sometimes try to take short-cuts that go straight downhill, but this can be harmful. Short-cutting trail switchbacks creates grooves that funnel water during rain storms and leads to heavy damage of the hill and trail.

1. How might water running in a groove damage a trail?
2. What's the best way to prevent this sort of damage? Why?

L2

Section 2 Section Focus Transparency **Nature's Hot Tub**

Imagine you're a Japanese macaque and it's cold and snowy. What do you do? If you're lucky, you live near a hot spring where you could find a warm refuge from the winter storm. The Japanese macaque lives further north than any other ape or monkey.

1. How can you tell the water in this spring is hot?
2. Where do you think the water in hot springs comes from?

L2

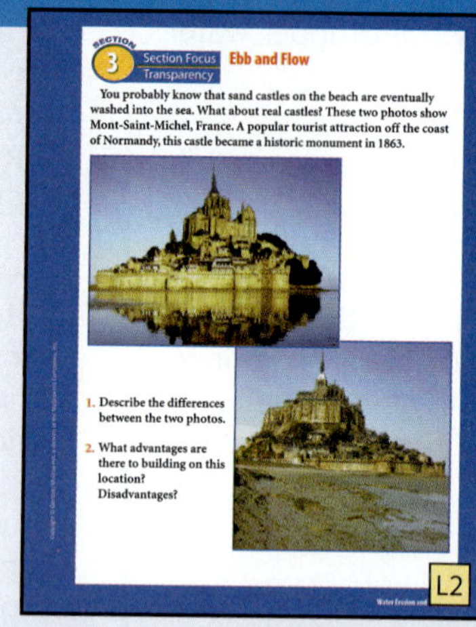
Section 3 Section Focus Transparency **Ebb and Flow**

You probably know that sand castles on the beach are eventually washed into the sea. What about real castles? These two photos show Mont-Saint-Michel, France. A popular tourist attraction off the coast of Normandy, this castle became a historic monument in 1863.

1. Describe the differences between the two photos.
2. What advantages are there to building on this location? Disadvantages?

L2

This is a representation of key blackline masters available in the Teacher Classroom Resources. See Resource Manager boxes within the chapter for additional information.

Assessment

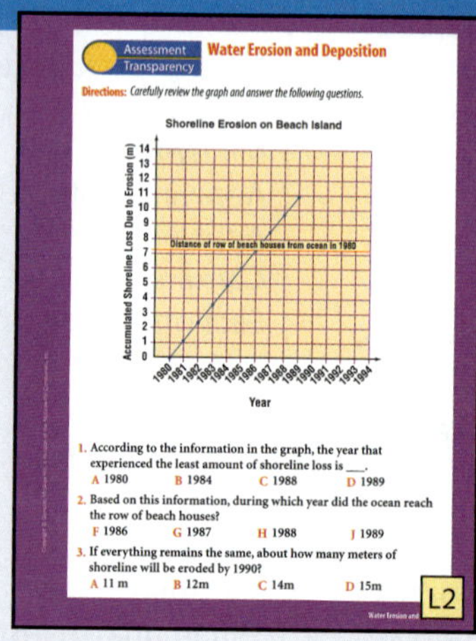
Assessment Transparency **Water Erosion and Deposition**

Directions: Carefully review the graph and answer the following questions.

1. According to the information in the graph, the year that experienced the least amount of shoreline loss is ___.
A 1980 B 1984 C 1988 D 1989
2. Based on this information, during which year did the ocean reach the row of beach houses?
F 1986 G 1987 H 1988 J 1989
3. If everything remains the same, about how many meters of shoreline will be eroded by 1990?
A 11 m B 12m C 14m D 15m

L2

Teaching

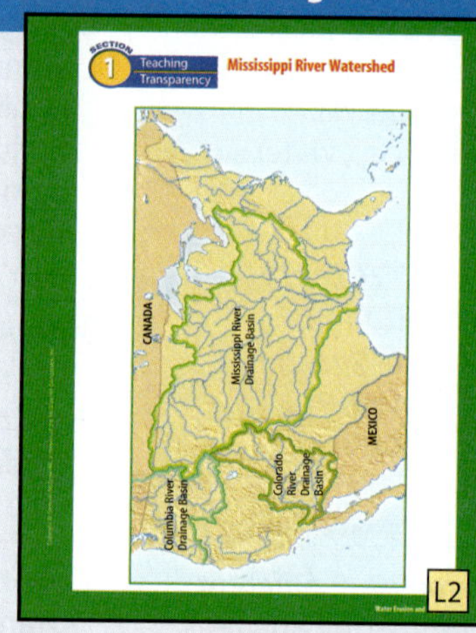
Section 1 Teaching Transparency **Mississippi River Watershed**

L2

Key to Teaching Strategies

The following designations will help you decide which activities are appropriate for your students.

- L1 Level 1 activities should be appropriate for students with learning difficulties.
- L2 Level 2 activities should be within the ability range of all students.
- L3 Level 3 activities are designed for above-average students.
- ELL ELL activities should be within the ability range of English Language Learners.
- COOP LEARN Cooperative Learning activities are designed for small group work.
- LS Multiple Learning Styles logos, as described on page 6T, are used throughout to indicate strategies that address different learning styles.
- P These strategies represent student products that can be placed into a best-work portfolio.
- PBL Problem-Based Learning activities apply real-world situations to learning.

Hands-on Activities

Student Text Lab Worksheet

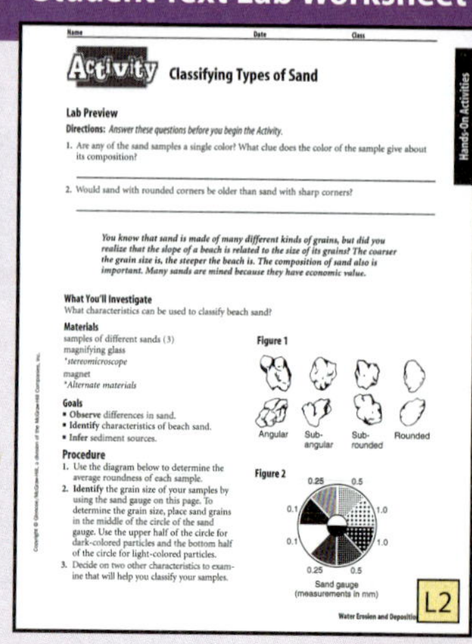
Activity **Classifying Types of Sand**

Lab Preview

Directions: *Answer these questions before you begin the Activity.*

1. Are any of the sand samples a single color? What clue does the color of the sample give about its composition?
2. Would sand with rounded corners be older than sand with sharp corners?

What You'll Investigate

Materials

Goals

Procedure

L2

Laboratory Activities

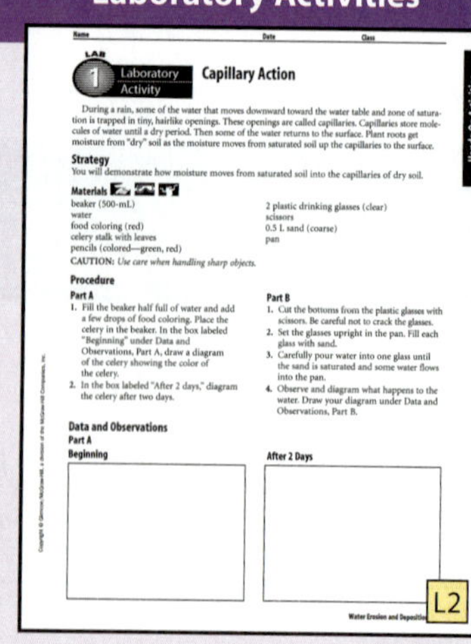
Lab 1 Laboratory Activity **Capillary Action**

Strategy

Materials

Procedure

Part A

Part B

Data and Observations

Part A

Beginning

After 2 Days

L2

Meeting Different Ability Levels

Content Outline

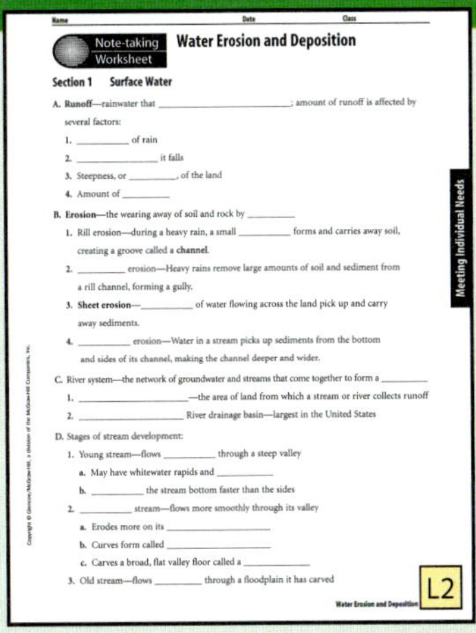

Reinforcement

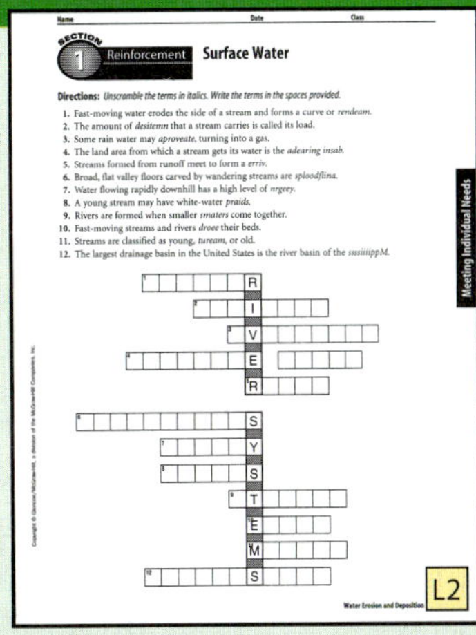

Enrichment

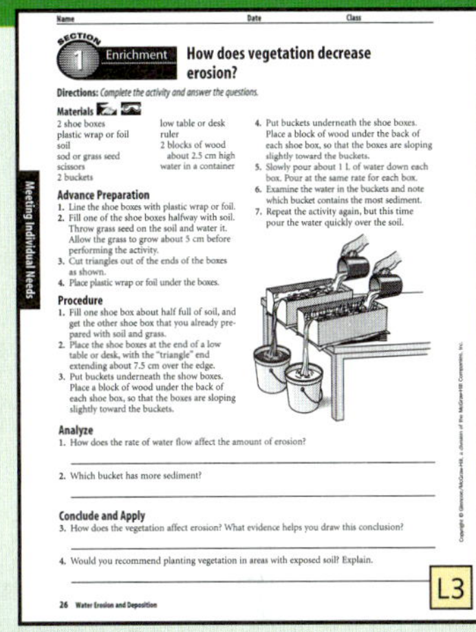

Directed Reading (English/Spanish)

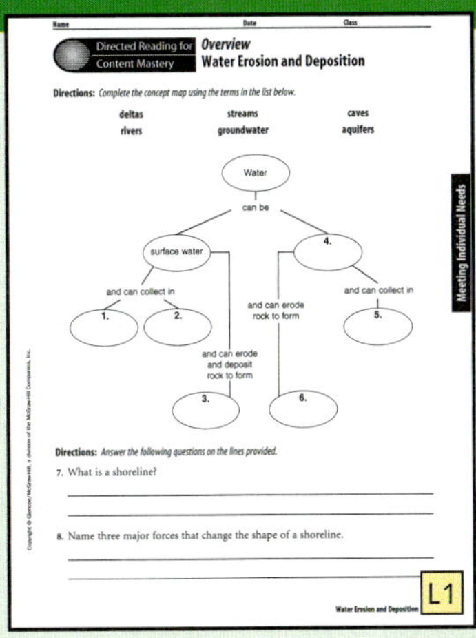

Study Guide

Reading Essentials

Assessment

Test Practice Workbook

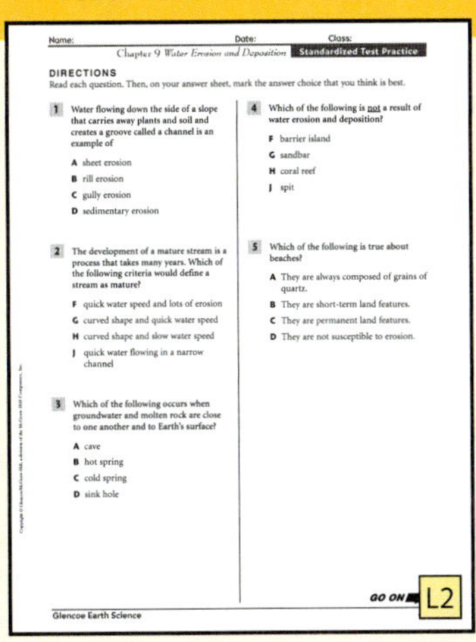

Chapter Review

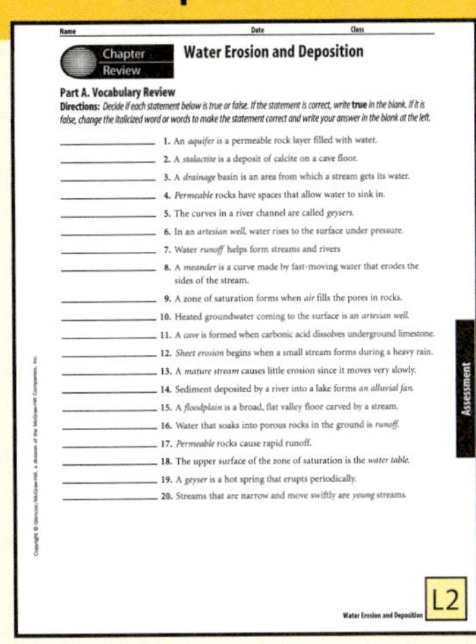

Chapter Tests

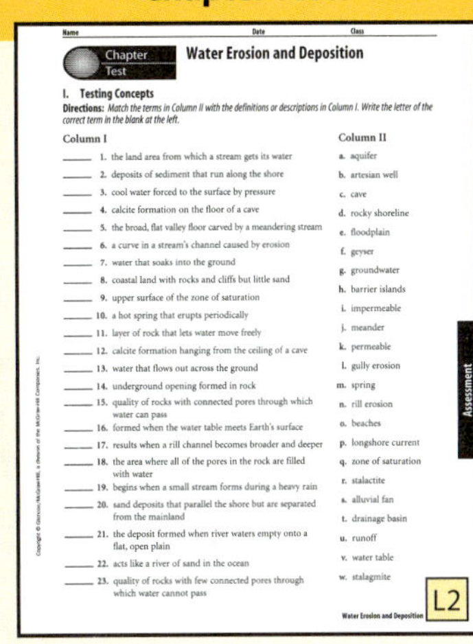

Science Content Background

section 1 Surface Water

Runoff

The average annual rainfall on the total land area of the United States is approximately 76 cm. Of this amount, 53 cm return to the atmosphere by evaporation and transpiration, less than 1 cm soaks into the ground, and 23 cm become runoff.

Effects of Gravity

Earth's gravity tries to pull everything on the planet toward its center. Every day, gravity pulls rocks, soil, and water down slopes. The weight of any object on Earth equals its mass times the acceleration of gravity, or $F_w = mg$.

Erosion

It generally takes between 80 and 400 years to form 1 cm of topsoil. In the United States, the amount of farmland soil lost to erosion each year far exceeds the amount of newly formed soil. Worldwide, depending on the region, topsoil is being lost 16 to 300 times faster than it can be replaced. With 1/3 of the world's topsoil already gone, the estimated additional loss each year exceeds 26 billion tons.

Drainage Basins

Drainage basins are separated from each other by high ground called divides. On a large scale, the Rocky Mountain and Appalachian Mountain divides form the boundaries of the Mississippi River drainage basin. When a river system develops in a smaller drainage basin, rills and gullies can erode a divide. With time, the divide will become narrower.

Teacher to Teacher

Rachel Shively,
Junior High Science Teacher
Aptakistic Junior High School
Buffalo Grove, Illinois

"To demonstrate water erosion, students take two plastic tubs (the longer the better) put dirt in one and sand in the other. With a finger, they make a 6 cm long indention in the soil and elevate the end with the indention approximately 10 cm. Supply two beakers of water to pour into the indention to see what happens. Then they do the same with the sand. They will either see a braided stream or meandering stream. Have them identify which one it is. Other supplies are then provided such as: wooden craft sticks, rocks, corrugated cardboard, yarn, etc. Have them scoop out the excess water and create barriers on the sides of their stream this time in the soil and sand. Then repeat the steps again. Which material appears to work the best? To further this investigation, you could have students investigate the materials used on their local waterways."

Rachel Shively

chapter content resources

Internet Resources

For additional content background, visit bookg.msscience.com to:

- access your book online
- find references to related articles in popular science magazines
- access Web links with related content background
- access current events with science journal topics

Print Resources

Rivers and Floodplains: Forms, Processes and Sedimentary Record, John Bridge Blackwell Science (UK), 2003

Erosion, Sherie Winner, Carolrhoda/Lerner, 1999

Soil and Water Conservation: Productivity and Environmental Protection, Fredrick R. Troeh, Roy L. Donahue, J. Arthur Hobbs, 1998

River systems develop specific drainage patterns. Dendritic drainage resembles a tree. Rectangular drainage forms when bedrock is fractured. Trellis drainage forms on areas covered with both hard and soft bedrock.

section 2

Groundwater

Groundwater Systems

Urban growth and agriculture irrigation compete for water use. In the United States, approximately two-thirds of the groundwater pumped out of aquifers is used for irrigation of crops. Groundwater provides about 40% of the public water supply in the United States.

Cave Formation

Rainwater dissolves carbon dioxide from the atmosphere and air trapped in soil to form carbonic acid according to the equation $H_2O + CO_2 \rightarrow H_2CO_3$. When the resulting acidic groundwater comes in contact with limestone, the carbonic acid reacts with calcite in the rocks to form soluble calcium hydrogen carbonate that may be carried away with the water. Although limestone is commonly said to dissolve in groundwater, the actual process involves a chemical reaction, not a solution process.

Bob Krist/Corbis

section 3

Ocean Shoreline

The Shore

During periods of calm weather, wave action is at a minimum. Waves have the most significant effect during a storm. Water is forced into cracks in cliffs and walls. This causes the air in these openings to become highly compressed. When the wave subsides, the air expands rapidly—in time, dislodging rock fragments and enlarging the preexisting fractures.

A sandy beach consists of four regions: dunes, berm, foreshore, and offshore. The offshore region is a zone with a gentle slope of fine sand. This region is exposed only when the tide is low. The foreshore is the zone of swash and backwash. The berm is a fairly flat area of beach that may be flooded during especially high tides or during storms. Dunes form on the dry sand behind the berm.

Barrier Islands

The origin of barrier islands is uncertain. Some may have originated as spits (small points of land) that were later severed from the mainland by wave erosion or by the general rise in sea level following the last episode of glaciation. Others may have been formed when turbulent water in the line of breakers heaped up sand that was scoured from the bottom. Still other barrier islands may be former sand dune ridges that developed along the shore when sea level was lower during the last glacial period. Sea level rose as the ice sheets melted and the area behind the beach-dune complex became flooded. Atlantic City, New Jersey; Galveston, Texas; and Kitty Hawk, North Carolina are built on barrier islands.

chapter 4

Chapter Vocabulary

runoff, p. 92
channel, p. 94
sheet erosion, p. 95
drainage basin, p. 96
meander, p. 97
groundwater, p. 103
permeable, p. 104
impermeable, p. 104
aquifer, p. 104
water table, p. 104
spring, p. 107
geyser, p. 107
cave, p. 107
longshore current, p. 110
beaches, p. 111

Science Journal Student answers will vary. They should include erosion by water, ice, or gravity.

INTERACTIVE CHALKBOARD with Image Bank

PowerPoint® Presentations

This CD-ROM is an editable Microsoft® PowerPoint® presentation that includes:

- a pre-made presentation for every chapter
- interactive graphics
- animations
- audio clips
- image bank
- all new section and chapter questions
- Standardized Test Practice
- transparencies
- pre-lab questions for all labs
- Foldables directions
- links to bookg.msscience.com

chapter 4

Water Erosion and Deposition

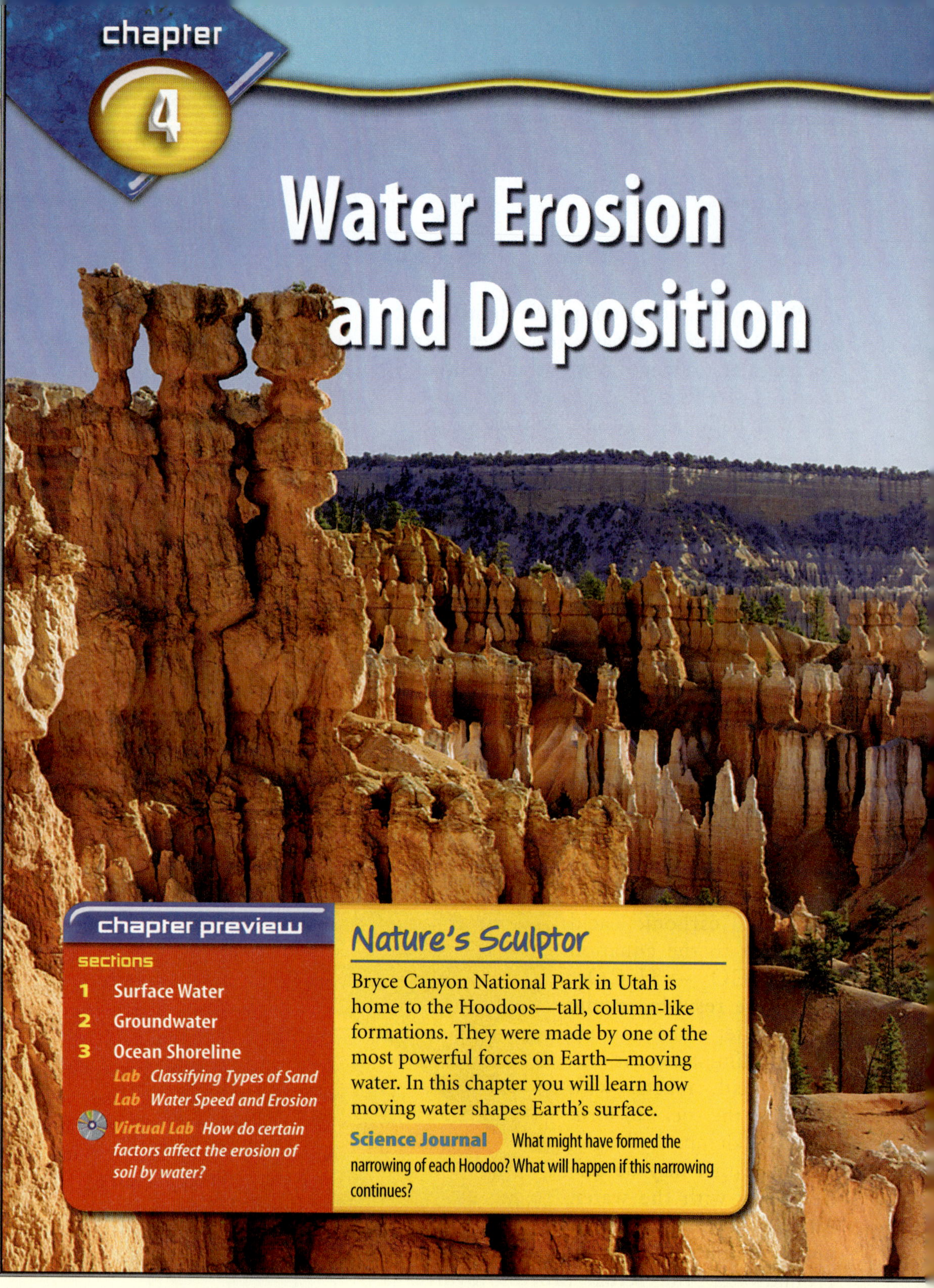

chapter preview

sections

1 Surface Water
2 Groundwater
3 Ocean Shoreline
Lab Classifying Types of Sand
Lab Water Speed and Erosion
Virtual Lab How do certain factors affect the erosion of soil by water?

Nature's Sculptor

Bryce Canyon National Park in Utah is home to the Hoodoos—tall, column-like formations. They were made by one of the most powerful forces on Earth—moving water. In this chapter you will learn how moving water shapes Earth's surface.

Science Journal What might have formed the narrowing of each Hoodoo? What will happen if this narrowing continues?

Theme Connection

Systems and Interactions Water forms a number of systems. Streams, rivers, and the ocean form surface systems, while groundwater forms a separate system. These systems interact with rocks and soil, producing erosion and deposition.

About the Photo

Caster of magical spells? A hoodoo is a natural column of rock left by erosion. Millions of years ago the area that today is the Bryce Canyon was uplifted by the Colorado Plateau. Over time, erosion cut deeply into rock layers forming deep slot canyons with tall fins of rock of unusual morphology.

Start-Up Activities

Launch LAB

Model how Erosion Works

Moving water has great energy. Sometimes rainwater falls softly and soaks slowly into soil. Other times it rushes down a slope with tremendous force and carries away valuable topsoil. What determines whether rain soaks into the ground or runs off and wears away the surface?

1. Place an aluminum pie pan on your desktop.
2. Put a pile of dry soil about 7 cm high into the pan.
3. Slowly drip water from a dropper onto the pile and observe what happens next.
4. Drip the water faster and continue to observe what happens.
5. Repeat steps 1 through 4, but this time change the slope of the hill by increasing the central pile. Start again with dry soil.
6. **Think Critically** In your Science Journal, write about the effect the water had on the different slopes.

FOLDABLES™ Study Organizer

Characteristics of Surface Water, Groundwater, and Shoreline Water Make the following Foldable to help you identify the main concepts relating to surface water, groundwater, and shoreline water.

STEP 1 **Fold** the top of a vertical piece of paper down and the bottom up to divide the paper into thirds.

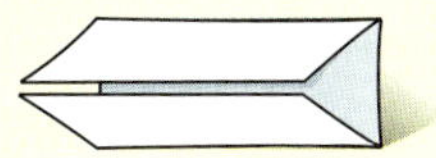

STEP 2 **Turn** the paper horizontally; **unfold and label** the three columns as shown.

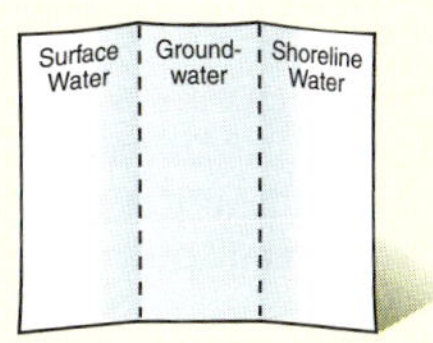

Read for Main Ideas As you read the chapter, list the concepts relating to surface water, groundwater, shoreline water.

Science Online — Preview this chapter's content and activities at bookg.msscience.com

G ◆ 91

Launch LAB

Purpose Students will model the soil erosion caused by running water.

Materials aluminum pie pan, dry soil, water, dropper

Safety Have students wear goggles and apron during this activity. Make provision for disposing of the sand.

Teaching Strategy To assist learning disabled students, substitute layers of colored sand for the mound of dry soil. As the trickle of water cuts a deeper groove into the sand, new colors will be eroded.

Think Critically

Students should observe more erosion when the "rain" fell faster, fell for longer periods, or fell on steeper slopes.

Assessment

Performance Ask students to infer how vegetation growing on the slope would affect their results. Use **Performance Assessment in the Science Classroom,** p. 89.

FOLDABLES Study Organizer — Dinah Zike Study Fold

Student preparation materials for this Foldable are available in the Chapter *FAST FILE* Resources.

Probeware Labs

This chapter's data-collection lab using Probeware technology is included on the *Video Labs CD-ROM*. See the *Probeware Lab Manual* pages 25–28 for student worksheets.

1 Motivate

INTERACTIVE CHALKBOARD
PowerPoint® Presentations

Bellringer

Section Focus Transparencies also are available on the Interactive Chalkboard CD-ROM. L2 ELL

Tie to Prior Knowledge

Water Color Have students recall what they have seen happen to the color of a river's water after a heavy rain. In this section, they will learn that the color is usually caused by eroded sediments carried by the water.

section 1

Surface Water

as you read

What You'll Learn

- **Identify** the causes of runoff.
- **Compare** rill, gully, sheet, and stream erosion.
- **Identify** three different stages of stream development.
- **Explain** how alluvial fans and deltas form.

Why It's Important

Runoff and streams shape Earth's surface.

Review Vocabulary
erosion: transport of surface materials by agents such as gravity, wind, water, or glaciers

New Vocabulary
- runoff
- channel
- sheet erosion
- drainage basin
- meander

Runoff

Picture this. You pour a glass of milk, and it overflows, spilling onto the table. You grab a towel to clean up the mess, but the milk is already running through a crack in the table, over the edge, and onto the floor. This is similar to what happens to rainwater when it falls to Earth. Some rainwater soaks into the ground and some evaporates, turning into a gas. The rainwater that doesn't soak into the ground or evaporate runs over the ground. Eventually, it enters streams, lakes, or the ocean. Water that doesn't soak into the ground or evaporate but instead flows across Earth's surface is called **runoff**. If you've ever spilled milk while pouring it, you've experienced something similar to runoff.

Factors Affecting Runoff What determines whether rain soaks into the ground or runs off? The amount of rain and the length of time it falls are two factors that affect runoff. Light rain falling over several hours probably will have time to soak into the ground. Heavy rain falling in less than an hour or so will run off because it cannot soak in fast enough, or it can't soak in because the ground cannot hold any more water.

Figure 1 In areas with gentle slopes and vegetation, little runoff and erosion take place. Lack of vegetation has led to severe soil erosion in some areas.

Section 1 Resource Manager

Chapter *FAST FILE* Resources
- Transparency Activity, pp. 40, 43–44
- Directed Reading for Content Mastery, pp. 15, 16
- Note-taking Worksheets, pp. 29–31
- Enrichment, p. 26
- MiniLAB, p. 3
- Reinforcement, p. 23

Earth Science Critical Thinking/Problem Solving, p.7

Life Science Critical Thinking/Problem Solving, p. 23

Cultural Diversity, p. 33

92 ◆ G CHAPTER 4 Water Erosion and Deposition

Other Factors Another factor that affects the amount of runoff is the steepness, or slope, of the land. Gravity, the attractive force between all objects, causes water to move down slopes. Water moves rapidly down steep slopes so it has little chance to soak into the ground. Water moves more slowly down gentle slopes and across flat areas. Slower movement allows water more time to soak into the ground.

Vegetation, such as grass and trees, also affects the amount of runoff. Just like milk running off the table, water will run off smooth surfaces that have little or no vegetation. Imagine a tablecloth on the table. What would happen to the milk then? Runoff slows down when it flows around plants. Slower-moving water has a greater chance to sink into the ground. By slowing down runoff, plants and their roots help prevent soil from being carried away. Large amounts of soil may be carried away in areas that lack vegetation, as shown in **Figure 1.**

INTEGRATE Physics

Effects of Gravity When you lie on the ground and feel as if you are being held in place, you are experiencing the effects of gravity. Gravity is the attracting force all objects have for one another. The greater the mass of an object is, the greater its force of gravity is. Because Earth has a much greater mass than any of the objects on it, Earth's gravitational force pulls objects toward its center. Water runs downhill because of Earth's gravitational pull. When water begins to run down a slope, it picks up speed. As its speed increases, so does its energy. Fast-moving water, shown in **Figure 2,** carries more soil than slow-moving water does.

INTEGRATE Career

Conservation Farmers sometimes have to farm on some kind of slope. The steeper the slope, the more erosion will occur. Not only is slope an important factor but other factors have to be considered as well. The Natural Resources Conservation Service, a government agency, studies these factors to determine soil loss from a given area. Find out what other factors this agency uses to determine soil loss.

Figure 2 During floods, the high volume of fast-moving water erodes large amounts of soil.

2 Teach

Fun Fact

When the surface of an area is paved or in some other way becomes less porous, the amount of runoff increases because the new surface keeps water from soaking in.

Text Question Answer

The milk would soak into the cloth.

INTEGRATE Career

Conservation The Natural Resources Conservation Services (NRCS) is a branch of the U.S. Department of Agriculture. The NRCS provides information necessary for understanding, managing, and conserving soil. Soil surveys include soil maps, data about physical and chemical properties of soils, and other information to meet the needs of farmers, community planners, engineers, and scientists. Aside from slope, other factors students might research include: types of soil, precipitation patterns, and vegetation cover.

Activity

Erosion on Maps Relief maps show elevation differences with contour lines or shading. Have students examine topographic or other relief maps of your area to find the areas of greatest potential erosion. L2 LS **Visual-Spatial**

Teacher FYI

Grand Canyon A river may erode a canyon, a steep-sided river valley that is usually deeper than it is wide. One of the world's largest canyons, the Grand Canyon, is a 1.6-kilometer-deep gash in Earth's surface eroded by the Colorado River.

Quick Demo

Runoff and Slope

Materials 2 cookie sheets, 1 brick, 2 empty beakers, 2 containers of water (equal amounts), stopwatch

Estimated Time 10 minutes

Procedure Place one cookie sheet flat and the other with the brick under one end. Place the 2 empty containers on the opposite ends of the brick to catch the water that runs-off. Pour the water over the cookie sheets, timing this procedure. At the end of the given time, see which beaker caught the most runoff. It should be the beaker under the raised cookie sheet.

Figure 3 Heavy rains can remove large amounts of sediment, forming deep gullies in the side of a slope.

Water Erosion

Suppose you and several friends walk the same way to school each day through a field or an empty lot. You always walk in the same footsteps as you did the day before. After a few weeks, you've worn a path through the field. When water travels down the same slope time after time, it also wears a path. The movement of soil and rock from one place to another is called erosion.

Rill and Gully Erosion You may have noticed a groove or small ditch on the side of a slope that was left behind by running water. This is evidence of rill erosion. Rill erosion begins when a small stream forms during a heavy rain. As this stream flows along, it has enough energy to erode and carry away soil. Water moving down the same path creates a groove, called a **channel**, on the slope where the water eroded the soil. If water frequently flows in the same channel, rill erosion may change over time into another type of erosion called gully erosion.

During gully erosion, a rill channel becomes broader and deeper. **Figure 3** shows gullies that were formed when water carried away large amounts of soil.

Sheet Erosion Water often erodes without being in a channel. Rainwater that begins to run off during a rainstorm often flows as thin, broad sheets before forming rills and streams. For example, when it rains over an area, the rainwater accumulates until it eventually begins moving down a slope as a sheet, like the water flowing off the hood of the car in **Figure 4.** Water also can flow as sheets if it breaks out of its channel.

Figure 4 When water accumulates, it can flow in sheets like the water seen flowing over the hood of this car.

Curriculum Connection

Industrial Technology Arrange for a local contractor or a city engineer to visit the class. Ask the speaker to explain how construction areas are evaluated in terms of runoff and drainage. In some areas, barriers must be placed around construction sites to prevent topsoil erosion. If possible, take students to visit a construction site to see the work firsthand.

Active Reading

Problem/Solution Journal Have students divide a sheet of paper in half, labeling the left side *Problems* and the right side *Consequences of Failing to Resolve the Problem.* Through writing, the student identifies a problem, brainstorms possible alternatives, chooses a probable solution, anticipates stumbling blocks, and proposes arguments.

Floodwaters spilling out of a river can flow as sheets over the surrounding flatlands. Streams flowing out of mountains fan out and may flow as sheets away from the foot of the mountain. **Sheet erosion** occurs when water that is flowing as sheets picks up and carries away sediments.

Stream Erosion Sometimes water continues to flow along a low place it has formed. As the water in a stream moves along, it picks up sediments from the bottom and sides of its channel. By this process, a stream channel becomes deeper and wider.

The sediment that a stream carries is called its load. Water picks up and carries some of the lightweight sediments, called the suspended load. Larger, heavy particles called the bed load just roll along the bottom of the stream channel, as shown in **Figure 5.** Water can even dissolve some rocks and carry them away in solution. The different-sized sediments scrape against the bottom and sides of the channel like a piece of sandpaper. Gradually, these sediments can wear away the rock by a process called abrasion.

Figure 5 This cross section of a stream channel shows the location of the suspended load and the bed load.
Describe *how the stream carries dissolved material.*

Activity

Model Runoff Direct student groups to investigate how different surfaces affect runoff. Each team needs a paint tray half-filled with soil, a piece of cardboard, a clump of grass, and a beaker of water. Have groups design and carry out an experiment using the cardboard to simulate a paved street and the grass to simulate a grassy field. Students should find that paved surfaces have more runoff than grassy surfaces. L2 ELL

COOP LEARN LS **Interpersonal**

Activity

Erosion Evidence Take students outside to look for evidence of water erosion around the school grounds. For example, they might find small channel scars on slopes or low, flat areas where fine sediments have collected. L2 ELL

Discussion

Rain Runoff Have students compare and contrast the runoff from paved streets and grassy fields during a thunderstorm and a gentle rain. During a thunderstorm, paved streets have much more runoff than grassy fields. In a gentle rain, water will likely puddle slightly and evaporate from paved streets and be absorbed by grassy fields. There will be little runoff from either surface during a gentle rain.

Caption Answer

Figure 5 They are transported downstream in the water.

Science Journal

Stream Erosion Have students write poems in their journals describing erosion in a stream channel like the one shown in **Figure 5**. Poems can describe lightweight particles carried in suspension and heavy particles rolled along the bottom of the channel. L2 LS **Auditory-Musical** P

Reading Check

Answer an area of land from which a stream or river collects runoff

Visual Learning

Figure 6 Have students trace this river's drainage system from source to mouth. Encourage them to use an atlas to name several of the major tributaries of the Mississippi. Some tributaries are the Ohio, Illinois, Cumberland, Missouri, Arkansas, and Tennessee Rivers. Have students discuss why the main river follows the path it does. The river flows through a lowland area or depression, which forms its bed.

Caption Answer

Figure 6 the Mississippi River

Teacher FYI

Protecting New Orleans The city of New Orleans is built on a part of the Mississippi River delta where sediments are compacting. As a result, the city and surrounding area are sinking, and the city is actually below sea level. To protect it from flooding, levees have been built along the river's banks.

Science Online

Topic: Drainage Basins
Visit bookg.msscience.com for Web links to information about drainage basins in your region.

Activity Locate the drainage basin in which you live. Make a poster showing the shape, or boundary, of the drainage basin. Also, include the main river, or trunk, and the major tributaries, or branches.

River System Development

Have you spent time near a river or stream in your community? Each day, probably millions of liters of water flow through that stream. Where does all the water come from? Where is it flowing to?

River Systems Streams are parts of river systems. The water comes from rills, gullies, and smaller streams located upstream. Just as the tree in **Figure 6** is a system containing twigs, branches, and a trunk, a river system also has many parts. Runoff enters small streams, which join together to form larger streams. Larger streams come together to form rivers. Rivers grow and carry more water as more streams join.

Drainage Basins A **drainage basin** is the area of land from which a stream or river collects runoff. Compare a drainage basin to a bathtub. Water that collects in a bathtub flows toward one location—the drain. Likewise, all of the water in a river system eventually flows to one location—the main river, or trunk. The largest drainage basin in the United States is the Mississippi River drainage basin shown in **Figure 6.**

Reading Check *What is a drainage basin?*

Figure 6 River systems can be compared with the structure of a tree.

The system of twigs, branches, and trunk that make up a tree is similar to the system of streams and rivers that make up a river system.

A large number of the streams and rivers in the United States are part of the Mississippi River drainage basin, or watershed.
State *what river represents the trunk of this system.*

Differentiated Instruction

Learning Disabled Have students look for and photograph places where water erosion has occurred in the community. Gather these photographs into a booklet that students can refer to when reviewing this material. L1

Curriculum Connection

Geography Have students use an atlas to explain how sediments on a Tennessee hillside might end up in the Mississippi delta. Local streams take sediments eroded from the hillside to the Mississippi River. When the river enters the Gulf of Mexico, it slows and deposits the sediments. L2 LS **Visual-Spatial**

Stages of Stream Development

Streams come in a variety of forms. Some are narrow and swift moving, and others are wide and slow moving. Streams differ because they are in different stages of development. These stages depend on the slope of the ground over which the stream flows. Streams are classified as young, mature, or old. **Figure 8** shows how the stages come together to form a river system.

The names of the stages of development aren't always related to the actual age of a river. The New River in West Virginia is one of the oldest rivers in North America. However, it has a steep valley and flows swiftly. As a result, it is classified as a young stream.

Figure 7 A meander is a broad bend in a river or stream. As time passes, erosion of the outer bank increases the bend.

Young Streams A stream that flows swiftly through a steep valley is a young stream. A young stream may have whitewater rapids and waterfalls. Water flowing through a steep channel with a rough bottom has a high level of energy and erodes the stream bottom faster than its sides.

Mature Streams The next stage in the development of a stream is the mature stage. A mature stream flows more smoothly through its valley. Over time, most of the rocks in the streambed that cause waterfalls and rapids are eroded by running water and the sediments it carries.

Erosion is no longer concentrated on the bottom in a mature stream. A mature stream starts to erode more along its sides, and curves develop. These curves form because the speed of the water changes throughout the width of the channel.

Water in a shallow area of a stream moves slower because it drags along the bottom. In the deeper part of the channel, the water flows faster. If the deep part of the channel is next to one side of the river, water will erode that side and form a slight curve. Over time, the curve grows to become a broad arc called a **meander** (mee AN dur), as shown in **Figure 7.**

The broad, flat valley floor formed by a meandering stream is called a floodplain. When a stream floods, it often will cover part or all of the floodplain.

Activity

Stream Speed Explain to students that erosion is greatest where water speed is greatest. Then take students to a local stream. Have them determine the speed at the middle of the stream and at the edge of the stream by measuring 10 m along the bank and timing how long it takes a fishing bob to float that distance. Speed equals distance divided by time ($s = d/t$). Have them determine where in the stream erosion is greatest.

Make a Model

Riverbed Have students work in groups to produce a 3-D model of a local riverbed. L2 ELL COOP LEARN

Differentiated Instruction

Challenge Have students research what happens to local runoff and draw a map of the local drainage basin. L3

NATIONAL GEOGRAPHIC

Visualizing Stream Development

Have students examine the pictures and read the captions. Then ask the following questions.

Which stage stream would be most likely to have a gravel streambed? Because older streams flow with less energy, finer sediment can be deposited in their bed. Turbulent, young streams often carry finer sediment in suspension and have gravel bottoms.

Why do old streams drop their loads of sediment near their mouths? Because when they reach the ocean, they slow down and the water no longer has the velocity to suspend sediments. Sediments drop to the bottom, building up deltas.

Where is the water flow fastest in a meander? How do you know? The water on the outside of the curve has the greater velocity. Proof is that on the outside of the curve, there is more erosion of the bank. Where the velocity is lower (on the inside of the curve), deposition of sediments take place.

Activity

Slope and Velocity Have students work in groups to create simple models using classroom materials that show the relationship between slope and velocity. Have them demonstrate their models to the class, and explain how they relate to stream development.

NATIONAL GEOGRAPHIC **VISUALIZING STREAM DEVELOPMENT**

Figure 8

Although no two streams are exactly alike, all go through three main stages—young, mature, and old—as they flow from higher to lower ground. A young stream, below, surging over steep terrain, moves rapidly. In a less steep landscape, right, a mature stream flows more smoothly. On nearly level ground, the stream–considered old–winds leisurely through its valley. The various stages of a stream's development are illustrated here.

A A young stream begins at a source—here, a melting mountain glacier. From its source, the stream flows swiftly downhill, cutting a narrow valley.

B A mature stream flows smoothly through its valley. Mature streams often develop broad curves called meanders.

C Old streams flow through broad, flat floodplains. Near its mouth, the stream gradually drops its load of silt. This sediment forms a delta, an area of flat, fertile land extending into the ocean.

NATIONAL GEOGRAPHIC

Teacher FYI

Floodplains Mature rivers are usually associated with a floodplain. On a floodplain where there are no levees, floodwaters spread over a wide area and deposit fine sediment.

Discussion

White-Water Rafting Ask students why most whitewater rafting is done in young rivers. Young rivers flow downhill with enough energy to form rapids. Older rivers have already eroded away the rocks and irregularities of their beds and flow with less energy. Therefore, they cannot provide the thrilling rides that young rivers can.

Fun Fact

One of the oldest rivers in the United States is the New River in West Virginia.

Reading Check

Answer Old streams flow smoothly through flat, broad flood plains, while young streams flow swiftly through steep valleys.

Teacher FYI

Water Scarcity In most places the problem isn't too much water, but too little. Many people live along the 300 rivers that cross national borders. Damming these rivers causes countries downstream to lose water. Some fear that water wars will occur. Others think that water will be recognized as a commodity that nations will trade.

Visual Learning

Figure 9 What may have caused this levee to break? Possible answers: The force of the flowing water may have eroded the levee; the water may have flowed over the top causing the levee's destruction.

Science Online

Topic: Classification of Rivers

Visit bookg.msscience.com for Web links to information about major rivers in the United States. Classify two of these streams as young, mature, or old.

Activity Write a paragraph about why you think the Colorado River is a young, mature, or old stream.

Old Streams The last stage in the development of a stream is the old stage. An old stream flows smoothly through a broad, flat floodplain that it has deposited. South of St. Louis, Missouri, the lower Mississippi River is in the old stage.

Major river systems, such as the Mississippi River, usually contain streams in all stages of development. In the upstream portion of a river system, you find whitewater streams moving swiftly down mountains and hills. At the bottom of mountains and hills, you find streams that start to meander and are in the mature stage of development. These streams meet at the trunk of the drainage basin and form a major river.

Reading Check *How do old streams differ from young streams?*

Too Much Water

Sometimes heavy rains or a sudden melting of snow can cause large amounts of water to enter a r ver system. What happens when a river system has too much water in it? The water needs to go somewhere, and out and over the banks is the only choice. A river that overflows its banks can bring disaster by flooding homes or washing away bridges or crops.

Figure 9 Flooding causes problems for people who live along major rivers. Floodwater broke through a levee during the Mississippi River flooding in 1993.

Dams and levees are built in an attempt to prevent this type of flooding. A dam is built to control the water flow downstream. It may be built of soil, sand, or steel and concrete. Levees are mounds of earth that are built along the sides of a river. Dams and levees are built to prevent rivers from overflowing their banks. Unfortunately, they do not stop the water when flooding is great. This was the case in 1993 when heavy rains caused the Mississippi River to flood parts of nine midwestern states. Flooding resulted in billions of dollars in property damage. **Figure 9** shows some of the damage caused by this flood.

As you have seen, floods can cause great amounts of damage. But at certain times in Earth's past, great floods have completely changed the surface of Earth in a large region. Such floods are called catastrophic floods.

LAB DEMONSTRATION

Purpose to show how levees help prevent flooding along rivers

Materials long paint tray, sand, water

Alternate Materials stream table

Preparation Fill the tray with sand, and make a shallow channel with flat banks in the sand.

Procedure Pour a stream of water through the channel. Slowly increase the volume until the stream overflows its banks. Rebuild the channel with levees and repeat.

Expected Outcome Students will see that with levees, overbank flooding takes longer.

Assessment

Why do some cities build levees along their rivers? Levees help prevent rivers from overflowing their banks. Why don't levees work when flooding is great? The water breaks through or flows over the levee.

100 ◆ G CHAPTER 4 Water Erosion and Deposition

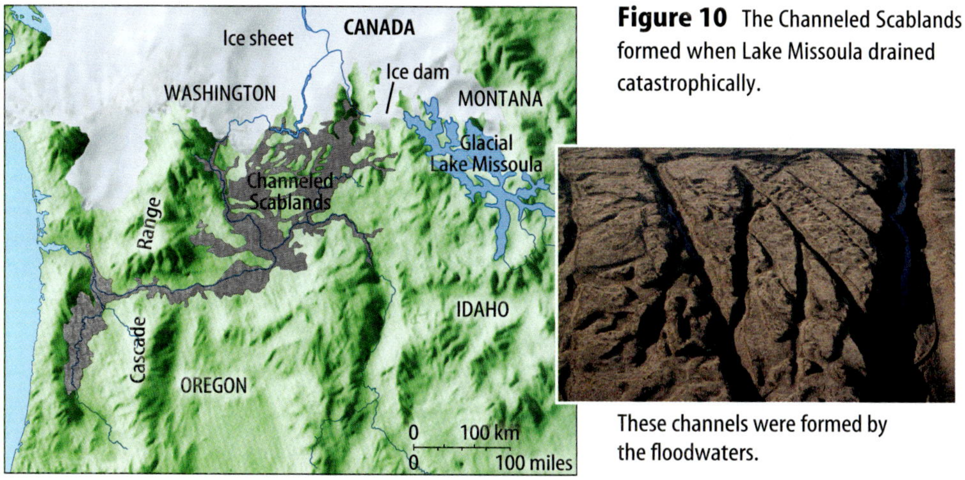

Figure 10 The Channeled Scablands formed when Lake Missoula drained catastrophically.

These channels were formed by the floodwaters.

Catastrophic Floods During Earth's long history, many catastrophic floods have dramatically changed the face of the surrounding area. One catastrophic flood formed the Channeled Scablands in eastern Washington State, shown here in **Figure 10.** A vast lake named Lake Missoula covered much of western Montana. A natural dam of ice formed this lake. As the dam melted or was eroded away, tremendous amounts of water suddenly escaped through what is now the state of Idaho into Washington. In a short period of time, the floodwater removed overlying soil and carved channels into the underlying rock, some as deep as 50 m. Flooding occurred several more times as the lake refilled with water and the dam broke loose again. Scientists say the last such flood occurred about 13,000 years ago.

Deposition by Surface Water

You know how hard it is to carry a heavy object for a long time without putting it down. As water moves throughout a river system, it loses some of its energy of motion. The water can no longer carry some of its sediment. As a result, it drops, or is deposited, to the bottom of the stream.

Some stream sediment is carried only a short distance. In fact, sediment often is deposited within the stream channel itself. Other stream sediment is carried great distances before being deposited. Sediment picked up when rill and gully erosion occur is an example of this. Water usually has a lot of energy as it moves down a steep slope. When water begins flowing on a level surface, it slows, loses energy, and deposits its sediment. Water also loses energy and deposits sediment when it empties into an ocean or lake.

Mini LAB

Observing Runoff Collection

Procedure

1. Put a plastic **rain gauge** into a narrow **drinking glass** and place the glass in the **sink.**
2. Fill a plastic **sprinkling can** with **water.**
3. Hold the sprinkling can one-half meter above the sink for 30 s.
4. Record the amount of water in the rain gauge.
5. After emptying the rain gauge, place a **plastic funnel** into the rain gauge and sprinkle again for 30 s.
6. Record the amount of water in the gauge.

Analysis

Explain how a small amount of rain falling on a drainage basin can have a big effect on a river or stream.

Identifying Misconceptions

Rock Cycle Because students can see the effects of erosion more readily, they may not consider deposition to be of equal importance. Remind students of the rock cycle. In this cycle, erosion is followed by sedimentation to form sedimentary rocks.

Mini LAB

Purpose Students use a model to infer how the effect of rain on a drainage basin affects a river or stream.

Materials plastic rain gauge, drinking glass, sink, sprinkling can, water, funnel

Teaching Strategy Use funnels with wide mouths for a more dramatic effect.

Analysis

The drainage basin funnels water from over a large area into a stream or river.

Assessment

Oral Ask students to predict what would happen to the amount of water in the gauge if they used a funnel with a wider mouth. The amount of water would increase. Use **Performance Assessment in the Science Classroom,** p. 89.

Cultural Diversity

Egypt's Nile No country depends on a river more than Egypt depends on the Nile. The Nile's floodplain provides fertile soil that has supported Egypt's agriculture and has been home to most of its people for thousands of years. Today, the Nile provides irrigation water, drinking water, transportation, and hydroelectric power for a nation which otherwise would be a desert wasteland.

Use Science Words

Word Origin Have students find out how deltas got their name. Both the Greek letter *delta* (Δ), and the deposit have a triangular shape.

3 Assess

DAILY INTERVENTION

Check for Understanding

Visual-Spatial Have students research how the border between the United States and Mexico has changed over time. Draw a map of the year 1890, 1990, and 2003 to show the changes because of the change in the Rio Grande.

Reteach

A River's Age Have students examine rivers shown on maps or in photographs. Ask students to classify these rivers as young, mature, or old and give reasons for their choices. L1 LS **Visual-Spatial**

Assessment

Process Have students refer to **Figure 11** and then draw pictures of alluvial fans and deltas. Use **Performance Assessment in the Science Classroom**, p. 127.

Virtual Labs

Water Erosion *How do certain factors affect the erosion of soil by water?*

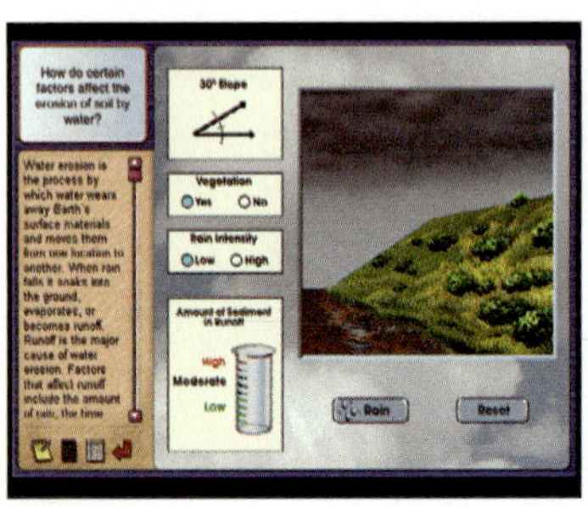

Figure 11 This satellite image of the Nile River Delta in Egypt shows the typical triangular shape. The green color shows areas of vegetation.

Agriculture is important on the Nile Delta.

Deltas and Fans Sediment that is deposited as water empties into an ocean or lake forms a triangular, or fan-shaped, deposit called a delta, shown in **Figure 11.** When the river waters empty from a mountain valley onto an open plain, the deposit is called an alluvial (uh LEW vee ul) fan. The Mississippi River exemplifies the topics presented in this section. Runoff causes rill and gully erosion. Sediment is picked up and carried into the larger streams that flow into the Mississippi River. As the Mississippi River flows, it cuts into its banks and picks up more sediment. Where the land is flat, the river deposits some of its sediment in its own channel. As the Mississippi enters the Gulf of Mexico, it slows, dropping much of its sediment and forming the Mississippi River delta.

section 1 review

Summary

Runoff

- Rainwater that doesn't soak into the ground or evaporate becomes runoff.
- Slope of land and vegetation affect runoff.

Water Erosion

- Water flowing over the same slope causes rills and gullies to form.

River System Development

- A drainage basin is an area of land from which a stream or river collects runoff.

Self Check

1. **Explain** how the slope of an area affects runoff.
2. **Compare and contrast** rill and gully erosion.
3. **Describe** the three stages of stream development.
4. **Think Critically** How is a stream's rate of flow related to the amount of erosion it causes? How is it related to the size of the sediments it deposits?

Applying Skills

5. **Compare and contrast** the formation of deltas and alluvial fans.

 Science Online bookg.msscience.com/self_check_quiz

section 1 review

1. the greater the slope, the greater the runoff
2. Both are types of erosion. Gully erosion is broader and deeper than rill erosion.
3. A youthful stream erodes its bottom more than its sides. A mature stream flows less swiftly, with erosion mostly along its sides. An old stream meanders slowly through a broad, flat floodplain.
4. The greater the rate of flow, the greater the erosion and the larger the sediment size a stream can carry.
5. Both are deposits of sediment formed when moving water slows down. Alluvial fans form when a river flows from mountains onto an open plain. Deltas form when rivers enter large bodies of water.

section 2

Groundwater

Groundwater Systems

What would have happened if the spilled milk in Section 1 ran off the table onto a carpeted floor? It probably would have quickly soaked into the carpet. Water that falls on Earth can soak into the ground just like the milk into the carpet.

Water that soaks into the ground becomes part of a system, just as water that stays above ground becomes part of a river system. Soil is made up of many small rock and mineral fragments. These fragments are all touching one another, as shown in **Figure 12,** but some empty space remains between them. Holes, cracks, and crevices exist in the rock underlying the soil. Water that soaks into the ground collects in these pores and empty spaces and becomes part of what is called **groundwater.**

How much of Earth's water do you think is held in the small openings in rock? Scientists estimate that 14 percent of all freshwater on Earth exists as groundwater. This is almost 30 times more water than is contained in all of Earth's lakes and rivers.

as you read

What You'll Learn

- **Recognize** the importance of groundwater.
- **Describe** the effect that soil and rock permeability have on groundwater movement.
- **Explain** how groundwater dissolves and deposits minerals.

Why It's Important

The groundwater system is an important source of your drinking water.

Review Vocabulary
pore: a small, or minute, opening in rock or soil

New Vocabulary
- groundwater
- permeable
- impermeable
- aquifer
- water table
- spring
- geyser
- cave

Figure 12 Soil has many small, connected pores that are filled with water when soil is wet.

Section 2 Resource Manager

Chapter *FAST FILE* Resources
- Transparency Activity, p. 41
- Directed Reading for Content Mastery, p. 17
- MiniLAB, p. 4
- Reinforcement, p. 24
- Enrichment, p. 27

Mathematics Skill Activities, p. 11

Physical Science Critical Thinking/Problem Solving, p. 12

1 Motivate

INTERACTIVE CHALKBOARD PowerPoint® Presentations

Bellringer

Section Focus Transparencies also are available on the Interactive Chalkboard CD-ROM. L2 ELL

Section 2 Section Focus Transparency **Nature's Hot Tub**

Imagine you're a Japanese macaque and it's cold and snowy. What do you do? If you're lucky, you live near a hot spring where you could find a warm refuge from the winter storm. The Japanese macaque lives further north than any other ape or monkey.

1. How can you tell the water in this spring is hot?
2. Where do you think the water in hot springs comes from?

L2

Tie to Prior Knowledge

Groundwater Have students recall watching a gentle rain fall onto and disappear into a field or lawn. Ask them where the water goes. It soaks into the soil to become a deposit of groundwater. If rainfall lasts long enough, the ground may become saturated and water will run off the lawn and possibly into a storm drain.

SECTION 2 Groundwater G ◆ 103

2 Teach

Mini LAB

Purpose Students measure and compare pore space. L2 ELL

LS **Logical-Mathematical**

Materials 2 plastic containers, sand, gravel, metric measuring cups, measuring cup, water

Teaching Strategies

- Have students predict which material will absorb more water before reaching capacity.
- Students must subtract the final volume reading on the measuring cup from the beginning volume reading to find the volume of water used.

Troubleshooting The sand and the gravel mixtures should be of relatively uniform grain size and perfectly dry.

Analysis

Sand and gravel are about equal in total amount of pore space. Gravel has fewer but larger pore spaces when compared to the smaller but more numerous spaces among the sand grains. Average results should show that it takes about the same amount of water to fill both materials.

Assessment

Content Have students draw a magnified image of the pore spaces and sediments in their two cups.

Reading Check

Answer Water flows through interconnected spaces in rock.

Mini LAB

Measuring Pore Space

Procedure

1. Use two identical, **clear-plastic containers.**
2. Put 3 cm of **sand** in one container and 3 cm of **gravel** in the other.
3. Pour **water** slowly into the containers and stop when the water just covers the top of the sediment.
4. Record the volume of water used in each.

Analysis

Which substance has more pore space—sand or gravel?

Try at Home

Permeability A groundwater system is similar to a river system. However, instead of having channels that connect different parts of the drainage basin, the groundwater system has connecting pores. Soil and rock are **permeable** (PUR mee uh bul) if the pore spaces are connected and water can pass through them. Sandstone is an example of a permeable rock.

Soil or rock that has many large, connected pores is permeable. Water can pass through it easily. However, if a rock or sediment has few pore spaces or they are not well connected, then the flow of groundwater is blocked. These materials are **impermeable,** which means that water cannot pass through them. Granite has few or no pore spaces at all. Clay has many small pore spaces, but the spaces are not well connected.

Reading Check *How does water move through permeable rock?*

Groundwater Movement How deep into Earth's crust does groundwater go? **Figure 13** shows a model of a groundwater system. Groundwater keeps going deeper until it reaches a layer of impermeable rock. When this happens, the water stops moving down. As a result, water begins filling up the pores in the rocks above. A layer of permeable rock that lets water move freely is an **aquifer** (AK wuh fur). The area where all of the pores in the rock are filled with water is the zone of saturation. The upper surface of this zone is the **water table.**

Figure 13 A stream's surface level is the water table. Below that is the zone of saturation.

Differentiated Instruction

Challenge Have students research what has happened to the Ogallala Aquifer as a result of overpumping. Students should make a map to show where the Ogallala Aquifer is located. Have the students make diagrams to illustrate their findings. L3 P

Water Table

Why are the zone of saturation and the water table so important? An average United States resident uses about 626 L of water per day. That's enough to fill nearly two thousand soft drink cans. Many people get their water from groundwater through wells that have been drilled into the zone of saturation. However, the supply of groundwater is limited. During a drought, the water table drops. This is why you should conserve water.

Applying Math — Calculate Rate of Flow

GROUNDWATER FLOW You and your family are hiking and the temperature is hot. You feel as if you can't walk one step farther. Luckily, relief is in sight. On the side of a nearby hill you see a stream, and you rush to splash some water on your face. Although you probably feel that it's taking you forever to reach the stream, your pace is quick when compared to how long it takes groundwater to flow through the aquifer that feeds the stream. The following problem will give you some idea of just how slowly groundwater flows through an aquifer.

The groundwater flows at a rate of 0.6 m/day. You've run 200 m to get some water from a stream. How long does it take the groundwater in the aquifer to travel the same distance?

Solution

1. *This is what you know:*
   - the distance that the groundwater has to travel: $d = 200$ m
   - the rate that groundwater flows through the aquifer: $r = 0.6$ m/day
2. *This is what you want to find:* time = t
3. *This is the equation you use:* $r \times t = d$ (rate × time = distance)
4. *Solve the equation for* t *and then substitute known values:* $t = \frac{d}{r} = \frac{(200 \text{ m})}{(0.6 \text{ m/day})} = 333.33$ days

Practice Problems

1. The groundwater in an aquifer flows at a rate of 0.5 m/day. How far does the groundwater move in a year?
2. How long does it take groundwater in the above aquifer to move 100 m?

Science Online — For more practice, visit bookg.msscience.com/math_practice

Discussion

Groundwater Pollution Given the way groundwater collects in aquifers, how might pollution collect in aquifers? It can also seep into the ground and collect there. Do you think pollution would be easier or harder to remove from groundwater or from surface water? It would be harder to get pollution out of groundwater because the system is fairly closed.

Fun Fact

The United States is the world's largest user of both surface and groundwater. It loses about half of its potential supply through evaporation, leaks, and other means.

Applying Math

National Math Standards

Correlation to Mathematics Objectives

1, 2, 9

Answers to Practice Problems

1. Solve as originally given: $d = (0.5 \text{ m/day})(365 \text{ days}) = 182.5$ m
2. $t = (100 \text{ m})/(0.5 \text{ m/day}) = 200$ days

Use an Analogy

Coffee Filter An aquifer is like a coffee filter. When water goes through an aquifer, the soil and rocks filter out the pollution just as a coffee filter keeps coffee grounds out of the coffee. However, if the soil is saturated with pollution, it can no longer hold the pollution. If the coffee filter is saturated, it may rip, allowing grounds into the coffee.

Science Journal

Groundwater Quality Have students research and describe in their journals ways pollutants from landfills, agriculture, and industry might affect groundwater quality and why this would be important. Pesticides, solvents, septic tank cleaners, and other wastes get into soil and permeable rocks from runoff. Some become dissolved in groundwater, which can cause harm to all organisms that use the water.

Teacher FYI

Alarming Rates In the United States, about half the drinking water and 40% of irrigation water comes from aquifers. But groundwater is being withdrawn at four times its replacement rate, with the greatest problems in the Plains from South Dakota to Texas, the Southwest, and California's Central Valley.

Quick Demo

A Well in a Bottle

Materials spray bottle, water

Estimated Time five minutes

Procedure Use a pump spray bottle to demonstrate how water is pumped from a well. Explain that the liquid in the bottle represents groundwater. The bottle represents the aquifer. The hollow tube represents the well. A pump attached to a well is used to extract water from an aquifer. LS **Visual-Spatial**

Discussion

Water Flavor Why does well water sometimes taste like iron or sulfur? Groundwater dissolves minerals in rocks it flows through. These minerals may flavor the water in an aquifer. Iron from pipes can also enter the water as it is transported through them.

Visual Learning

Figure 14 Why would the ground level here sink or collapse after pumping? There is empty pore space left after the water is removed. The weight of the overlying rock layers or sediment can no longer be supported and the ground above collapses. What type of rock probably underlies an area where something like this occurs? It is most likely a loosely packed rock that has many pores.

Caption Answer

Figure 15 The elevation of the recharge area and friction.

Reading Check

Answer Gravity and pressure of the water in the higher part of the aquifer force the water to move through the interconnected pores of the rock.

Figure 14 The years on the pole show how much the ground level dropped in the San Joaquin Valley, California, between 1925 and 1977.

Figure 15 The pressure of water in a sloping aquifer keeps an artesian well flowing.
Describe *what limits how high water can flow in an artesian well.*

Wells A good well extends deep into the zone of saturation, past the top of the water table. Groundwater flows into the well, and a pump brings it to the surface. Because the water table sometimes drops during very dry seasons, even a good well can go dry. Then time is needed for the water table to rise, either from rainfall or through groundwater flowing from other areas of the aquifer.

Where groundwater is the main source of drinking water, the number of wells and how much water is pumped out are important. If a large factory were built in such a town, the demand on the groundwater supply would be even greater. Even in times of normal rainfall, the wells could go dry if water were taken out at a rate greater than the rate at which it can be replaced.

In areas where too much water is pumped out, the land level can sink from the weight of the sediments above the now-empty pore spaces. **Figure 14** shows what occurred when too much groundwater was removed in a region of California.

One type of well doesn't need a pump to bring water to the surface. An artesian well is a well in which water rises to the surface under pressure. Artesian wells are less common than other types of wells because of the special conditions they require.

As shown in **Figure 15,** the aquifer for an artesian well needs to be located between two impermeable layers that are sloping. Water enters at the high part of the sloping aquifer. The weight of the water in the higher part of the aquifer puts pressure on the water in the lower part. If a well is drilled into the lower part of the aquifer, the pressurized water will flow to the surface. Sometimes, the pressure is great enough to force the water into the air, forming a fountain.

Reading Check *How does water move through permeable rock?*

Active Reading

Bubble Map Using a bubble map helps students start ideas flowing about a given topic. Words are clustered to describe a topic or idea that is studied. Students can use a bubble map for a prewriting, to generate ideas before writing in their Journals, or to review for a test. Have students design a Bubble Map for a water table.

Sample map:

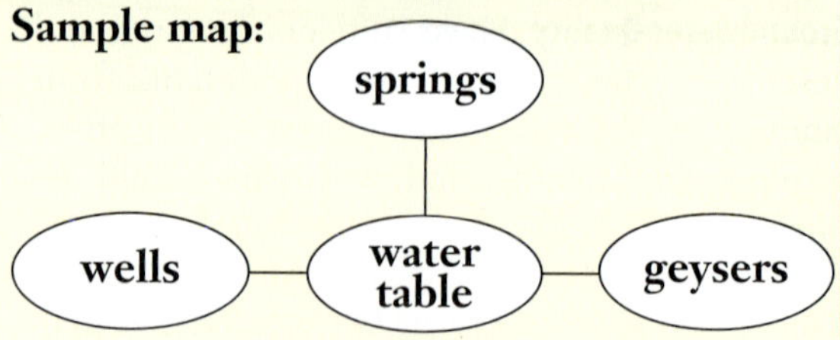

Springs In some places, the water table is so close to Earth's surface that water flows out and forms a **spring.** Springs are found on hillsides or other places where the water table meets a sloping surface. Springs often are used as a source of freshwater.

The water from most springs is a constant, cool temperature because soil and rock are good insulators and protect the groundwater from changes in temperature on Earth's surface. However, in some places, magma rises to within a few kilometers of Earth's surface and heats the surrounding rock. Groundwater that comes in contact with these hot rocks is heated and can come to the surface as a hot spring.

Geysers When water is put into a teakettle to boil, it heats slowly at first. Then some steam starts to come out of the cap on the spout, and suddenly the water starts boiling. The teakettle starts whistling as steam is forced through the cap. A similar process can occur with groundwater. One of the places where groundwater is heated is in Yellowstone National Park in Wyoming. Yellowstone has hot springs and geysers. A **geyser** is a hot spring that erupts periodically, shooting water and steam into the air. Groundwater is heated to high temperatures, causing it to expand underground. This expansion forces some of the water out of the ground, taking the pressure off the remaining water. The remaining water boils quickly, with much of it turning to steam. The steam shoots out of the opening like steam out of a teakettle, forcing the remaining water out with it. Yellowstone's famous geyser, Old Faithful, pictured in **Figure 16,** shoots between 14,000 and 32,000 L of water and steam into the air about once every 80 min.

INTEGRATE Chemistry

Acid Rain Effects Acid rain occurs when gases released by burning oil and coal mix with water in the air. Infer what effect acid rain can have on a statue made of limestone.

Figure 16 Yellowstone's famous geyser, Old Faithful, used to erupt once about every 76 min. An earthquake on January 9, 1998, slowed Old Faithful's "clock" by 4 min to an average of one eruption about every 80 min. The average height of the geyser's water is 40.5 m.

The Work of Groundwater

Although water is the most powerful agent of erosion on Earth's surface, it also can have a great effect underground. Water mixes with carbon dioxide gas to form a weak acid called carbonic acid. Some of this carbon dioxide is absorbed from the air by rainwater or surface water. Most carbon dioxide is absorbed by groundwater moving through soil. One type of rock that is dissolved easily by this acid is limestone. Acidic groundwater moves through natural cracks and pores in limestone, dissolving the rock. Gradually, the cracks in the limestone enlarge until an underground opening called a **cave** is formed.

INTEGRATE Chemistry

Acid Rain Effects The acid in acid rain reacts chemically to erode the limestone in statues.

Inquiry Lab

Purpose To observe and explore groundwater pollution, have students construct a model of an aquifer.

Possible Materials 1-quart jar, 1 cup gravel, 1 cup sand, 1 coffee filter, a straw, 1 syringe, water, red food coloring

Estimated Time 30 minutes

Teaching Strategies

- Have students place gravel into the bottom of a jar, place a coffee filter with a hole the size of the straw on top of the gravel, and pour sand on top of the coffee filter.
- Next, have them place the drinking straw into the jar through the hole of the filter, fill the jar half full of water, and drop a few drops of food coloring into the water.
- Finally, students should stick the syringe in the straw and suck out water to see if it is colored with food coloring.
- Allow students to explore questions that arise from this lab.

For additional inquiry activities, see *Science Inquiry Labs*.

Differentiated Instruction

Hearing Impaired Have students research artesian wells and write a report of how they work.

Science Journal

Yellowstone Have students write short reports on geysers and other natural features in Yellowstone National Park. Have them include the mechanism that causes these eruptions. Geysers are produced by the heating of groundwater within fractures of hot rock. L2 LS **Linguistic**

Activity

Permeability Groundwater travels through the soil and rocks. The rate of movement depends on the permeability of the soil and rocks. Have students fill a can that has small holes on the bottom, with sand. Fill a second similar can with rocks. Pour water in each one and see which can the water comes out faster. L1

Make a Model

Cave Formations After reading about cave formation and researching the topic in the library or on-line, have students make models of limestone caves that include labeled stalactites, stalagmites, and other dripstone features. Display completed models in the classroom. LS **Visual-Spatial**

3 Assess

DAILY INTERVENTION

Check for Understanding

Linguistic Have students research why groundwater is such an important resource. Have the students write a short report.

Reteach

Groundwater Components Direct students to **Figure 13**. Have students identify each component of the groundwater system and describe how all the components interact.

Assessment

Portfolio Have groups of students make models that show how the position of the water table affects a well, a hillside spring, a stream, a lake, and a hot spring. Use **Performance Assessment in the Science Classroom**, p. 123.

Figure 17 Water dissolves rock to form caves and also deposits material to form spectacular formations, such as these in Carlsbad Caverns in New Mexico.

Cave Formation You've probably seen a picture of the inside of a cave, like the one shown in **Figure 17,** or perhaps you've visited one. Groundwater not only dissolves limestone to make caves, but it also can make deposits on the insides of caves.

Water often drips slowly from cracks in the cave walls and ceilings. This water contains calcium ions dissolved from the limestone. If the water evaporates while hanging from the ceiling of a cave, a deposit of calcium carbonate is left behind. Stalactites form when this happens over and over. Where drops of water fall to the floor of the cave, a stalagmite forms. The words *stalactite* and *stalagmite* come from Greek words that mean "to drip."

Sinkholes If underground rock is dissolved near the surface, a sinkhole may form. A sinkhole is a depression on the surface of the ground that forms when the roof of a cave collapses or when material near the surface dissolves. Sinkholes are common features in places like Florida and Kentucky that have lots of limestone and enough rainfall to keep the groundwater system supplied with water. Sinkholes can cause property damage if they form in a populated area.

In summary, when rain falls and becomes groundwater, it might dissolve limestone and form a cave, erupt from a geyser, or be pumped from a well to be used at your home.

section 2 review

Summary

Groundwater Systems

- Water that soaks into the ground and collects in pore spaces is called groundwater.
- 14 percent of all freshwater on Earth exists as groundwater.
- Groundwater systems have connecting pores.
- The zone of saturation is the area where all pores in the rock are filled with water.

Water Table

- The supply of groundwater is limited.

Self Check

1. **Describe** how the permeability of soil and rocks affects the flow of groundwater.
2. **Describe** why a well might go dry.
3. **Explain** how caves form.
4. **Think Critically** Why would water in wells, geysers, and hot springs contain dissolved materials?

Applying Skills

5. **Compare and contrast** wells, geysers, and hot springs.

 Science Online bookg.msscience.com/self_check_quiz

section 2 review

1. The more permeable a soil or rock is, the better groundwater flows through it.
2. If water is removed from an aquifer faster than it can be replaced by nature, a well may go dry.
3. Caves form when water dissolves underground rock.
4. because groundwater dissolves materials as it flows through rock and soil.
5. All bring water to Earth's surface. In most wells, water must be pumped. Water is hot in geysers and hot springs. A geyser is a hot spring that periodically erupts steam and hot water.

section 3

Ocean Shoreline

The Shore

Picture yourself sitting on a beautiful, sandy beach like the one shown in **Figure 18.** Nearby, palm trees sway in the breeze. Children play in the quiet waves lapping at the water's edge. It's hard to imagine a place more peaceful. Now, picture yourself sitting along another shore. You're on a high cliff watching waves crash onto boulders far below. Both of these places are shorelines. An ocean shoreline is where land meets the ocean.

The two shorelines just described are different even though both experience surface waves, tides, and currents. These actions cause shorelines to change constantly. Sometimes you can see these changes from hour to hour. Why are shorelines so different? You'll understand why they look different when you learn about the forces that shape shorelines.

as you read

What **You'll Learn**

- **Identify** the different causes of shoreline erosion.
- **Compare and contrast** different types of shorelines.
- **Describe** some origins of sand.

Why **It's Important**

Constantly changing shorelines impact the people who live and work by them.

Review Vocabulary

tides: the alternating rise and fall of sea level caused by the gravitational attraction of the Moon and the Sun

New Vocabulary

- longshore current
- beach

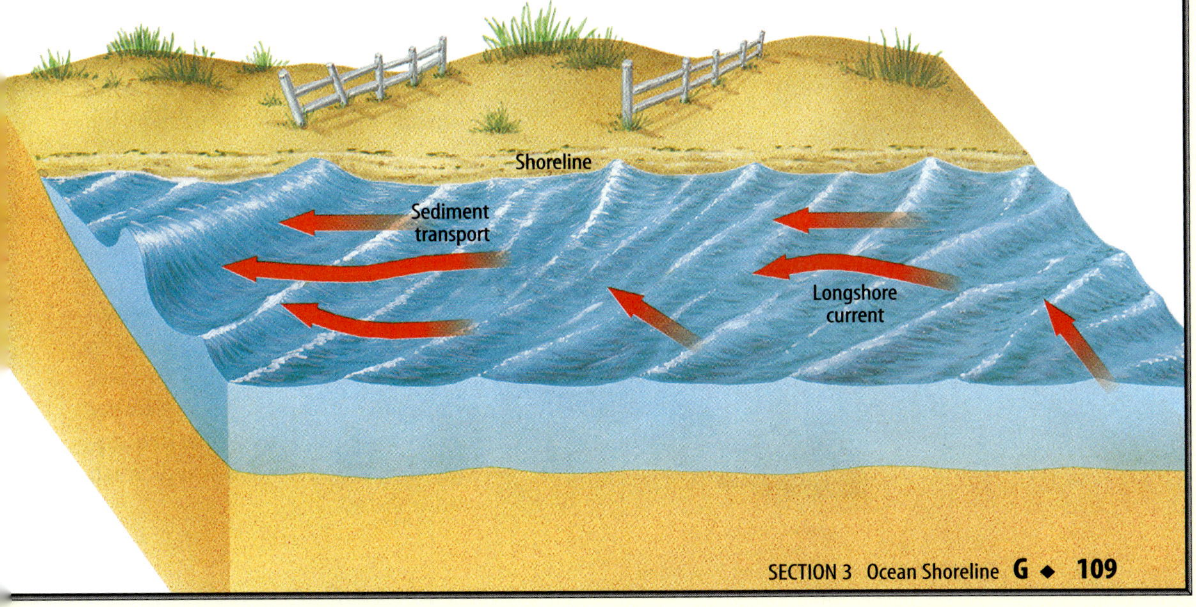

Figure 18 Waves, tides, and currents cause shorelines to change constantly. Waves approaching the shoreline at an angle create a longshore current.
Describe *the effects longshore currents have on a shoreline.*

Section 3 Resource Manager

Chapter *Fast File* Resources

Transparency Activity, p. 42

Directed Reading for Content Mastery, pp. 17, 18

Enrichment, p. 28

Lab Worksheets, pp. 5–6, 7–8

Reinforcement, p. 25

Home and Community Involvement, p. 19

Lab Management and Safety, p. 74

1 Motivate

Bellringer

Section Focus Transparencies also are available on the Interactive Chalkboard CD-ROM. L2 ELL

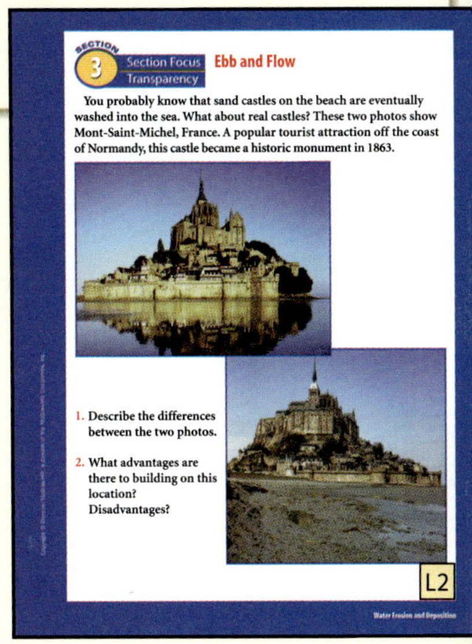

Section 3 Section Focus Transparency — Ebb and Flow

You probably know that sand castles on the beach are eventually washed into the sea. What about real castles? These two photos show Mont-Saint-Michel, France. A popular tourist attraction off the coast of Normandy, this castle became a historic monument in 1863.

1. Describe the differences between the two photos.
2. What advantages are there to building on this location? Disadvantages?

L2

Tie to Prior Knowledge

Shoreline Structures Many students have visited shorelines. Those who have not have probably seen them on television. Have students recall the shoreline structures they have seen.

Caption Answer

Figure 18 Longshore currents move sand and sediment from one place in the shoreline and carries it further down the shore

SECTION 3 Ocean Shoreline G ◆ 109

2 Teach

Teacher FYI

Tides Tides are the alternate rising and falling of the level of the ocean caused by the gravitational attraction of the Moon and Sun on Earth. In most places, there are two high tides and two low tides each day. Tidal range—the difference between high and low tide—varies. In some places, such as the shores of the Hawaiian and Caribbean Islands, the range is usually less than 1 m. In Nova Scotia's Bay of Fundy, on the other hand, the tidal range is more than 15 m.

Reading Check

Answer by waves colliding with shorelines at slight angles

Shoreline Forces When waves constantly pound against the shore, they break rocks into ever-smaller pieces. Currents move many metric tons of sediment along the shoreline. The sediment grains grind against each other like sandpaper. The tide goes out carrying sediment to deeper water. When the tide returns, it brings new sediment with it. These forces are always at work, slowly changing the shape of the shoreline. Water is always in motion along the shore.

The three major forces at work on the shoreline are waves, currents, and tides. Winds blowing across the water make waves. Waves, crashing against a shoreline, are a powerful force. They can erode and move large amounts of material in a short time. Waves usually collide with a shore at slight angles. This creates a **longshore current** of water that runs parallel to the shoreline. Longshore currents, shown in **Figure 18,** carry many metric tons of loose sediments and act like rivers of sand in the ocean.

Reading Check *How does a longshore current form?*

Tides create currents that move at right angles to the shore. These are called tidal currents. Outgoing tides carry sediments away from the shore, and incoming tides bring new sediments toward the shore. Tides work with waves to shape shorelines. You've seen the forces that affect all shorelines. Now you will see the differences that make one shore a flat, sandy beach and another shore a steep, rocky cliff.

Figure 19 Along a rocky shoreline, the force of pounding waves breaks rock fragments loose, then grinds them into smaller and smaller pieces.

Rocky Shorelines

Rocks and cliffs are the most common features along rocky shorelines like the one in **Figure 19.** Waves crash against the rocks and cliffs. Sediments in the water grind against the cliffs, slowly wearing the rock away. Then rock fragments broken from the cliffs are ground up by the endless motion of waves. They are transported as sediment by longshore currents.

Softer rocks become eroded before harder rocks do, leaving islands of harder rocks. This takes thousands of years, but remember that the ocean never stops. In a single day, about 14,000 waves crash onto shore.

Curriculum Connection

Music Suggest that students find music (tapes, CDs, videos, sheet music) about the coast or the beach. Have students play the music in class, or write the lyrics down and share them with classmates. Use the lyrics to spark a discussion of how people think about the shore and how it is used. L2 LS **Musical-Auditory**

Differentiated Instruction

Challenge The shape of coastal cliffs depends on the type of rock and the strength of waves. Basalt, granite, and other hard rocks result in tall, straight cliffs. Erosion of softer rock usually forms a more sloping landscape where erosion occurs faster. Have students research a well-known coastal cliff and write a short report describing the type of rock and the effects of ocean waves upon it. L3 **Linguistic** P

This quartz sand from a Texas beach is clear and glassy.

Some Hawaiian beaches are composed of black basalt sand.

Figure 20 Beach sand varies in size, color, and composition.

Sandy Beaches

Smooth, gently sloping shorelines are different from steep, rocky shorelines. Beaches are the main feature here. **Beaches** are deposits of sediment that are parallel to the shore.

Beaches are made up of different materials. Some are made of rock fragments from the shoreline. Many beaches are made of grains of quartz, and others are made of seashell fragments. These fragments range in size from stones larger than your hand to fine sand. Sand grains range from 0.06 mm to 2 mm in diameter. Why do many beaches have particles of this size? Waves break rocks and seashells down to sand-sized particles like those shown in **Figure 20.** The constant wave motion bumps sand grains together. This bumping not only breaks particles into smaller pieces but also smooths off their jagged corners, making them more rounded.

Reading Check *How do waves affect beach particles?*

Sand in some places is made of other things. For example, Hawaii's black sands are made of basalt, and its green sands are made of the mineral olivine. Jamaica's white sands are made of coral and shell fragments.

Fun Fact

Some beach sands squeak, bark, and whistle. Grains of these sands are smooth, polished, and of similar size. These characteristics may allow them to slip over one another with little friction, producing sound.

Quick Demo

Sand in a Glass

Material glass, sand, water

Estimated Time 10 minutes

Procedure Use a glass of water and sand to demonstrate deposition. Swirl the glass, and then stop the motion. Students will see that moving water keeps materials suspended, but when movement stops, the sand settles out. LS **Visual-Spatial**

Reading Check

Answer They break them down into sand-sized particles and make them more rounded.

Discussion

Taking a Risk What factors would you consider if building a house along the shoreline? You would have to look at waves, currents, and tides to see how they would affect that particular shoreline.

Differentiated Instruction

English-Language Learners Have the students compare groundwater levels of their native country to groundwater levels where they live in the United States. ELL

Curriculum Connection

Economics Many coastal cities depend on tourism. When beaches disappear because of erosion, tourists do not visit. For this reason, many cities have spent a lot of money to preserve their beaches. Have students research costs involved in dredging and building seawalls. Sand trapped by one city is "denied" to another. Dredging and building seawalls are expensive and temporary measures. One strong storm can remove all the dredged sand and undermine seawalls.

SECTION 3 Ocean Shoreline G ◆ 111

Visual Learning

Figure 21 How would buildings such as these be affected by hurricanes? They would be in danger from coastal flooding and storm surges. Is development right along the coast a good idea? Accept all opinions as a basis for class discussion.

3 Assess

DAILY INTERVENTION

Check for Understanding

Visual-Spatial Have students research the history of New Orleans. Have them make graphs showing how, over time, it has dropped below sea level.

Reteach

Coastline Shaping Obtain photographs or slides of various kinds of coastlines. Have students describe what caused the coastline features shown in each photograph. L1 LS **Visual-Spatial**

Assessment

Performance Have students make a sequence concept map that shows how an arch of rock can form along a rocky shoreline. Use **Performance Assessment in the Science Classroom,** p. 161

Figure 21 Shorelines change constantly. Human development is often at risk from shoreline erosion.

Sand Erosion and Deposition

Longshore currents carry sand along beaches to form features such as barrier islands, spits, and sandbars. Storms and wind also move sand. Thus, beaches are fragile, short-term land features that are damaged easily by storms and human activities such as some types of construction. Communities in widely separated places such as Long Island, New York; Malibu, California; and Padre Island, Texas, have problems because of beach erosion.

Barrier Islands Barrier islands are sand deposits that lay parallel to the shore but are separated from the mainland. These islands start as underwater sand ridges formed by breaking waves. Hurricanes and storms add sediment to them, raising some to sea level. When a barrier island becomes large enough, the wind blows the loose sand into dunes, keeping the new island above sea level. As with all seashore features, barrier islands are short term, lasting from a few years to a few centuries.

The forces that build barrier islands also can erode them. Storms and waves carry sediments away. Beachfront development, as in **Figure 21,** can be affected by shoreline erosion.

section 3 review

Summary

The Shore

- An ocean shoreline is where land meets the ocean.
- The forces that shape shorelines are waves, currents, and tides.

Rocky Shorelines

- Rocks and cliffs are the most common features along rocky shorelines.

Sandy Beaches

- Beaches are made up of different materials. Some are made of rock fragments and others are made of seashell fragments.

Sand Erosion and Deposition

- Longshore currents, storms, and wind move sand.
- Beaches are fragile, short-term land features.

Self Check

1. **Identify** major forces that cause shoreline erosion.
2. **Compare and contrast** the features you would find along a steep, rocky shoreline with the features you would find along a gently sloping, sandy shoreline.
3. **Explain** how the type of shoreline could affect the types of sediments you might find there.
4. **List** several materials that beach sand might be composed of. Where do these materials come from?
5. **Think Critically** How would erosion and deposition of sediment along a shoreline be affected if the longshore current was blocked by a wall built out into the water?

Applying Math

6. **Solve One-Step Equations** If 14,000 waves crash onto a shore daily, how many waves crash onto it in a year? How many crashed onto it since you were born?

Science Online bookg.msscience.com/self_check_quiz

section 3 review

1. waves and currents
2. rocky: boulders, cliffs, caves; sandy: flat beaches, barrier islands
3. Sediment size is larger along rocky shores than along gentle, sandy slopes.
4. Materials include quartz, seashell and coral fragments, olivine, and basalt. They originate from rock fragments from the shoreline or from the shells of sea organisms.
5. Sediments would be trapped by the wall instead of being deposited farther down the beach.
6. 5,110,000

LAB

Classifying Types of Sand

Sand is made of different kinds of grains, but did you realize that the slope of a beach is related to the size of its grains? The coarser the grain size is, the steeper the beach is. The composition of sand also is important. Many sands are mined because they have economic value.

Real-World Question

What characteristics can be used to classify different types of beach sand?

Goals

- **Observe** differences in sand.
- **Identify** characteristics of beach sand.
- **Infer** sediment sources.

Materials

samples of different sands (3)
magnifying lens
**stereomicroscope*
magnet
**Alternate materials*

Safety Precautions

Procedure

1. **Design** a five-column data table to compare the three sand samples. Use column one for the samples and the others for the characteristics you will be examining.

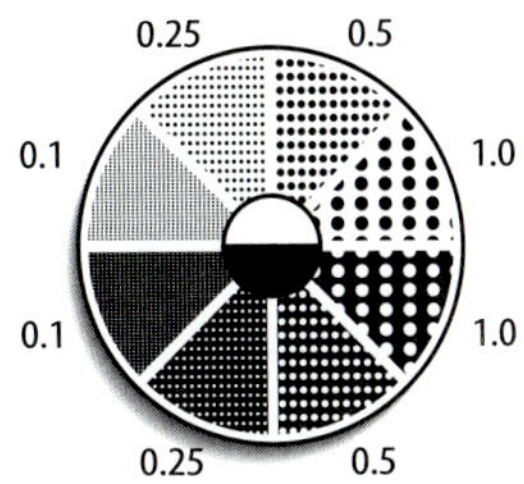

Sand gauge
(measurements in mm)

2. **Use the diagram** in the previous column to determine the average roundness of each sample.
3. **Identify** the grain size of your samples by using the sand gauge above. To determine the grain size, place sand grains in the middle of the circle of the sand gauge. Use the upper half of the circle for dark-colored particles and the bottom half for light-colored particles.
4. **Decide** on two other characteristics to examine that will help you classify your samples.

Conclude and Apply

1. **Compare and contrast** some characteristics of beach sand.
2. **Describe** why there are variations in the characteristics of different sand samples.
3. **Explain** what your observations tell you about the sources of the three samples.

Communicating Your Data

Compare your results with those of other students.

BENCH TESTED

Real-World Question

Purpose Students observe and classify beach sand. L2 LS **Visual-Spatial**

Process Skills classify, observe and infer, compare and contrast, interpret data, make tables

Procedure

Time Required 45 minutes

Alternate Materials stereomicroscope

Safety Precautions

Teaching Strategy Beach sands can be obtained from scientific supply distributors, hardware stores, and garden centers. Contact teachers in coastal areas for possible trading of sand samples.

Troubleshooting Advise students to avoid mixing the sand samples. They will need to examine only a small quantity of sand.

Conclude and Apply

1. Possible characteristics: color, texture, shape, grain size, luster, magnetism, and composition
2. Samples will vary according to the length of time exposed to weathering and erosion, the type of erosion (wind or water), and the characteristics of parent material.
3. Answer depends on the characteristics of the samples. Possible answer: all three samples were from different sources. Students may infer the specific origin of a sample to be quartz-rich rocks or sea shells.

Assessment

Performance Show students pictures of different beaches. Have them use the results of this activity and the information that steeper beaches have coarser-sized sand grains to hypothesize as to the grain size likely to be found on each beach. Use **Performance Assessment in the Science Classroom,** p. 93.

Communicating Your Data

Students should discuss why their conclusions did or did not agree with those of classmates. If possible, make overhead transparencies of several student's tables. Project them for the class to use while discussing and critiquing different classification systems.

BENCH TESTED

Real-World Question

Purpose Students will determine how water speed affects its ability to erode.

Process Skills observe, collect data, interpret data, identify and manipulate variables and controls, infer, predict

Time Required one class period

Procedure

Materials Disposable paint trays are available at paint and hardware stores.

Alternate Materials Disposable wallpaper trays can be substituted for paint roller pans.

Safety Precautions Remind students to immediately clean up any water that spills on the floor. Have students hold the rubber tubing while turning the faucet on and increasing water speed. This will prevent possible mishaps and spills. Have students wash their hands after handling sand.

LAB

Water Speed and Erosion

Goals

- **Assemble** an apparatus for measuring the effect of water speed on erosion.
- **Observe and measure** the ability of water traveling at different speeds to erode sand.

Materials

paint roller pan
**disposable wallpaper trays*
sand
1-L beaker
rubber tubing (20 cm)
metric ruler
water
stopwatch
fine-mesh screen
wood block
**Alternate materials*

Safety Precautions

Wash your hands after you handle the sand. Immediately clean up any water that spills on the floor.

Real-World Question

What would it be like to make a raft and use it to float on a river? Would it be easy? Would you feel like Tom Sawyer? Probably not. You'd be at the mercy of the current. Strong currents create fast rivers. But does fast moving water affect more than just floating rafts and other objects? How does the speed of a stream or river affect its ability to erode?

Procedure

1. Copy the data table on the following page.
2. Place the screen in the sink. Pour moist sand into your pan and smooth out the sand. Set one end of the pan on the wood block and hang the other end over the screen in the sink. Excess water will flow onto the screen in the sink.
3. Attach one end of the hose to the faucet and place the other end in the beaker. Turn on the water so that it trickles into the beaker. Time how long it takes for the trickle of water to fill the beaker to the 1-L mark. Divide 1 L by your time in seconds to calculate the water speed. Record the speed in your data table.
4. Without altering the water speed, hold the hose over the end of the pan that is resting on the wood block. Allow the water to flow into the sand for 2 min. At the end of 2 min, turn off the water.
5. **Measure** the depth and length of the eroded channel formed by the water. Count the number of branches formed on the channel. Record your measurements and observations in your data table.

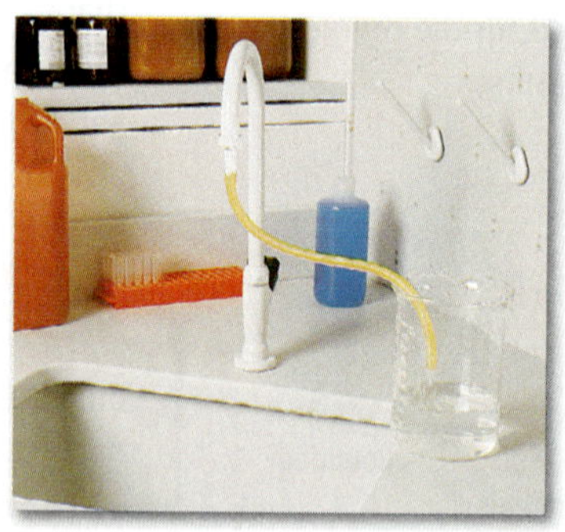

Differentiated Instruction

Visually Impaired After other group members have measured the depth and length of the eroded channels, have these students feel the channels made by different water speeds so they can make a comparison.

Alternative Inquiry Lab

Obstacles and Erosion How would vegetation and other obstacles affect the erosion rate? Encourage students to design an experiment to explore this question or other questions that arise from the lab.

Water Speed and Erosion

Water Speed (Liters per Second)	Depth of Channel	Length of Channel	Number of Channel Branches

6. Empty the excess water from the tray and smooth out the sand. Repeat steps 3 through 5 two more times, increasing your water speed each time.

Conclude and Apply

1. **Identify** the constants and variables in your experiment.
2. **Observe** Which water speed created the deepest and longest channel?
3. **Observe** Which water speed created the greatest number of branches?
4. **Infer** the effect that water speed has on erosion.
5. **Predict** how your results would have differed if one end of the pan had been raised higher.
6. **Infer** how streams and rivers can shape Earth's surface.

Communicating Your Data

Write a pamphlet for people buying homes near rivers or streams that outlines the different effects that water erosion could have on their property.

Teaching Strategies

- Have students read through the procedure completely and prepare the data table before beginning.
- Set up stream tables in advance to save time.

Tie to Prior Knowledge Most students have observed how gullies form when water runs down a bare slope so they are aware of what to expect.

Analyze Your Data

Expected Outcome Most results will reflect an increase in the amount of sand eroded as the water speed increased.

Error Analysis Have each group compare their results with those of another group. If the data disagree, have students list possible reasons for the discrepancy.

Conclude and Apply

1. constants: amount of sand, elevation of the slope, the amount of time water ran on the slope; variables: water speed, amount of water that ran on the slope, amount of sand eroded by water.
2. the fastest water speed
3. the fastest water speed
4. The faster the water moves, the greater the erosion that occurs.
5. The steeper the slope, the faster the water flows and the more it erodes.
6. When water flows through streams and rivers, it erodes soil and carries it to other areas. These processes change the shape of the land.

Assessment

Performance Have students design and carry out a procedure to test their prediction in question 5 of Conclude and Apply. Use **Performance Assessment in the Science Classroom,** p. 95.

Communicating Your Data

Students can use a word processing program and graphics program to prepare their pamphlets.

LAB G ◆ 115

TIME

Content Background

Explain that most of the erosion caused by the sea occurs where the breaking waves interact with the land. As waves approach the shore, the bottom of each wave touches the shallow ocean bottom near the shore and is slowed down while the top of the wave continues unhindered and bends toward the shore. This results in an unequal distribution of the energy stored in the wave. As a consequence, the wave crests as it breaks onto the shore. Erosion occurs primarily by abrasion. Sediments are carried out to sea by the retreating water and the shoreline retreats.

Beach nourishment is one method commonly used to help stabilize eroded coastlines. Other methods include the construction of jetties (structures that extend perpendicular from the shore to trap sand), building of breakwaters (long dikes built offshore to slow waves down), and using concrete slabs to create erosion-resistant shores or embankments.

Two factors that should be considered when determining which method to use are natural features of the shoreline and cost.

Activity

Shoreline Erosion Have students search magazines to obtain pictures of different kinds of beach environments. Organize these pictures into three categories: shores resistant to erosion, shores easily eroded, and shores unaffected by erosion. Explain why each picture belongs in that category. The third category, shores unaffected by erosion, should be empty since all beaches are affected by erosion.

TIME SCIENCE AND Society

SCIENCE ISSUES THAT AFFECT YOU!

Is there hope for America's coastlines or is beach erosion a "shore" thing?

Sands in Time

Water levels are rising along the coastline of the United States. Serious storms and the building of homes and businesses along the shore are leading to the erosion of anywhere from 70 percent to 90 percent of the U.S. coastline. A report from the Federal Emergency Management Agency (FEMA) confirms this. The report says that one meter of United States beaches will be eaten away each year for the next 60 years. Since 1965, the federal government has spent millions of dollars replenishing more than 1,300 eroding sandy shores around the country. And still beaches continue to disappear.

The slowly eroding beaches are upsetting to residents and officials of many communities, who depend on their shore to earn money from visitors. Some city and state governments are turning to beach nourishment—a process in which sand is taken from the seafloor and dumped on beaches. The process is expensive, however. The state of Delaware, for example, is spending 7,000,000 dollars to bring in sand for its beaches.

This beach house will collapse as its underpinnings are eroded.

Other methods of saving eroding beaches are being tried. In places along the Great Lakes shores and coastal shores, one company has installed fabrics underwater to slow currents. By slowing currents, sand is naturally deposited and kept in place.

Another shore-saving device is a synthetic barrier that is shaped like a plastic snowflake. A string of these barriers is secured just offshore. They absorb the energy of incoming waves. Reducing wave energy can prevent sand from being eroded from the beach. New sand also might accumulate because the barriers slow down the currents that flow along the shore.

Many people believe that communities along the shore must restrict the beachfront building of homes, hotels, and stores. Since some estimates claim that by the year 2025, nearly 75 percent of the U.S. population will live in coastal areas, it's a tough solution. Says one geologist, "We can retreat now and save our beaches or we can retreat later and probably ruin the beaches in the process."

Debate **Using the facts in this article and other research you have done in your school media center or through Web links at msscience.com, make a list of methods that could be used to save beaches. Debate the issue with your classmates.**

Science Online
For more information, visit bookg.msscience.com/time

Debate Help students prepare their positions by constructing a table to compare the pros and cons (columns) of the methods (rows) used for saving beaches. Other factors that might be compared include natural shoreline features, land ownership (public vs. private), land use, erosion history, storm and flooding history, cost effectiveness, and environmental impacts.

Resources for Teachers and Students

Evaluation of Erosion Hazards Report, 2000. (FEMA), 500 C Street, Washington DC, 20472

Beach Nourishment and Protection. Committee on Beach Nourishment and Protection, National Research Council, 1995

Against the Tide: The Battle for America's Beaches. Cornelia Dean, Columbia University Press, 1999

Reviewing Main Ideas

Section 1 Surface Water

1. Rainwater that does not soak into the ground is pulled down the slope by gravity. This water is called runoff.
2. Runoff can erode sediment. Factors such as steepness of slope and number and type of plants affect the amount of erosion. Rill, gully, and sheet erosion are types of surface water erosion caused by runoff.
3. Runoff generally flows into streams that merge with larger rivers until emptying into a lake or ocean. Major river systems usually contain several different types of streams.
4. Young streams flow through steep valleys and have rapids and waterfalls. Mature streams flow through gentler terrain and have less energy. Old streams are often wide and meander across their floodplains.

Section 2 Groundwater

1. When water soaks into the ground, it becomes part of a vast groundwater system.
2. Although rock may seem solid, many types are filled with connected spaces called pores. Such rocks are permeable and can contain large amounts of groundwater.

Section 3 Ocean Shoreline

1. Ocean shorelines are always changing.
2. Waves and currents have tremendous amounts of energy which break up rocks into tiny fragments called sediment. Over time, the deposition and relocation of sediment can change beaches, sandbars, and barrier islands.

Visualizing Main Ideas

Copy and complete the following concept map on caves.

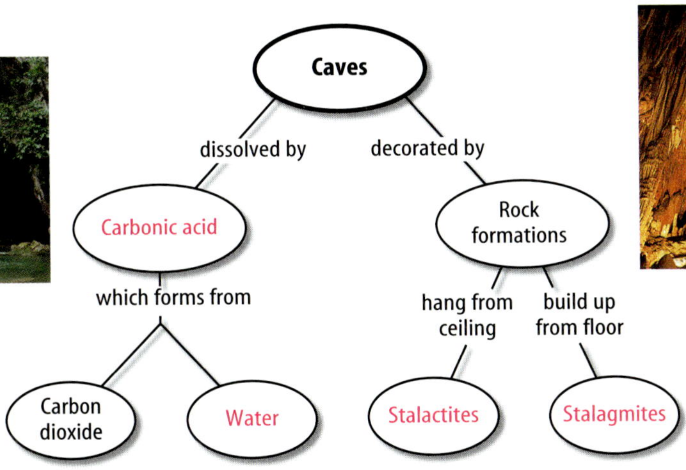

Reviewing Main Ideas

Summary statements can be used by students to review the major concepts of the chapter.

Visualizing Main Ideas

See student page.

Science Online

Visit bookg.msscience.com
/self_check_quiz
/interactive_tutor
/vocabulary_puzzlemaker
/chapter_review
/standardized_test

Assessment Transparency

For additional assessment questions, use the *Assessment Transparency* located in the transparency book.

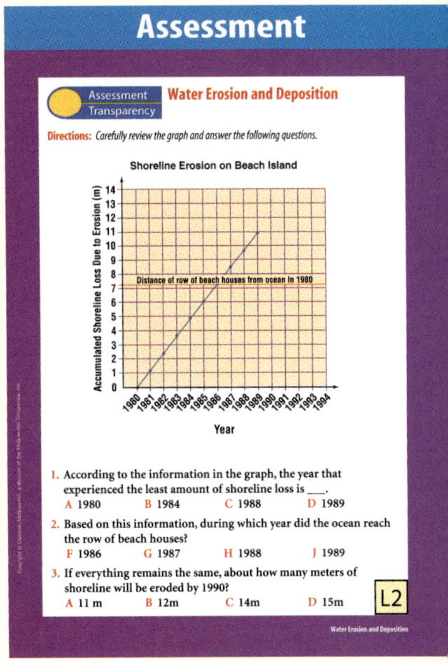

Assessment

Assessment Transparency **Water Erosion and Deposition**

Directions: *Carefully review the graph and answer the following questions.*

Shoreline Erosion on Beach Island
Accumulated Shoreline Loss Due to Erosion (m)
0 1 2 3 4 5 6 7 8 9 10 11 12 13 14
Distance of row of beach houses from ocean in 1980
1980 1981 1982 1983 1984 1985 1986 1987 1988 1989 1990 1991 1992 1993 1994
Year

1. According to the information in the graph, the year that experienced the least amount of shoreline loss is ___.
   A 1980 B 1984 C 1988 D 1989
2. Based on this information, during which year did the ocean reach the row of beach houses?
   F 1986 G 1987 H 1988 J 1989
3. If everything remains the same, about how many meters of shoreline will be eroded by 1990?
   A 11 m B 12m C 14m D 15m

L2

Water Erosion and Deposition

CHAPTER STUDY GUIDE G ◆ 117

chapter 4 Review

Using Vocabulary

1. Runoff is water that flows across Earth's surface. Sheet erosion is runoff that flows without being in a channel.
2. A channel is a groove made by flowing water. A drainage basin is the land area from which a stream collects runoff.
3. An aquifer is a layer of permeable rock that lets water move freely, albeit slowly, underground. A cave is an enlarged underground opening where the water has worn away or dissolved the rock.
4. A spring is where the water table intersects Earth's surface, allowing water to flow out. A geyser is a hot spring that erupts periodically.
5. Permeable materials let water pass through. Materials that are impermeable do not.
6. Sheet erosion moves sediments as water flows in sheets over flat land. A meander is a broad bend in a stream.
7. Groundwater is water that collects in underground pores in soil and rocks. The water table is the upper surface of the underground saturated zone.
8. Permeable substances allow water to pass through. An aquifer is a layer of permeable rock through which water moves slowly, but freely.
9. A longshore current runs along a shore. A beach is a deposit of sediments parallel to the shore.
10. A meander is a curve in a stream's channel. A channel is a groove where water flows.

Checking Concepts

11. B
12. C
13. D
14. D
15. A
16. A
17. B
18. D
19. A
20. B

chapter 4 Review

Using Vocabulary

aquifer p. 104
beach p. 111
cave p. 107
channel p. 94
drainage basin p. 96
geyser p. 107
groundwater p. 103
impermeable p. 104
longshore current p. 110
meander p. 97
permeable p. 104
runoff p. 92
sheet erosion p. 95
spring p. 107
water table p. 104

Explain the difference between the vocabulary words in each of the following sets.

1. runoff—sheet erosion
2. channel—drainage basin
3. aquifer—cave
4. spring—geyser
5. permeable—impermeable
6. sheet erosion—meander
7. groundwater—water table
8. permeable—aquifer
9. longshore current—beach
10. meander—channel

Checking Concepts

Choose the word or phrase that best answers the question.

11. Where are beaches most common?
   A) rocky shorelines
   B) flat shorelines
   C) aquifers
   D) young streams

12. What is the network formed by a river and all the smaller streams that contribute to it?
   A) groundwater system
   B) zone of saturation
   C) river system
   D) water table

13. Why does water rise in an artesian well?
   A) a pump
   B) erosion
   C) heat
   D) pressure

14. Which term describes rock through which fluids can flow easily?
   A) impermeable
   B) meanders
   C) saturated
   D) permeable

15. Identify an example of a structure created by deposition.
   A) beach
   B) rill
   C) cave
   D) geyser

16. Which stage of development are mountain streams in?
   A) young
   B) mature
   C) old
   D) meandering

17. What forms as a result of the water table meeting Earth's surface?
   A) meander
   B) spring
   C) aquifer
   D) stalactite

18. What contains heated groundwater that reaches Earth's surface?
   A) water table
   B) cave
   C) aquifer
   D) hot spring

19. What is a layer of permeable rock that water flows through?
   A) an aquifer
   B) a pore
   C) a water table
   D) impermeable

20. Name the deposit that forms when a mountain river runs onto a plain.
   A) subsidence
   B) an alluvial fan
   C) infiltration
   D) water diversion

Science Online bookg.msscience.com/vocabulary_puzzlemaker

Use the ExamView® Pro Testmaker CD-ROM to:

- create multiple versions of tests
- create modified tests with one mouse click for inclusion students
- edit existing questions and add your own questions
- build tests aligned with state standards using built-in State Curriculum Tags
- change English tests to Spanish with one mouse click and vice versa

Thinking Critically

21. **Concept Map** Copy and complete the concept map below using the following terms: *developed meanders, gentle curves, gentle gradient, old, rapids, steep gradient, wide floodplain,* and *young.*

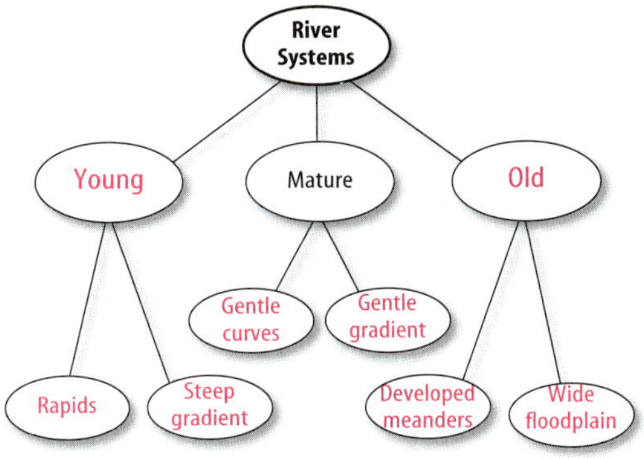

22. **Describe** what determines whether a stream erodes its bottom or its sides.
23. **Interpret Data** The rate of water flowing out of the Brahmaputra River in India, the La Plata River in South America, and the Mississippi River in North America are given in the table below. Infer which river carries the most sediment.

River Flow Rates

River	Flow (m^3/s)
Brahmaputra River, India	19,800
La Plata River, South America	79,300
Mississippi River, North America	175,000

24. **Explain** why the Mississippi River has meanders along its course.
25. **Outline** Make an outline that explains the three stages of stream development.
26. **Form Hypotheses** Hypothesize why most of the silt in the Mississippi delta is found farther out to sea than the sand-sized particles are.
27. **Infer** Along what kind of shoreline would you find barrier islands?
28. **Explain** why you might be concerned if developers of a new housing project started drilling wells near your well.
29. **Use Variables, Constants, and Controls** Explain how you could test the effect of slope on the amount of runoff produced.

Performance Activities

30. **Poster** Research a beach that interests you. Make a poster that shows different features you would find at a beach.

Applying Math

Use the illustration below to answer questions 31–32.

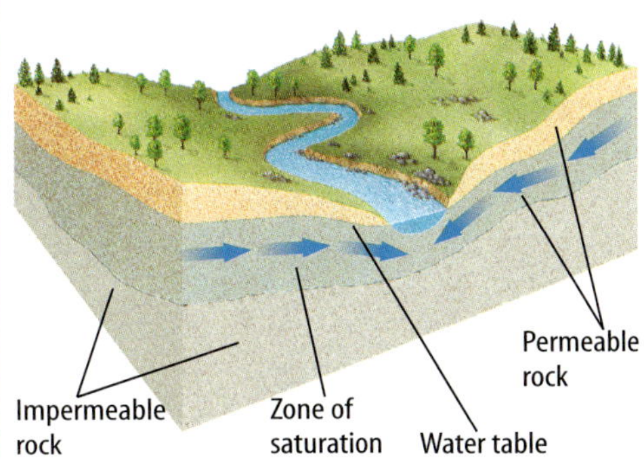

31. **Flow Distance** The groundwater in an aquifer flows at a rate of 0.2 m/day. How far does the groundwater move in one week?
32. **Flow Time** If groundwater in an aquifer flows at a rate of 0.4 m/day, how long does it take groundwater to move 24 m?

Assessment Resources

Reproducible Masters

Chapter *Fast File* Resources
- Chapter Review, pp. 33–34
- Chapter Tests, pp. 35–38
- Assessment Transparency Activity, p. 45

Glencoe Science Web site
- Chapter Review Test
- Standardized Test Practice

Glencoe Technology
- Assessment Transparency
- Exam*View*® Pro Testmaker
- MindJogger Videoquiz
- Interactive Chalkboard

chapter 4 Review

Thinking Critically

21. See student page.
22. The energy of the moving water, which is related to a stream's stage of development, determines erosion. Young streams erode downward; old streams erode their sides.
23. The Mississippi carries more because it has more energy of motion. Its flow is much greater than the other two rivers.
24. The Mississippi River is in the old stage of development. Meanders are one characteristic of this stage.
25. Students' outlines should include details of the young, mature, and old stages.
26. Silt is smaller than sand. As water slows at the river's mouth, it drops larger sediments (sand) first.
27. along flat, sandy shores
28. If a lot of water is drawn from the same aquifer, the supply could be depleted by overuse.
29. Answer should include (1) a way to allow a fixed amount of water to flow in the same way over the same type of soil surface, the difference being the angle of slope, and (2) a method to collect and measure the amount of run-off produced.

Performance Activities

30. Sands vary according to their parent material and the length of time they have been exposed to weathering and erosion. Use **Performance Assessment in the Science Classroom,** p. 145.

Applying Math

National Math Standards

1, 2, 9

31. distance = (0.2 m/day)(7 days) = 1.4 m
32. time = (24 m)/(0.4 m/day) = 60 days

FAST FILE

Answer Sheet A practice answer sheet can be found at bookg.msscience.com/answer_sheet.

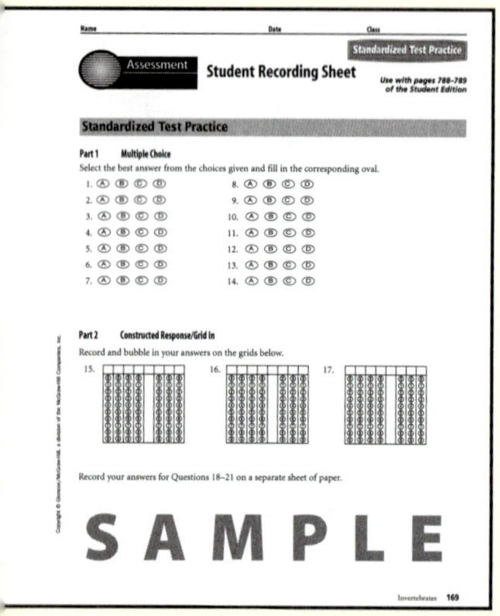
Name Date Class

Standardized Test Practice

Assessment Student Recording Sheet

Use with pages 788–789 of the Student Edition

Standardized Test Practice

Part 1 Multiple Choice

Select the best answer from the choices given and fill in the corresponding oval.

1. Ⓐ Ⓑ Ⓒ Ⓓ 8. Ⓐ Ⓑ Ⓒ Ⓓ
2. Ⓐ Ⓑ Ⓒ Ⓓ 9. Ⓐ Ⓑ Ⓒ Ⓓ
3. Ⓐ Ⓑ Ⓒ Ⓓ 10. Ⓐ Ⓑ Ⓒ Ⓓ
4. Ⓐ Ⓑ Ⓒ Ⓓ 11. Ⓐ Ⓑ Ⓒ Ⓓ
5. Ⓐ Ⓑ Ⓒ Ⓓ 12. Ⓐ Ⓑ Ⓒ Ⓓ
6. Ⓐ Ⓑ Ⓒ Ⓓ 13. Ⓐ Ⓑ Ⓒ Ⓓ
7. Ⓐ Ⓑ Ⓒ Ⓓ 14. Ⓐ Ⓑ Ⓒ Ⓓ

Part 2 Constructed Response/Grid In

Record and bubble in your answers on the grids below.

15. 16. 17.

Record your answers for Questions 18–21 on a separate sheet of paper.

SAMPLE

Part 1 Multiple Choice

1. D	4. C	7. C	10. C
2. A	5. B	8. B	11. A
3. A	6. A	9. A	12. A

Part 2 Short Response

13. It causes water to move downhill. The steeper the grade the faster the water moves and greater the erosion.
14. Lighter sediments are carried by the water in the suspended load of the stream. Heavier particles called the bed load roll along the bottom of the stream channel.
15. During heavy rains the floodplain can flood and cause damage to homes, business and crops. People need to be aware of this situation so that they can seek shelter and take measures to protect themselves and their property.
16. An aquifer is a layer of permeable rock where water flows freely. The area where all of the pores in the rock are filled with water is the zone of saturation. The upper surface of this zone is the water table.

chapter 4 Standardized Test Practice

Part 1 Multiple Choice

Record your answers on the answer sheet provided by your teacher or on a sheet of paper.

1. Which is erosion over a large, flat area?
   A. gully **C.** runoff
   B. rill **D.** sheet
2. Which is erosion where a rill becomes broader and deeper?
   A. gully **C.** runoff
   B. rill **D.** sheet
3. Which is the area of land from which a stream collects runoff?
   A. drainage basin **C.** runoff
   B. gully **D.** stream channel
4. Which type of soil or rock allows water to pass through them?
   A. impermeable **C.** permeable
   B. nonporous **D.** underground

Refer to the figure below to answer question 5.

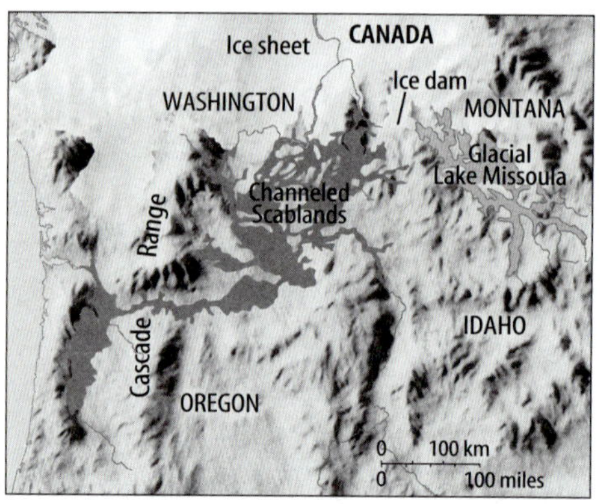

5. Which formed the Channeled Scablands?
   A. deposition **C.** rill erosion
   B. floodwaters **D.** sheet erosion
6. Which dissolves limestone to form caves?
   A. carbonic acid **C.** stalactites
   B. hydrochloric acid **D.** stalagmites
7. Which forms on the ceilings of caves as water drips through cracks?
   A. aquifer **C.** stalactite
   B. geyser **D.** stalagmite
8. Which are piles of sand found on barrier islands?
   A. deltas **C.** geysers
   B. dunes **D.** streams
9. Which creates springs and geysers?
   A. groundwater **C.** rills
   B. gullies **D.** runoff

Refer to the figure below to answer question 10 and 11.

10. Wh:ch feature is shown?
    A. artesian well **C.** geyser
    B. aquifer **D.** waterfall
11. Which provides the water?
    A. groundwater **C.** stream
    B. runoff **D.** surface water
12. Which is a layer of permeable rock through which water moves freely?
    A. aquifer **C.** geyser
    B. clay **D.** granite

Test-Taking Tip

Correct Answer Bubbles For each question, double check that you are filling in the correct answer bubble for the question number you are working on.

17. The chemicals can soak into the ground and soil water. This water then could travel through permeable rock and become part of the groundwater system.
18. gully erosion
19. Water moving down the same path on a slope creates a small channel, called a rill. If water frequently flows in the same channel, rill erosion can change into broader and deeper gully erosion.
20. amount of rain, length of time, slope, and vegetation cover
21. flows smoothly and can develop broad curves called meanders

Part 3 Open Ended

22. Rocks, steep cliffs, and very little sediment are the most common features along rocky shorelines. Smooth, gently sloping shorelines with lots of sediment are characteristic of sandy beaches.

Part 2 Short Response/Grid In

Record your answers on the answer sheet provided by your teacher or on a sheet of paper.

13. How does gravity affect water erosion?
14. Describe the different types of load in a stream.
15. What will happen to homes and businesses located in a floodplain?
16. Explain how the water table and the zone of saturation are related.
17. How would harmful chemicals in the soil enter into the groundwater system?

Refer to the picture below to answer questions 18 and 19.

18. What type of erosion is shown?
19. What led to the erosion shown?
20. What factors affect runoff?
21. What are the characteristics of a mature stream?

Part 3 Open Ended

Record your answers on a sheet of paper.

22. Compare and contrast rocky shorelines and sandy beaches.
23. What are the Outer Banks in North Carolina and how were they formed?
24. What can humans do to try to control flood waters?
25. Explain how waves produce longshore currents.

Refer to the figure below to answer question 26.

26. What stream formation is shown in the diagram above? Explain how differing speeds of water in a stream can cause this type of stream formation.
27. Explain how a drainage basin works. Compare and contrast a drainage basin to the gutter drainage system of a rooftop.
28. Compare and contrast a pumped well and an artesian well.
29. Compare and contrast alluvial fans and deltas.

Science Online bookg.msscience.com/standardized_test

23. Barrier islands are sand deposits that lay parallel to the shore but are separate from the mainland. These islands start as underwater sand ridges but storms and hurricanes raise them to sea level by adding sediment to them. They are short term seashore features that can be destroyed by the same forces that built them.
24. Dams and levees are built to protect homes, crops and bridges from flood waters. Dams control the water flow downstream. They can be made of soil, sand, or steel and concrete. Levees are mounds of soil that are built along the sides of a river.
25. The three major forces at work along the shoreline are waves, currents and tides. Powerful waves crash against a shoreline and can move large amounts of material in a short amount of time. Waves also form a longshore current that runs parallel to the shoreline and carries tons of loose sediment. Tides create currents that move at right angles to the shore called tidal currents. Outgoing tides carry sediment away from the shore and incoming tides bring new sediment towards the shore.
26. a mature stream; A mature stream starts to erode more on the sides causing curves to develop. Water in the shallow areas of a stream moves slower than in the deeper areas. If the deeper part of the channel is next to one side of the river, water will erode that side and form a slight bend. Over time, the curve becomes a broad arc called a meander.
27. A drainage basin is the area of land from which a stream or river collects runoff. The runoff flows downhill and inward toward the main river, or trunk. Similar to a river drainage basin, runoff flows downhill toward the gutter of a rooftop drainage system. However, a rooftop drainage system is like an inverted river drainage basin with the runoff flowing outward toward the gutters.
28. Both are a source of groundwater from an aquifer. Water from a pumped well must be forced to the surface. A pump is not needed for an artesian well, because pressure within the aquifer forces the water to the surface.
29. Both are a deposition caused by a stream or river. Sediment that is deposited as water empties into an ocean or lake forms a fan-shaped, deposit called a delta. When the stream waters empty from a mountain valley onto an open plain, the deposit is called an alluvial fan.

Rubrics

For more help evaluating open-ended assessment questions, see the rubric on p. 10T.

chapter 5 Organizer

Section/Objectives	Standards		Labs/Features
	National	State/Local	
Chapter Opener	See pp. 9T–10T for a Key to Standards.		**Launch Lab:** Clues to Life's Past, p. 123 **Foldables,** p. 123
Section 1 Fossils 2 sessions 1 block 1. **List** the conditions necessary for fossils to form. 2. **Describe** several processes of fossil formation. 3. **Explain** how fossil correlation is used to determine rock ages. 4. **Determine** how fossils can be used to explain changes in Earth's surface, life forms, and environment.	National Content Standards: UCP.1, UCP.2, UCP.3, UCP.5, A.1, A.2, F.5		**MiniLAB:** Predicting Fossil Preservation, p. 125 **Integrate Social Studies,** p. 127 **Integrate Life Science,** p. 130
Section 2 Relative Ages of Rocks 2 sessions 1 block 5. **Describe** methods used to assign relative ages to rock layers. 6. **Interpret** gaps in the rock record. 7. **Give** an example of how rock layers can be correlated with other rock layers.	National Content Standards: UCP.1, UCP.2, UCP.3, UCP.5, A.1, A.2		**Science Online,** p. 133 **Visualizing Unconformities,** p. 135 **Science Online,** p. 136 **Lab:** Relative Ages, p. 138
Section 3 Absolute Ages of Rocks 3 sessions 1.5 blocks 8. **Identify** how absolute age differs from relative age. 9. **Describe** how the half-lives of isotopes are used to determine a rock's age.	National Content Standards: UCP.1, UCP.2, UCP.3, UCP.5, A.1, A.2, B.1, F.5, G.3		**MiniLAB:** Modeling Carbon-14 Dating, p. 140 **Science Online,** p. 142 **Applying Science:** When did the Iceman die?, p. 142 **Lab:** Trace Fossils, p. 144 **Oops! Accidents in Science:** The World's Oldest Fish Story, p. 146

Glencoe Exclusive! TeacherWorks™ All-In-One Planner and Resource Center

Lab Materials	Reproducible Resources	Section Assessment	Technology
Launch Lab: 500-mL jar with lid, plaster of paris, water, several small shells, paper or plastic bowl	**Chapter *FAST FILE* Resources** Foldables Worksheet, p. 17 Directed Reading Overview, p. 19 Note-taking Worksheets, pp. 33–34	Glencoe's Assessment Advantage	TeacherWorks includes: • Interactive Teacher Edition • Lesson Planner with calendar • Access to all program blacklines • Correlations to standards • Web links
	Chapter *FAST FILE* Resources Transparency Activity, p. 44 MiniLAB, p. 3 Enrichment, p. 30 Reinforcement, p. 27 Directed Reading, p. 20 Transparency Activity, pp. 47–48 Lab Activities, pp. 9–12 **Science Inquiry Labs,** p. 30	**Portfolio** Integrate Social Studies, p. 127 **Performance** MiniLAB, p. 125 Applying Skills, p. 131 **Content** Section Review, p. 131	Section Focus Transparency Teaching Transparency Virtual Labs CD-ROM Guided Reading Audio Program Interactive Chalkboard CD-ROM
Lab: no materials needed *Need materials?* Contact Science Kit at 1-800-828-7777 or www.sciencekit.com on the Internet.	**Chapter *FAST FILE* Resources** Transparency Activity, p. 45 Enrichment, p. 31 Reinforcement, p. 28 Directed Reading, p. 21 Lab Worksheets, pp. 5–6 Lab Activities, pp. 13–15 **Reading and Writing Skill Activities,** p. 21	**Performance** Applying Skills, p. 137 **Content** Section Review, p. 137	Section Focus Transparency Virtual Labs CD-ROM Guided Reading Audio Program Interactive Chalkboard CD-ROM
MiniLAB: 80 red jelly beans, 80 green jelly beans **Lab:** construction paper, rigid plastic, plaster of paris, sturdy cardboard, pipe cleaners, wire, scissors, toothpicks, clay, glue	**Chapter *FAST FILE* Resources** Transparency Activity, p. 46 MiniLAB, p. 4 Enrichment, p. 32 Reinforcement, p. 29 Directed Reading, p. 21, 22 Lab Worksheets, pp. 7–8 **Lab Management and Safety,** p. 39	**Portfolio** Science Journal, p. 142 **Performance** MiniLAB, p. 140 Applying Science, p. 142 Applying Math, p. 143 **Content** Section Review, p. 143	Section Focus Transparency Virtual Labs CD-ROM Guided Reading Audio Program Interactive Chalkboard CD-ROM Video Lab

End of Chapter Assessment

Glencoe's Assessment Advantage

Blackline Masters	Technology	Professional Series
Chapter *FAST FILE* Resources Chapter Review, pp. 37–38 Chapter Tests, pp. 39–42 **Standardized Test Practice,** pp. 23–26	MindJogger Videoquiz Virtual Labs CD-ROM Exam*View*® Pro Testmaker TeacherWorks CD-ROM Interactive Chalkboard CD-ROM	**Performance Assessment in the Science Classroom (PASC)**

chapter 5 Clues to Earth's Past

Transparencies

Section Focus

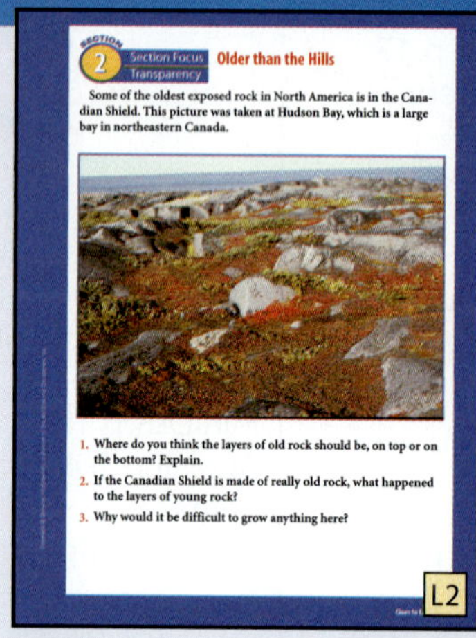

This is a representation of key blackline masters available in the Teacher Classroom Resources. See Resource Manager boxes within the chapter for additional information.

Assessment

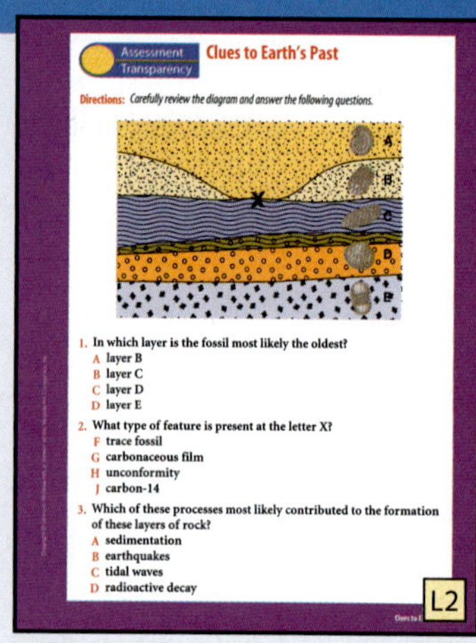

Teaching

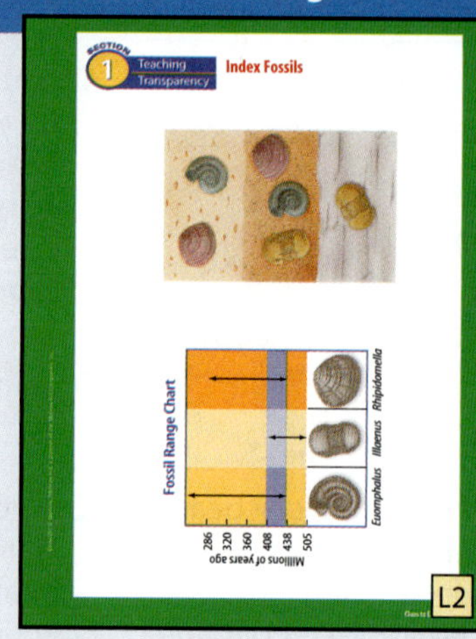

Key to Teaching Strategies

The following designations will help you decide which activities are appropriate for your students.

- **L1** Level 1 activities should be appropriate for students with learning difficulties.
- **L2** Level 2 activities should be within the ability range of all students.
- **L3** Level 3 activities are designed for above-average students.
- **ELL** ELL activities should be within the ability range of English Language Learners.
- **COOP LEARN** Cooperative Learning activities are designed for small group work.
- **LS** Multiple Learning Styles logos, as described on page 6T, are used throughout to indicate strategies that address different learning styles.
- **P** These strategies represent student products that can be placed into a best-work portfolio.
- **PBL** Problem-Based Learning activities apply real-world situations to learning.

Hands-on Activities

Student Text Lab Worksheet

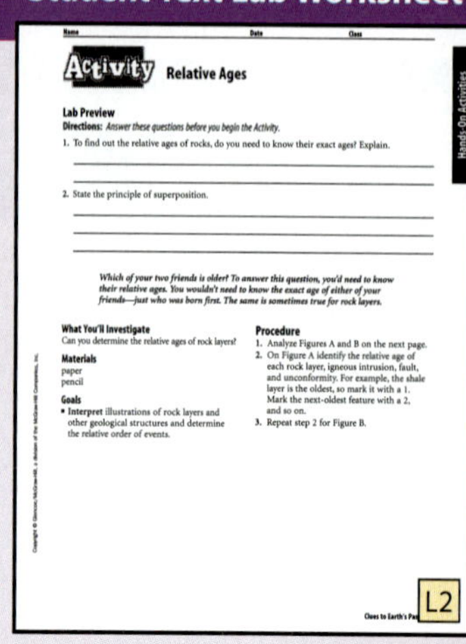

Laboratory Activities

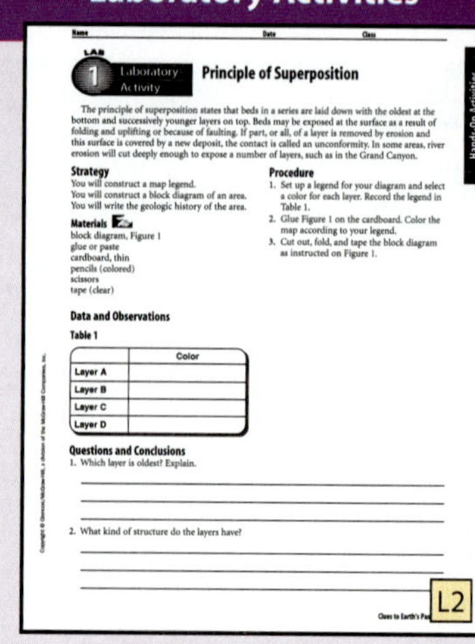

Meeting Different Ability Levels

Content Outline

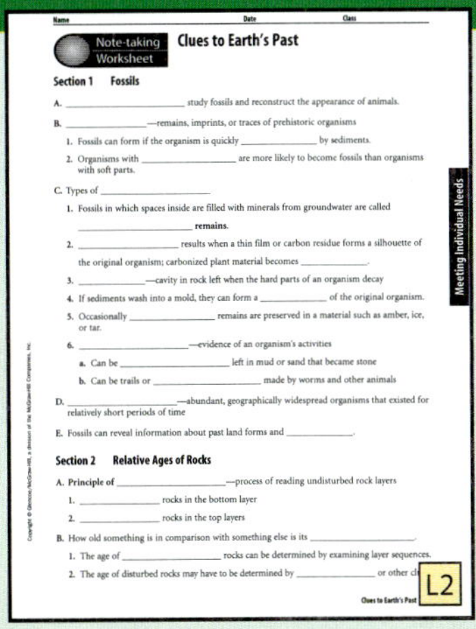

Reinforcement

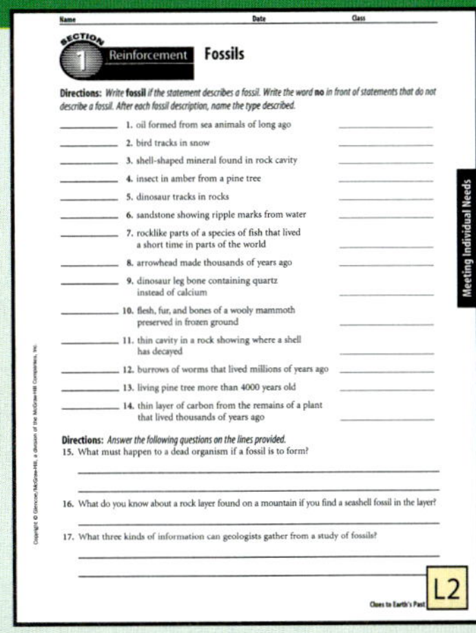

Enrichment

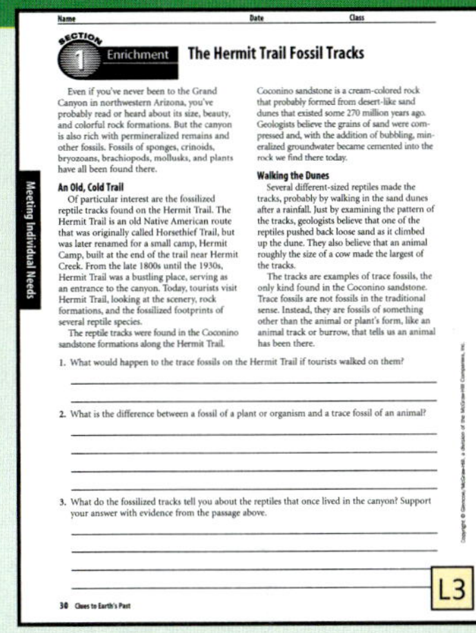

Directed Reading (English/Spanish)

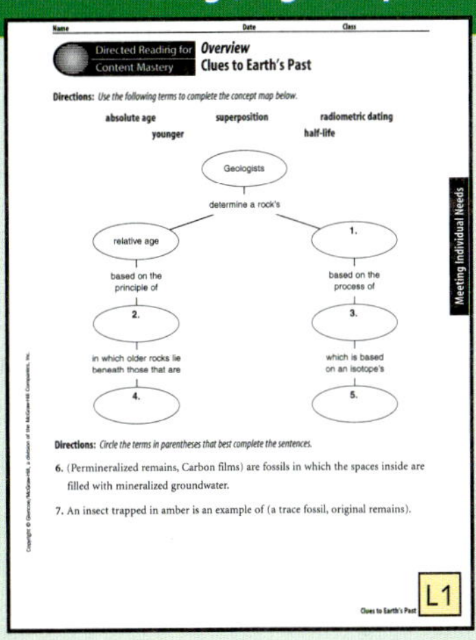

Study Guide

Reading Essentials

Assessment

Test Practice Workbook

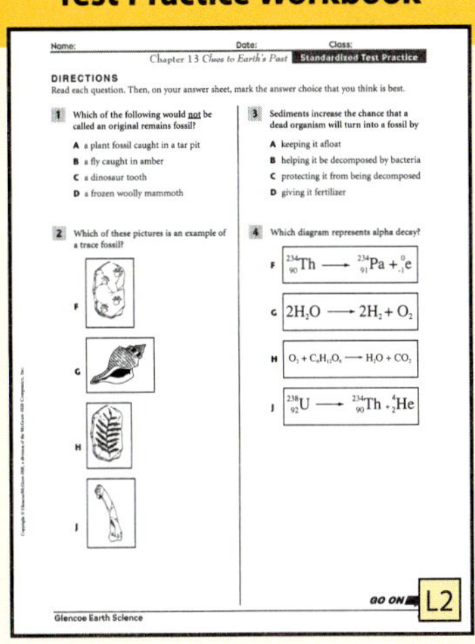

Chapter Review

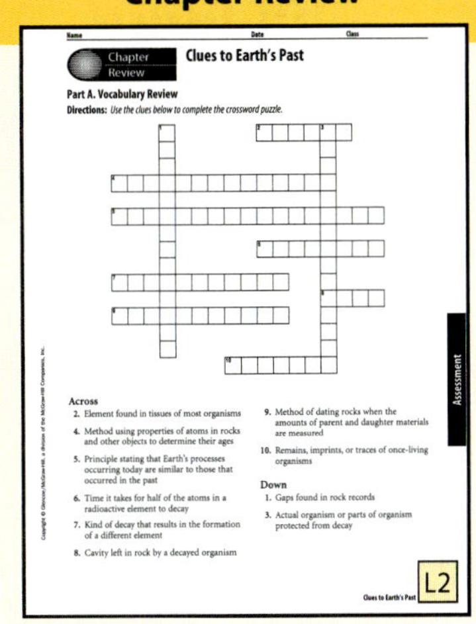

Chapter Tests

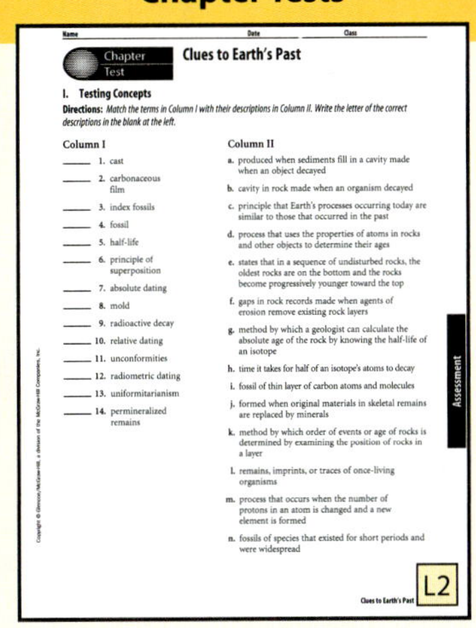

Science Content Background

section 1 Fossils

Trace Fossils

Scientists hypothesize that certain reptiles swallowed stones to aid in digestion. When the organism died and decayed, the polished stones, called gastroliths, were left behind. Because gastroliths are traces of the original organism and not the organism itself, they are called trace fossils. Another trace fossil is a coprolite, or fossilized fecal material. Jewelry has been made from cut and polished coprolites.

Fossil Formation

Some fossils consist of the original parts of the organism. An insect trapped in amber and the skin and hair of some woolly mammoths have been found intact because they were protected from bacteria which would cause decay. Most fossils are formed as dead organisms are buried by layers of sediment that harden to form sedimentary rocks.

section 2 Relative Ages of Rocks

Relative Dating

The principle of faunal succession states that fossil organisms occur in rocks in a definite and determinable order. Thus, rocks formed during a particular interval of geological time can be recognized by their fossils. A geological column is a composite diagram that shows the rocks of an area in the sequence in which they occur. Correlation of geological columns from many locations enables geologists to construct geologic maps. Geologic maps show the relative age, distribution, and structural features of surface rocks.

section 3 Absolute Ages of Rocks

Isotopes

When daughter isotopes are part of the original makeup of a rock, the calculated age of that rock can be wrong. However, for some isotopes, geologists are able to distinguish between original daughter isotopes and those that formed during radioactive decay. In order to cross-check their work in radiometric dating, geologists use several parent isotopes for the same age measurement. Because the different parent isotopes decay at different rates, using them for the same age measurement can ensure accuracy or expose errors.

chapter content resources

Internet Resources

For additional content background, visit bookg.msscience.com to:

- access your book online
- find references to related articles in popular science magazines
- access Web links with related content background
- access current events with science journal topics

Print Resources

Roadside Geology Series, by various authors, Mountain Press Publishing Company, various dates

Fossils, Frank H. Rhodes, Raymond Perlman, and Herbert Spencer Zim, St. Martin's Press, 2001

Janice VanCleave's Earth Science for Every Kid, Janice Pratt VanCleave, Barbara Clark, John Wiley & Sons, 1991

IDENTIFYING Misconceptions

Find Out What Students Think

Students may think that . . .

The present is not the key to the past.
Students are often confused about uniformitarianism, which is one of the fundamental principles of geology. The principle of uniformitarianism states that Earth processes occurring today are similar to those that occurred in the past. Thus, our understanding of past Earth processes is based on those that are presently taking place.

Discussion

Have students imagine that they are geologists. During a field study, they observe that deposits of sediment in a mudflat contain cracks and ripple marks. Back in the lab, you observe several sedimentary rocks that also contain these features. Have students discuss what they can infer from these observations, and how observations such as these help scientists explain past events. Students should infer that the same processes that cause the ripple marks to form today also caused them to form in the past.

Promote Understanding

Activity

- Organize the class into small groups. Have groups examine samples of a layered sedimentary rock, such as shale. Ask each student to record his or her observations of the shale and then share these observations with their group members.
- Provide each group with a jar of water into which you have put sediment containing fine particles of clay. After students make sure that the lids are tightly sealed, have them gently shake the jar and observe the contents as they settle. Have students compare the rock sample and the layered sediments and determine how many of their original observations "fit" the sediments in the jar. Students should recognize that many of their observations are the same for both samples.
- Have each student group present their list of common characteristics to illustrate the principle of uniformitarianism.

Assess

After completing the chapter, see *Identifying Misconceptions* in the Study Guide at the end of the chapter.

chapter 5

Chapter Vocabulary

fossil, p. 125
permineralized remains, p. 126
carbon film, p. 126
mold, p. 127
cast, p. 127
index fossil, p. 129
principle of superposition, p. 132
relative age, p. 133
unconformity, p. 134
absolute age, p. 139
radioactive decay, p. 139
half-life, p. 140
radiometric dating, p. 141
uniformitarianism, p. 143

Science Journal Answers will vary, but should make sense for the local environment. If near water, include mollusks. Insects in amber also are likely. Point out that animal bones are unlikely to survive for fossilization. Leaf impressions or pollen grains may be possible.

INTERACTIVE CHALKBOARD with Image Bank

PowerPoint® Presentations

This CD-ROM is an editable Microsoft® PowerPoint® presentation that includes:

- a pre-made presentation for every chapter
- interactive graphics
- animations
- audio clips
- image bank
- all new section and chapter questions
- Standardized Test Practice
- transparencies
- pre-lab questions for all labs
- Foldables directions
- links to bookg.msscience.com

chapter 5

Clues to Earth's Past

chapter preview

sections

1. Fossils
2. Relative Ages of Rocks
   Lab *Relative Ages*
3. Absolute Ages of Rocks
   Lab *Trace Fossils*

Virtual Lab *How can fossil and rock data determine when an organism lived?*

Reading the Past

The pages of Earth's history, much like the pages of human history, can be read if you look in the right place. Unlike the pages of a book, the pages of Earth's past are written in stone. In this chapter you will learn how to read the pages of Earth's history to understand what the planet was like in the distant past.

Science Journal List three fossils that you would expect to find a million years from now in the place you live today.

122 ◆ G

Theme Connection

Stability and Change The evolution of organisms throughout geologic time as shown by the fossil record is a major focus of the chapter. The changes in organisms on Earth can be used to obtain relative ages of rocks and to date geologic events.

About the Photo

Grand Canyon Aside from being a famous tourist attraction, the Grand Canyon is studied extensively by geologists. Since it cuts deeply into Earth's crust, it exposes rocks that are millions of years old. This allows scientists access to rocks that formed in Earth's distant past.

Start-Up Activities

Launch Lab

Clues to Life's Past

Fossil formation begins when dead plants or animals are buried in sediment. In time, if conditions are right, the sediment hardens into sedimentary rock. Parts of the organism are preserved along with the impressions of parts that don't survive. Any evidence of once-living things contained in the rock record is a fossil.

1. Fill a small jar (about 500 mL) one-third full of plaster of paris. Add water until the jar is half full.
2. Drop in a few small shells.
3. Cover the jar and shake it to model a swift, muddy stream.
4. Now model the stream flowing into a lake by uncovering the jar and pouring the contents into a paper or plastic bowl. Let the mixture sit for an hour.
5. Crack open the hardened plaster to locate the model fossils.
6. **Think Critically** Remove the shells from the plaster and study the impressions they made. In your Science Journal, list what the impressions would tell you if found in a rock.

FOLDABLES™ Study Organizer

Age of Rocks Make the following Foldable to help you understand how scientists determine the age of a rock.

STEP 1 **Fold** a sheet of paper in half lengthwise.

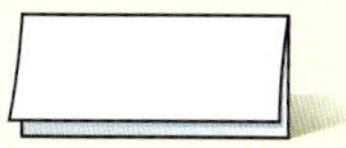

STEP 2 **Fold** paper down 2.5 cm from the top. (Hint: From the tip of your index finger to your middle knuckle is about 2.5 cm.)

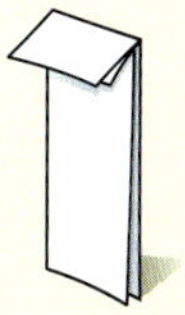

STEP 3 **Open and draw** lines along the 2.5-cm fold. **Label** as shown.

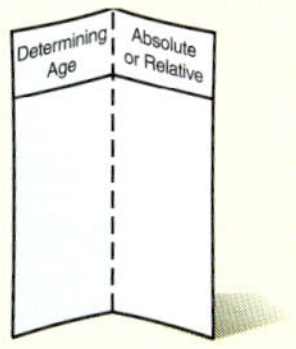

Summarize in a Table As you read the chapter, in the left column, list four different ways in which one could determine the age of a rock. In the right column, note whether each method gives an absolute or a relative age.

Science Online — Preview this chapter's content and activities at bookg.msscience.com

Launch Lab

Purpose Use this Launch Lab to introduce students to fossil formation. L2 LS **Kinesthetic**

Preparation Put dry plaster of paris into jars before class starts.

Materials 500-mL jar with lid, plaster of paris, water, several small shells, paper or plastic bowl

Teaching Strategy Have students wear gloves when working with the plaster of paris. You may wish to do this activity as a demonstration.

Think Critically

Answers will vary but might include the size of the organism, the size of the population, or how deep impressions might reveal different information than surface ones.

Assessment

Process Ask your students to infer what other types of fossils the remains of organisms could form. organisms trapped in amber, organisms trapped in tar beds, frozen animals such as wooly mammoths, petrified wood, molds, and casts Use **Performance Assessment in the Science Classroom**, p. 89

FOLDABLES™ Study Organizer — Dinah Zike Study Fold

Student preparation materials for this Foldable are available in the Chapter *Fast File* Resources.

section 1

Fossils

1 Motivate

Bellringer

Section Focus Transparencies also are available on the Interactive Chalkboard CD-ROM. L2 ELL

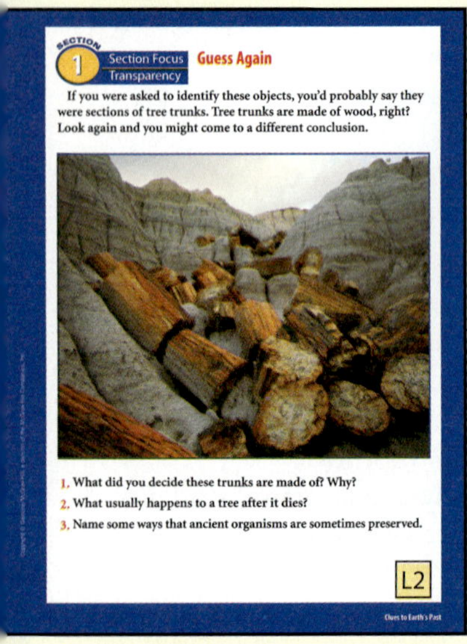
Section 1 Section Focus Transparency **Guess Again**

If you were asked to identify these objects, you'd probably say they were sections of tree trunks. Tree trunks are made of wood, right? Look again and you might come to a different conclusion.

1. What did you decide these trunks are made of? Why?
2. What usually happens to a tree after it dies?
3. Name some ways that ancient organisms are sometimes preserved.

L2

Tie to Prior Knowledge

Fossilization Ask students to list things that can become fossils. In this section, they will learn that, under certain conditions, the remains of almost any organism or evidence of its activities can become a fossil.

as you read

What You'll Learn

- **List** the conditions necessary for fossils to form.
- **Describe** several processes of fossil formation.
- **Explain** how fossil correlation is used to determine rock ages.
- **Determine** how fossils can be used to explain changes in Earth's surface, life forms, and environments.

Why It's Important

Fossils help scientists find oil and other sources of energy necessary for society.

Review Vocabulary

paleontologist: a scientist who studies fossils

New Vocabulary

- fossil
- permineralized remains
- carbon film
- mold
- cast
- index fossil

Traces of the Distant Past

A giant crocodile lurks in the shallow water of a river. A herd of *Triceratops* emerges from the edge of the forest and cautiously moves toward the river. The dinosaurs are thirsty, but danger waits for them in the water. A large bull *Triceratops* moves into the river. The others follow.

Does this scene sound familiar to you? It's likely that you've read about dinosaurs and other past inhabitants of Earth. But how do you know that they really existed or what they were like? What evidence do humans have of past life on Earth? The answer is fossils. Paleontologists, scientists who study fossils, can learn about extinct animals from their fossil remains, as shown in **Figure 1.**

Figure 1 Scientists can learn how dinosaurs looked and moved using fossil remains. A skeleton can then be reassembled and displayed in a museum.

Section 1 Resource Manager

Chapter *Fast File* Resources

- Transparency Activity, pp. 44, 47–48
- Directed Reading for Content Mastery, pp. 19, 20
- Note-taking Worksheets, pp. 33–35
- MiniLAB, p. 3
- Enrichment, p. 30
- Lab Activities, pp. 9–12
- Reinforcement, p. 27

Earth Science Critical Thinking/Problem Solving, p. 12

Science Inquiry Labs, p. 30

Formation of Fossils

Figure 2 These fossil shark teeth are hard parts. Soft parts of animals do not become fossilized as easily.

Fossils are the remains, imprints, or traces of prehistoric organisms. Fossils have helped scientists determine approximately when life first appeared, when plants and animals first lived on land, and when organisms became extinct. Fossils are evidence of not only when and where organisms once lived, but also how they lived.

For the most part, the remains of dead plants and animals disappear quickly. Scavengers eat and scatter the remains of dead organisms. Fungi and bacteria invade, causing the remains to rot and disappear. If you've ever left a banana on the counter too long, you've seen this process begin. In time, compounds within the banana cause it to break down chemically and soften. Microorganisms, such as bacteria, cause it to decay. What keeps some plants and animals from disappearing before they become fossils? Which organisms are more likely to become fossils?

Conditions Needed for Fossil Formation Whether or not a dead organism becomes a fossil depends upon how well it is protected from scavengers and agents of physical destruction, such as waves and currents. One way a dead organism can be protected is for sediment to bury the body quickly. If a fish dies and sinks to the bottom of a lake, sediment carried into the lake by a stream can cover the fish rapidly. As a result, no waves or scavengers can get to it and tear it apart. The body parts then might be fossilized and included in a sedimentary rock like shale. However, quick burial alone isn't always enough to make a fossil.

Organisms have a better chance of becoming fossils if they have hard parts such as bones, shells, or teeth. One reason is that scavengers are less likely to eat these hard parts. Hard parts also decay more slowly than soft parts do. Most fossils are the hard parts of organisms, such as the fossil teeth in **Figure 2.**

Types of Preservation

Perhaps you've seen skeletal remains of *Tyrannosaurus rex* towering above you in a museum. You also have some idea of what this dinosaur looked like because you've seen illustrations. Artists who draw *Tyrannosaurus rex* and other dinosaurs base their illustrations on fossil bones. What preserves fossil bones?

Mini LAB

Predicting Fossil Preservation

Procedure

1. Take a brief walk outside and observe your neighborhood.
2. Look around and notice what kinds of plants and animals live nearby.

Analysis

1. Predict what remains from your time might be preserved far into the future.
2. Explain what conditions would need to exist for these remains to be fossilized.

Try at Home

2 Teach

Mini LAB

Purpose Students consider what conditions are necessary for living things to become fossils. L2

LS **Logical-Mathematical**

Teaching Strategy Introduce this activity by asking students how we know about conditions that existed long ago on Earth's surface. Scientists study evidence of past life preserved in Earth's crust.

Analysis

1. The objects named must be the remains of living things or evidence of their activities, such as footprints or burrows. Possible answers: leaves, sticks, bones, shells
2. To best be preserved, the remains should have hard parts and be buried quickly.

Assessment

Oral Have students explain why shark teeth are often found as fossils but impressions of jellyfish are rarely found. Shark teeth are hard, and thus likely to be preserved. Jellyfish have no hard parts and thus are unlikely to be preserved. Use **Performance Assessment in the Science Classroom,** p. 89.

Try at Home

Differentiated Instruction

Challenge Have students investigate the process of fossilization. Scientists often study the decay of today's organisms to understand which animals from the past could become fossils and to help interpret the marks seen on those fossils. Have the students prepare a poster that discusses how current organisms can provide information on ancient fossils. L3

Caption Answer

Figure 3 During the permineralization process, the pore spaces inside the parts of a dead organism are replaced by minerals carried in solution. This preserves the original shape of the bone.

IDENTIFYING Misconceptions

Today's Fossils Students might think that fossils are not forming today. Ask them what happens to organisms that die today and settle to the bottom of a lake or ocean. Explain that sediments bury these remains quickly. Once buried, they begin the process of fossilization.

Teacher FYI

Handling Fossils Some fossils are fragile. In order for them to be moved from the rock outcrop in which they were found, they must be strengthened and protected. The fossils are first soaked in glue. Next, strips of wet newspaper are molded onto the fossils. Finally, a plaster mixture is used to coat burlap strips, which are applied to the fossils, producing a kind of cast that protects and supports the fossils. The fossils are now ready to be moved.

Figure 3 Opal and various minerals have replaced original materials and filled the hollow spaces in this permineralized dinosaur bone. **Explain** *why this fossil retained the shape of the original bone.*

Mineral Replacement Most hard parts of organisms such as bones, teeth, and shells have tiny spaces within them. In life, these spaces can be filled with cells, blood vessels, nerves, or air. When the organism dies and the soft materials inside the hard parts decay, the tiny spaces become empty. If the hard part is buried, groundwater can seep in and deposit minerals in the spaces. **Permineralized remains** are fossils in which the spaces inside are filled with minerals from groundwater. In permineralized remains, some original material from the fossil organism's body might be preserved—encased within the minerals from groundwater. It is from these original materials that DNA, the chemical that contains an organism's genetic code, can sometimes be recovered.

Sometimes minerals replace the hard parts of fossil organisms. For example, a solution of water and dissolved silica (the compound SiO_2) might flow into and through the shell of a dead organism. If the water dissolves the shell and leaves silica in its place, the original shell is replaced.

Often people learn about past forms of life from bones, wood, and other remains that became permineralized or replaced with minerals from groundwater, as shown in **Figure 3,** but many other types of fossils can be found.

Figure 4 Graptolites lived hundreds of millions of years ago and drifted on currents in the oceans. These organisms often are preserved as carbon films.

Carbon Films The tissues of organisms are made of compounds that contain carbon. Sometimes fossils contain only carbon. Fossils usually form when sediments bury a dead organism. As sediment piles up, the organism's remains are subjected to pressure and heat. These conditions force gases and liquids from the body. A thin film of carbon residue is left, forming a silhouette of the original organism called a **carbon film. Figure 4** shows the carbonized remains of graptolites, which were small marine animals. Graptolites have been found in rocks as old as 500 million years.

126 ◆ G CHAPTER 5 Clues to Earth's Past

Active Reading

Jigsaw Place students in five cooperative groups. Each group should be responsible for studying and summarizing one of the five types of fossil formation discussed in this section. Once finished, each group is to direct a discussion on the type of fossil it was responsible for. Instruct groups to include examples of their types of fossil formation. permineralized remains—dinosaur bone; carbon film—leaf impression; molds and casts—filled-in cavity; original remains—insect in amber; trace fossil—worm burrow or coprolite L2 COOP LEARN IN **Interpersonal**

Coal In swampy regions, large volumes of plant matter accumulate. Over millions of years, these deposits become completely carbonized, forming coal. Coal is an important fuel source, but since the structure of the original plant is usually lost, it cannot reveal as much about the past as other kinds of fossils.

Reading Check *In what sort of environment does coal form?*

Molds and Casts In nature, impressions form when seashells or other hard parts of organisms fall into a soft sediment such as mud. The object and sediment are then buried by more sediment. Compaction, together with cementation, which is the deposition of minerals from water into the pore spaces between sediment particles, turns the sediment into rock. Other open pores in the rock then let water and air reach the shell or hard part. The hard part might decay or dissolve, leaving behind a cavity in the rock called a **mold.** Later, mineral-rich water or other sediment might enter the cavity, form new rock, and produce a copy or **cast** of the original object, as shown in **Figure 5.**

INTEGRATE Social Studies

Coal Mining Many of the first coal mines in the United States were located in eastern states like Pennsylvania and West Virginia. In your Science Journal, discuss how the environments of the past relate to people's lives today.

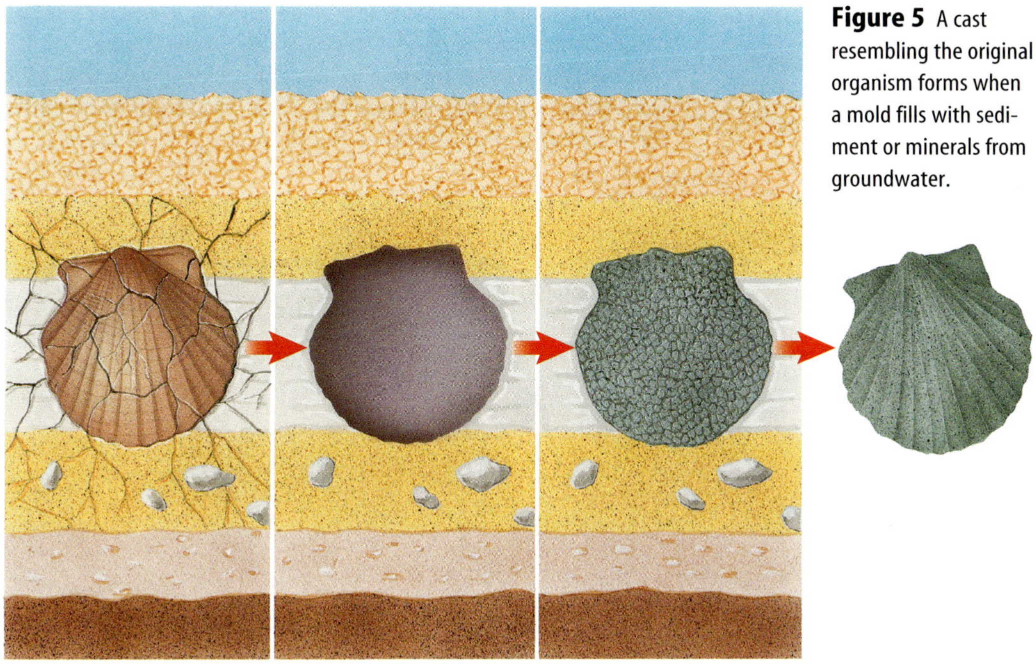

Figure 5 A cast resembling the original organism forms when a mold fills with sediment or minerals from groundwater.

The fossil begins to dissolve as water moves through spaces in the rock layers.

The fossil has been dissolved away. The harder rock once surrounding it forms a mold.

Sediment washes into the mold and is deposited, or mineral crystals form.

A cast results.

Reading Check

Answer Coal forms in swampy regions.

INTEGRATE Social Studies

Coal Mining Answers will vary; accept anything reasonable

Career Have students investigate how geologists find coal deposits. They also should learn what training is required. Have students prepare a recruitment poster to attract students to the field. P

Visual Learning

Figure 5 After students have studied the figure, provide them with a mold and cast of the same organism. Which of your fossils is a mold and which is a cast? The mold is the hollow or depression in the original rock, and the cast will be the bump that fits into the hollow.

Use an Analogy

Molds in Nature Have students discuss an analogy between molds used in making ceramic statues and those formed in nature. Use modeling clay and a small statue to demonstrate why the internal structure of an organism is not preserved when casts are formed. L2

Fun Fact

Scientists have extracted DNA from frozen mammoths and compared it with modern elephants to determine their relationships.

Curriculum Connection

Geography Have students research and find on a map where replacement fossils have been found in the United States. As part of the task, have students explain what the name "Petrified Forest" refers to. This name refers to the Petrified Forest National Monument in Arizona, where there is wood in which much of the original matter has been replaced by minerals. L2 LS **Visual-Spatial**

Quick Demo

Making an Impression

Materials modeling clay, shell, leaf, tooth

Estimated Time five minutes

Procedure Prepare clay as a flat piece the size of a piece of notebook paper and about an inch thick. Press the leaf and other objects into the clay. Show students that these impressions can become fossils if quickly covered and the clay hardens into rock.

Use Science Words

Word Origin Trace fossils are not the original organism; they are just a trace of its activities. Ask students to find words that also use *trace* and explain their meanings. Possible answers: tracing—a graphic record; tracer—person who looks for clues of missing people; tracery—lacy, decorative openwork in a window

Discussion

Fossil Information What information can a frozen mammoth provide that other fossils in this chapter cannot? Possible answers: frozen specimens often have preserved skin, hair, internal organs, and DNA

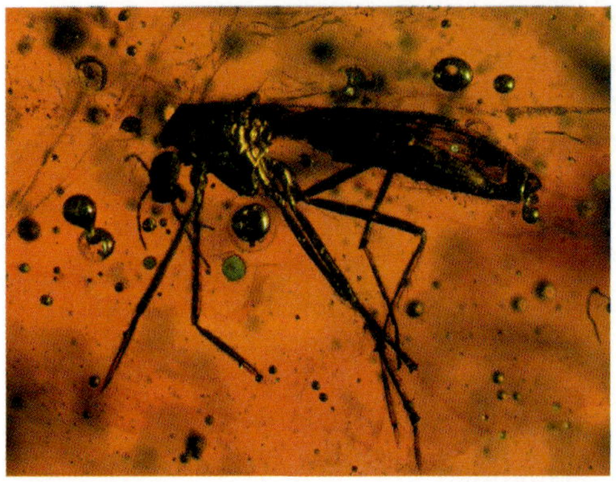

Figure 6 The original soft parts of this mosquito have been preserved in amber for millions of years.

Original Remains Sometimes conditions allow original soft parts of organisms to be preserved for thousands or millions of years. For example, insects can be trapped in amber, a hardened form of sticky tree resin. The amber surrounds and protects the original material of the insect's exoskeleton from destruction, as shown in **Figure 6.** Some organisms, such as the mammoth, have been found preserved in frozen ground in Siberia. Original remains also have been found in natural tar deposits, such as the La Brea tar pits in California.

Trace Fossils Do you have a handprint in plaster that you made when you were in kindergarten? If so, it's a record that tells something about you. From it, others can guess your size and maybe your weight at that age. Animals walking on Earth long ago left similar tracks, such as those in **Figure 7.** Trace fossils are fossilized tracks and other evidence of the activity of organisms. In some cases, tracks can tell you more about how an organism lived than any other type of fossil. For example, from a set of tracks at Davenport Ranch, Texas, you might be able to learn something about the social life of sauropods, which were large, plant-eating dinosaurs. The largest tracks of the herd are on the outer edges and the smallest are on the inside. These tracks led some scientists to hypothesize that adult sauropods surrounded their young as they traveled—perhaps to protect them from predators. A nearby set of tracks might mean that another type of dinosaur, an allosaur, was stalking the herd.

Figure 7 Tracks made in soft mud, and now preserved in solid rock, can provide information about animal size, speed, and behavior.

The dinosaur track below is from the Glen Rose Formation in north-central Texas.

The tracks to the right are located on a Navajo reservation in Arizona.

Cultural Diversity

The Copper Man of the Alps On September 19, 1991, a German couple hiking in the Alps came across a 5,000-year-old traveler, naturally mummified in a small depression covered by glacial ice. The Iceman is now known to be one of the oldest and best-preserved mummified humans. Plant remains on his clothing suggest that he came from lowland South Tyrol. He wore finely stitched skin clothing, a woven grass cape, and boots. His deerskin quiver held two arrows tipped with flint and 12 unfinished arrows. He carried two antibiotic fungi, a backpack, a half-finished longbow, high-quality flint, and a flint dagger. One of his most intriguing possessions was an axe, originally thought to be bronze but now known to be copper. It is flanged in a style not found anywhere nearby until nearly 800 years later.

128 ◆ G CHAPTER 5 Clues to Earth's Past

Trails and Burrows Other trace fossils include trails and burrows made by worms and other animals. These, too, tell something about how these animals lived. For example, by examining fossil burrows you can sometimes tell how firm the sediment the animals lived in was. As you can see, fossils can tell a great deal about the organisms that have inhabited Earth.

Reading Check *How are trace fossils different from fossils that are the remains of an organism's body?*

Index Fossils

One thing you can learn by studying fossils is that species of organisms have changed over time. Some species of organisms inhabited Earth for long periods of time without changing. Other species changed a lot in comparatively short amounts of time. It is these organisms that scientists use as index fossils.

Index fossils are the remains of species that existed on Earth for relatively short periods of time, were abundant, and were widespread geographically. Because the organisms that became index fossils lived only during specific intervals of geologic time, geologists can estimate the ages of rock layers based on the particular index fossils they contain. However, not all rocks contain index fossils. Another way to approximate the age of a rock layer is to compare the spans of time, or ranges, over which more than one fossil appears. The estimated age is the time interval where fossil ranges overlap, as shown in **Figure 8.**

Figure 8 The fossils in a sequence of sedimentary rock can be used to estimate the ages of each layer. The chart shows when each organism inhabited Earth.
Explain *why it is possible to say that the middle layer of rock was deposited between 440 million and 410 million years ago.*

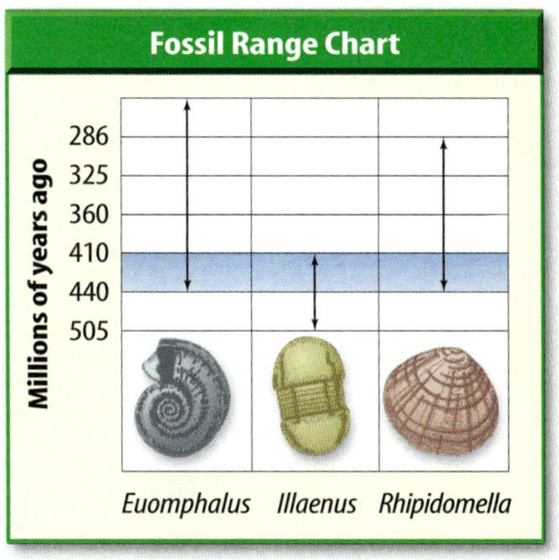

Reading Check

Answer Trace fossils are fossilized tracks and other evidence of an organism's activity. They are not parts of the original organism.

Make a Model

Fossil Models Encourage pairs of students to produce a model of a fossil that depicts one of the five methods of formation. Display student models and actual samples in the classroom. L2 ELL COOP LEARN LS **Kinesthetic**

Caption Answer

Figure 8 The three fossils have ranges that overlap between 440 and 410 million years. The age of the layer cannot be dated more accurately than that.

Activity

Index Fossils Have students prepare a list of index fossils for North America, including the periods in which they are found. Give them maps of various geologic formations in the United States. Have students show the class which index fossils might be found in their formations.

Science Journal

Traces of the Past Have students research and write a summary of what can be learned about extinct animals from studying gastroliths and coprolites. Possible answer: Gastroliths are stones swallowed to aid in digestion; coprolites are petrified feces. Both might provide clues about an organism's diet. L2 LS **Linguistic**

SECTION 1 Fossils G ◆ 129

INTEGRATE Life Science

Ancient Ecology Possible answers: Study of fossils might indicate predator-prey relationships, migratory patterns, and climatic changes. Accept reasonable answers.

Research Have students investigate the Burgess Shale, dinosaur footprints, or dinosaur nests. Let them put together a presentation that discusses the ecological insights available from these discoveries. If possible, have them connect these past organisms to current animals with similar behaviors.

Text Question Answer

Because different organisms live at different depths, fossils can be general indicators of water depth in a region.

Visual Learning

Figure 9 How does the position of North America explain the presence of fern fossils in shale beds of the Pennsylvanian Period? North America was closer to the equator during the Pennsylvanian Period. Much of the region was covered by swamps, in which ferns thrived. As ferns died, they fell into the swamp and were covered by sediments.

INTEGRATE Life Science

Ancient Ecology
Ecology is the study of how organisms interact with each other and with their environment. Some paleontologists study the ecology of ancient organisms. Discuss the kinds of information you could use to determine how ancient organisms interacted with their environment.

Fossils and Ancient Environments

Scientists can use fossils to determine what the environment of an area was like long ago. Using fossils, you might be able to find out whether an area was land or whether it was covered by an ocean at a particular time. If the region was covered by ocean, it might even be possible to learn the depth of the water. What clues about the depth of water do you think fossils could provide?

Fossils also are used to determine the past climate of a region. For example, rocks in parts of the eastern United States contain fossils of tropical plants. The environment of this part of the United States today isn't tropical. However, because of the fossils, scientists know that it was tropical when these plants were living. **Figure 9** shows that North America was located near the equator when these fossils formed.

Figure 9 The equator passed through North America 310 million years ago. At this time, warm, shallow seas and coal swamps covered much of the continent, and ferns like the *Neuropteris,* below, were common.

Differentiated Instruction

Learning Disabled To help students better understand how fossils can indicate ancient environments, show them samples of rocks that contain fossils of organisms that once lived in a shallow sea. Ask students to explain the environment that existed when the organisms were alive.
L1 LS **Visual-Spatial**

Shallow Seas How would you explain the presence of fossilized crinoids—animals that lived in shallow seas—in rocks found in what is today a desert? **Figure 10** shows a fossil crinoid and a living crinoid. When the fossil crinoids were alive, a shallow sea covered much of western and central North America. The crinoid hard parts were included in rocks that formed from the sediments at the bottom of this sea. Fossils provide information about past life on Earth and also about the history of the rock layers that contain them. Fossils can provide information about the ages of rocks and the climate and type of environment that existed when the rocks formed.

Figure 10 The crinoid on the left lived in warm, shallow seas that once covered part of North America. Crinoids like the one on the right typically live in warm, shallow waters in the Pacific Ocean.

section 1 review

Summary

Formation of Fossils

- Fossils are the remains, imprints, or traces of past organisms.
- Fossilization is most likely if the organism had hard parts and was buried quickly.

Fossil Preservation

- Permineralized remains have open spaces filled with minerals from groundwater.
- Thin carbon films remain in the shapes of dead organisms.
- Hard parts dissolve to leave molds.
- Trace fossils are evidence of past activity.

Index Fossils

- Index fossils are from species that were abundant briefly, but over wide areas.
- Scientists can estimate the ages of rocks containing index fossils.

Fossils and Ancient Environments

- Fossils tell us about the environment in which the organisms lived.

Self Check

1. **Describe** the typical conditions necessary for fossil formation.
2. **Explain** how a fossil mold is different from a fossil cast.
3. **Discuss** how the characteristics of an index fossil are useful to geologists.
4. **Describe** how carbon films form.
5. **Think Critically** What can you say about the ages of two widely separated layers of rock that contain the same type of fossil?

Applying Skills

6. **Communicate** what you learn about fossils. Visit a museum that has fossils on display. Make an illustration of each fossil in your Science Journal. Write a brief description, noting key facts about each fossil and how each fossil might have formed.
7. **Compare and contrast** original remains with other kinds of fossils. What kinds of information would only be available from original remains? Are there any limitations to the use of original remains?

Science Online bookg.msscience.com/self_check_quiz

3 Assess

DAILY INTERVENTION

Check for Understanding

Visual-Spatial Give groups of students stacks of magazines in chronological order. Insert an undated item in the middle of the stack. Have each group determine the relative age of the item based on the magazine dates.

Reteach

Fossils Have students form groups of four. Have them press shells or bones into mixtures of sand and water, soft clay and water, and gravel and water. The recorder should report which sediment allowed the best imprint to be formed. soft clay and water L2 COOP LEARN

LS **Kinesthetic and Interpersonal**

Assessment

Performance Have students devise concept maps summarizing the major concepts of this section. Have pairs exchange maps and make suggestions as to how to improve the maps. Use **Performance Assessment in the Science Classroom,** p. 161.

section 1 review

1. hard parts that are quickly buried
2. A buried object decays, leaving a cavity, or mold. Later, sediments fill the cavity, harden, and produce a cast.
3. They lived for a relatively short period of time, were abundant, and were geographically widespread. Geologists can define the ages of rock layers based on the particular index fossils they contain.
4. Pressure and heat force out gases and liquids; only a film of carbon residue is left.
5. Both rock layers formed when the fossil-producing organism lived; they may be similar in age if the organism existed for a short time.
6. Encourage students to set up their observations as a field notebook.
7. Original remains include structures like hair and organs that are usually absent in fossils. Original remains only exist in a few places, like glaciers or tar pits.

1 Motivate

INTERACTIVE CHALKBOARD PowerPoint® Presentations

Bellringer

Section Focus Transparencies also are available on the Interactive Chalkboard CD-ROM. L2 ELL

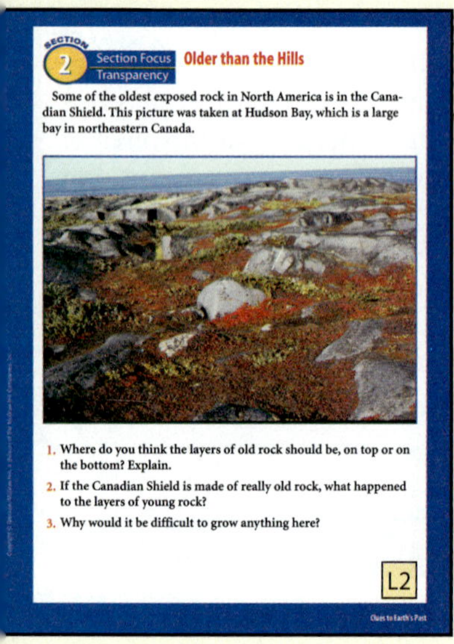
Section 2 Section Focus Transparency **Older than the Hills**

Some of the oldest exposed rock in North America is in the Canadian Shield. This picture was taken at Hudson Bay, which is a large bay in northeastern Canada.

1. Where do you think the layers of old rock should be, on top or on the bottom? Explain.
2. If the Canadian Shield is made of really old rock, what happened to the layers of young rock?
3. Why would it be difficult to grow anything here?

L2

Clues to Earth's Past

Tie to Prior Knowledge

Daily Superposition Students may be surprised to find that they observe superposition every day. Discuss how they know which layers of bricks were laid first as they observe a wall, or where they would look for last week's paper in a stack of newspapers set aside to be recycled. Explain that scientists use these same principles when they determine the age of rock layers.

Text Question Answer

Sediments accumulate to form rocks; the oldest layer is on the bottom.

section 2

Relative Ages of Rocks

as you read

What You'll Learn

- **Describe** methods used to assign relative ages to rock layers.
- **Interpret** gaps in the rock record.
- **Give** an example of how rock layers can be correlated with other rock layers.

Why It's Important

Being able to determine the age of rock layers is important in trying to understand a history of Earth.

Review Vocabulary
sedimentary rock: rock formed when sediments are cemented and compacted or when minerals are precipitated from solution

New Vocabulary
- principle of superposition
- relative age
- unconformity

Superposition

Imagine that you are walking to your favorite store and you happen to notice an interesting car go by. You're not sure what kind it is, but you remember that you read an article about it. You decide to look it up. At home you have a stack of magazines from the past year, as seen in **Figure 11.**

You know that the article you're thinking of came out in the January edition, so it must be near the bottom of the pile. As you dig downward, you find magazines from March, then February. January must be next. How did you know that the January issue of the magazine would be on the bottom? To find the older edition under newer ones, you applied the principle of superposition.

Oldest Rocks on the Bottom According to the **principle of superposition,** in undisturbed layers of rock, the oldest rocks are on the bottom and the rocks become progressively younger toward the top. Why is this the case?

Figure 11 The pile of magazines illustrates the principle of superposition. According to this principle, the oldest rock layer (or magazine) is on the bottom.

Section 2 Resource Manager

Chapter *FAST FILE* Resources
- Transparency Activity, p. 45
- Directed Reading for Content Mastery, p. 21
- Lab Activities, pp. 13–15
- Lab Worksheets, pp. 5–6

Reading and Writing Skill Activities, p. 21
Earth Science Critical Thinking/Problem Solving, pp. 11 and 14
Cultural Diversity, p. 13

Rock Layers Sediment accumulates in horizontal beds, forming layers of sedimentary rock. The first layer to form is on the bottom. The next layer forms on top of the previous one. Because of this, the oldest rocks are at the bottom. However, forces generated by mountain formation sometimes can turn layers over. When layers have been turned upside down, it's necessary to use other clues in the rock layers to determine their original positions and relative ages.

Relative Ages

Now you want to look for another magazine. You're not sure how old it is, but you know it arrived after the January issue. You can find it in the stack by using the principle of relative age.

The **relative age** of something is its age in comparison to the ages of other things. Geologists determine the relative ages of rocks and other structures by examining their places in a sequence. For example, if layers of sedimentary rock are offset by a fault, which is a break in Earth's surface, you know that the layers had to be there before a fault could cut through them. The relative age of the rocks is older than the relative age of the fault. Relative age determination doesn't tell you anything about the age of rock layers in actual years. You don't know if a layer is 100 million or 10,000 years old. You only know that it's younger than the layers below it and older than the fault cutting through it.

Other Clues Help Determination of relative age is easy if the rocks haven't been faulted or turned upside down. For example, look at **Figure 12.** Which layer is the oldest? In cases where rock layers have been disturbed you might have to look for fossils and other clues to date the rocks. If you find a fossil in the top layer that's older than a fossil in a lower layer, you can hypothesize that layers have been turned upside down by folding during mountain building.

Science Online

Topic: Relative Dating

Visit bookg.msscience.com for Web links to information about relative dating of rocks and other materials.

Activity Imagine yourself at an archaeological dig. You have found a rare artifact and want to know its age. Make a list of clues you might look for to provide a relative date and explain how each would allow you to approximate the artifact's age.

Figure 12 In a stack of undisturbed sedimentary rocks, the oldest rocks are at the bottom. This stack of rocks can be folded by forces within Earth.
Explain *how you can tell if an older rock is above a younger one.*

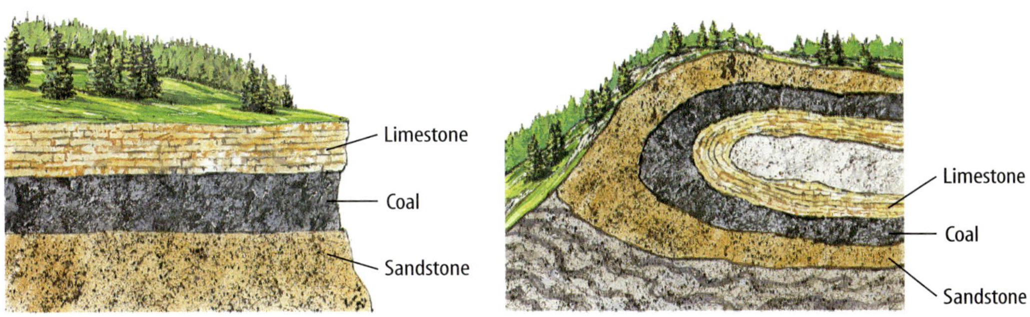

2 Teach

Quick Demo

Relative Ages

Materials 6 different colors of clay, knife

Estimated Time 10 minutes

Procedure Place layers of five different colored clays on top of a desk. Use a knife to carefully cut a fault through all of the layers. Place a sixth layer on top of the faulted ones. Ask students to determine the relative ages of each layer and the fault. L2

LS **Visual-Spatial**

Fun Fact

An inclusion forms when one type of rock is included in another. The rock with inclusions is younger than the rock from which the inclusions came. For example, if granite inclusions are found within a layer of sedimentary rock, the sedimentary rock formed after the granite.

Caption Answer

Figure 12 Possible answer: It might contain fossils that can be used to date the rock layer.

Differentiated Instruction

Challenge Have students research how index fossils can be used in relative dating. What index fossils are found in your state? Students can prepare a chart showing the rock layers in your area and the index fossils that might be in those layers. L3

SECTION 2 Relative Ages of Rocks G ◆ 133

Reading Check

Answer Unconformities form when rock is eroded, and new sediment is deposited on top of the erosion surface.

Visual Learning

Figure 13 Use several books from a set of encyclopedias to reinforce the appearance of an angular unconformity shown in Figure D. Place volumes 1, 2, 3, and 4 in numerical order, with volume 1 on the bottom. Tilt the books at an angle and then horizontally place volumes 5 and 6 on top of the tilted books. LS **Visual-Spatial**

Discussion

Unconformities What does an unconformity indicate about past geological processes of an area? At some point, erosion removed some rock, or time passed without any sediment deposition and rock formation, or both. Explain that angular unconformities form when rock layers are uplifted, tilted, and eroded. Ask students to explain what causes the forces that produce angular unconformities. Plate tectonics generates compression, tension, and shear forces, which in turn cause rock layers to be folded, tilted, or faulted. The agents of erosion produce forces that wear away exposed rock. L2 LS **Logical-Mathematical**

Figure 13 An angular unconformity results when horizontal layers cover tilted, eroded layers.

A Sedimentary rocks are deposited originally as horizontal layers.

B The horizontal rock layers are tilted as forces within Earth deform them.

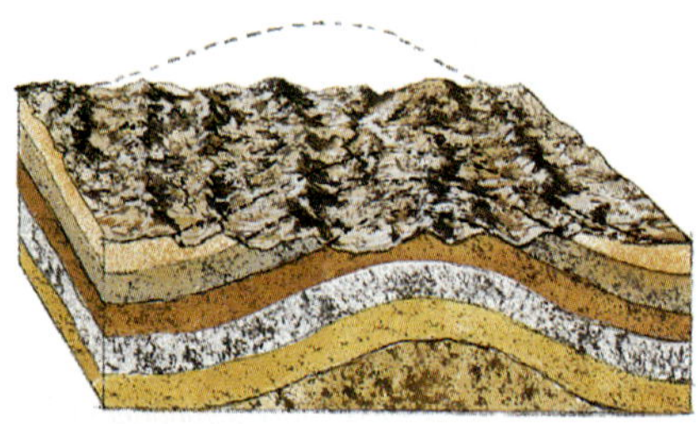

C The tilted layers erode.

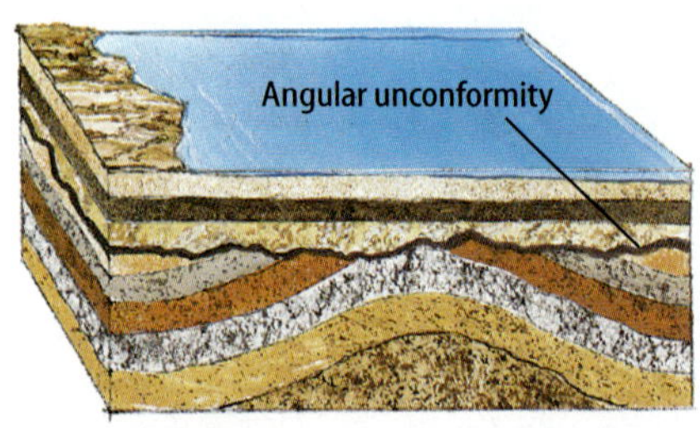

D An angular unconformity results when new layers form on the tilted layers as deposition resumes.

Unconformities

A sequence of rock is a record of past events. But most rock sequences are incomplete—layers are missing. These gaps in rock sequences are called **unconformities** (un kun FOR muh teez). Unconformities develop when agents of erosion such as running water or glaciers remove rock layers by washing or scraping them away.

Reading Check *How do unconformities form?*

Angular Unconformities Horizontal layers of sedimentary rock often are tilted and uplifted. Erosion and weathering then wear down these tilted rock layers. Eventually, younger sediment layers are deposited horizontally on top of the tilted and eroded layers. Geologists call such an unconformity an angular unconformity. **Figure 13** shows how angular unconformities develop.

Disconformity Suppose you're looking at a stack of sedimentary rock layers. They look complete, but layers are missing. If you look closely, you might find an old surface of erosion. This records a time when the rocks were exposed and eroded. Later, younger rocks formed above the erosion surface when deposition of sediment began again. Even though all the layers are parallel, the rock record still has a gap. This type of unconformity is called a disconformity. A disconformity also forms when a period of time passes without any new deposition occurring to form new layers of rock.

Nonconformity Another type of unconformity, called a nonconformity, occurs when metamorphic or igneous rocks are uplifted and eroded. Sedimentary rocks are then deposited on top of this erosion surface. The surface between the two rock types is a nonconformity. Sometimes rock fragments from below are incorporated into sediments deposited above the nonconformity. All types of unconformities are shown in **Figure 14.**

Differentiated Instruction

Visually Impaired To help students participate in the discussion of unconformities, mold models of each type of unconformity from clay. Allow students to handle the models and to remove each of the layers. L1 ELL LS **Kinesthetic**

NATIONAL GEOGRAPHIC **VISUALIZING UNCONFORMITIES**

Figure 14

An unconformity is a gap in the rock record caused by erosion or a pause in deposition. There are three major kinds of unconformities—nonconformity, angular unconformity, and disconformity.

▲ In a nonconformity, horizontal layers of sedimentary rock overlie older igneous or metamorphic rocks. A nonconformity in Big Bend National Park, Texas, is shown above.

▲ An angular unconformity develops when new horizontal layers of sedimentary rock form on top of older sedimentary rock layers that have been folded by compression. An example of an angular unconformity at Siccar Point in southeastern Scotland is shown above.

▼ A disconformity develops when horizontal rock layers are exposed and eroded, and new horizontal layers of rock are deposited on the eroded surface. The disconformity shown below is in the Grand Canyon.

NATIONAL GEOGRAPHIC

Visualizing Unconformities

Have students examine the pictures and read the captions. Then ask the following questions.

How do geologists know there is a gap in the rock record in the disconformity shown at the Grand Canyon? The sequence of rock layers does not match the complete sequence found in surrounding areas. Some layers are missing and somewhere younger layers abut older layers. Fossils are often used to date the layers.

Could an unconformity occur that has metamorphic rock on top of sedimentary rock? Not likely. The forces of heat and pressure responsible for the metamorphism would likely occur throughout the rock layers.

Activity

Modeling Unconformities Have students work in groups of three, with each student modeling one of the three types of unconformities shown in the feature. Have them use several colors of modeling clay so the models are vivid and obvious. Have students explain their models to each other.

Inquiry Lab

Modeling Relative Dating

Purpose To understand how various geologic features allow us to determine the ages of rocks.

Possible Materials modeling clay

Estimated Time one class period

Teaching Strategies

- Have students make models involving multiple rock layers with unconformities and faults.
- Students can then present their models to the class, including the relative ages of all features.
- Allow students to investigate other features that contribute to relative dating, such as index fossils.

For additional inquiry activities, see *Science Inquiry Labs*.

Reading Check

Answer If the fossils are the same type or can be shown to be the same age, it is likely that the rocks are part of the same layer.

Science Online

Topic: Correlating with Index Fossils

Visit bookg.msscience.com for Web links to information about using index fossils to match up layers of rock.

Activity Make a chart that shows the rock layers of both the Grand Canyon and Capitol Reef National Park in Utah. For each layer that appears in both parks, list an index fossil you could find to correlate the layers.

Matching Up Rock Layers

Suppose you're studying a layer of sandstone in Bryce Canyon in Utah. Later, when you visit Canyonlands National Park, Utah, you notice that a layer of sandstone there looks just like the sandstone in Bryce Canyon, 250 km away. Above the sandstone in the Canyonlands is a layer of limestone and then another sandstone layer. You return to Bryce Canyon and find the same sequence—sandstone, limestone, and sandstone. What do you infer? It's likely that you're looking at the same layers of rocks in two different locations. **Figure 15** shows that these rocks are parts of huge deposits that covered this whole area of the western United States. Geologists often can match up, or correlate, layers of rocks over great distances.

Evidence Used for Correlation It's not always easy to say that a rock layer exposed in one area is the same as a rock layer exposed in another area. Sometimes it's possible to walk along the layer for kilometers and prove that it's continuous. In other cases, such as at the Canyonlands area and Bryce Canyon as seen in **Figure 16,** the rock layers are exposed only where rivers have cut through overlying layers of rock and sediment. How can you show that the limestone sandwiched between the two layers of sandstone in Canyonlands is likely the same limestone as at Bryce Canyon? One way is to use fossil evidence. If the same types of fossils were found in the limestone layer in both places, it's a good indication that the limestone at each location is the same age, and, therefore, one continuous deposit.

Reading Check *How do fossils help show that rocks at different locations belong to the same rock layer?*

Figure 15 These rock layers, exposed at Hopi Point in Grand Canyon National Park, Arizona, can be correlated, or matched up, with rocks from across large areas of the western United States.

Teacher FYI

Formation Information The basic unit of layered rocks is the formation, a mappable unit of rock. A formation is named for basic characteristics such as rock type, color, and fossil content—all of which are readily observed. A group consists of more than one formation, stacked on top of one another. A member is a subdivision of a formation. It is designated by a specific characteristic within it, such as an abundance of fossils or a particular mineral.

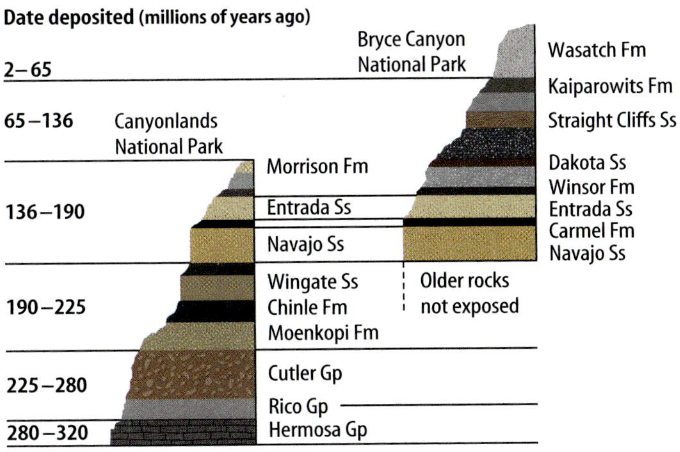

Figure 16 Geologists have named the many rock layers, or formations, in Canyonlands and in Bryce Canyon, Utah. They also have correlated some formations between the two canyons.
List *the labeled layers present at both canyons.*

Can layers of rock be correlated in other ways? Sometimes determining relative ages isn't enough and other dating methods must be used. In Section 3, you'll see how the numerical ages of rocks can be determined and how geologists have used this information to estimate the age of Earth.

section 2 review

Summary

Superposition

- Superposition states that in undisturbed rock, the oldest layers are on the bottom.

Relative Ages

- Rock layers can be ranked by relative age.

Unconformities

- Angular unconformities are new layers deposited over tilted and eroded rock layers.
- Disconformities are gaps in the rock record.
- Nonconformities divide uplifted igneous or metamorphic rock from new sedimentary rock.

Matching Up Rock Layers

- Rocks from different areas may be correlated if they are part of the same layer.

Self Check

1. **Discuss** how to find the oldest paper in a stack of papers.
2. **Explain** the concept of relative age.
3. **Illustrate** a disconformity.
4. **Describe** one way to correlate similar rock layers.
5. **Think Critically** Explain the relationship between the concept of relative age and the principle of superposition.

Applying Skills

6. **Interpret data** to determine the oldest rock bed. A sandstone contains a 400-million-year-old fossil. A shale has fossils that are over 500 million years old. A limestone, below the sandstone, contains fossils between 400 million and 500 million years old. Which rock bed is oldest? Explain.

Science Online bookg.msscience.com/self_check_quiz

Caption Answer

Figure 16 Both canyons have layers Navajo Ss and Entrada Ss.

3 Assess

DAILY INTERVENTION

Check for Understanding

Visual-Spatial Obtain a geologic map and correlation chart for the area that includes your state. Invite students to illustrate the correlation between any two points in the state.

Reteach

Geologic History Draw a geologic column on transparency acetate. Include layers of limestone, sandstone, shale, and metamorphic rock, a fault, a disconformity (indicated by a jagged line), and an igneous intrusion. Project the illustration, and have students explain the history of the area. L2 ELL LS **Visual-Spatial**

Assessment

Performance Illustrate Question 6 of the Section Review on the board. Then have students interpret the relative ages of the rock layers. Assume that the rocks are undisturbed and that shale underlies the limestone. Use **Performance Assessment in the Science Classroom,** p. 89.

section 2 review

1. Applying the principle of superposition would help in finding the papers.
2. Relative age describes the age of one rock layer in relation to another; it does not describe age in actual years.
3. a stack of sedimentary rock layers in which one or more layers is missing
4. Possible answers: match rock types, look for presence of matching fossils, compare relative positions in the geologic column
5. The principle of superposition states that older rock is on the bottom; therefore, the relative age of the bottom layer is older.
6. The shale; its fossils are older than those found in the other layers.

BENCH TESTED

Real-World Question

Purpose Students determine the relative order of events by interpreting geologic cross sections. L2 ELL LS **Logical-Mathematical**

Process Skills infer, interpret data, formulate models, recognize and use spatial relationships, sequence, interpret scientific illustrations

Time Required 40–45 minutes

Procedure

Teaching Strategy If students are having trouble, reproduce the cross sections on an overhead. Then work as a class to identify the relative ages of the different layers.

Conclude and Apply

1. Disconformity; yes, other rock layers could have been eroded.
2. The rocks moved upward.
3. as a result of a fault
4. A fault cuts through the intrusion but stops at the unconformity. Thus the fault occurred after the intrusion but while erosion was occurring. Therefore, the intrusion is older.
5. The one on the left is older than the one on the right. The one on the right intrudes the sandstone and is therefore younger. The one offset by the fault is older than the sandstone.
6. The sandstone and the upper shale may have been thicker. They exhibit erosional disconformities at their upper surfaces.

LAB

Relative Ages

Which of your two friends is older? To answer this question, you'd need to know their relative ages. You wouldn't need to know the exact age of either of your friends—just who was born first. The same is sometimes true for rock layers.

A

B

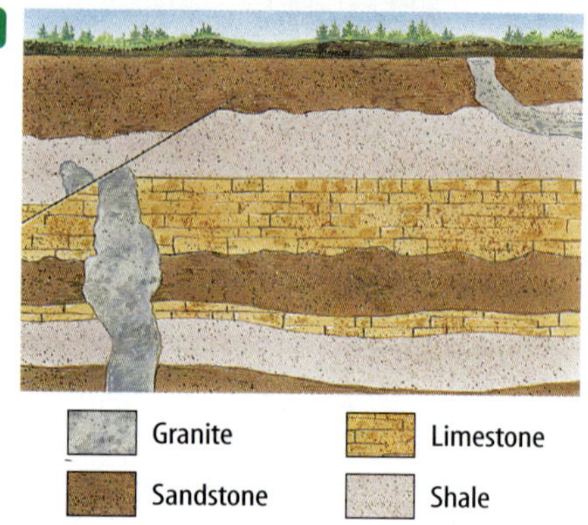

Real-World Question

Can you determine the relative ages of rock layers?

Goals

- **Interpret** illustrations of rock layers and other geological structures and determine the relative order of events.

Materials

paper pencil

Procedure

1. **Analyze Figures A** and **B.**
2. Make a sketch of **Figure A.** On it, identify the relative age of each rock layer, igneous intrusion, fault, and unconformity. For example, the shale layer is the oldest, so mark it with a 1. Mark the next-oldest feature with a 2, and so on.
3. Repeat step 2 for **Figure B.**

Conclude and Apply

Figure A

1. **Identify** the type of unconformity shown. Is it possible that there were originally more layers of rock than are shown?
2. **Describe** how the rocks above the fault moved in relation to rocks below the fault.
3. **Hypothesize** how the hill on the left side of the figure formed.

Figure B

4. Is it possible to conclude if the igneous intrusion on the left is older or younger than the unconformity nearest the surface?
5. **Describe** the relative ages of the two igneous intrusions. How did you know?
6. **Hypothesize** which two layers of rock might have been much thicker in the past.

Communicating Your Data

Compare your results with other students' results. **For more help, refer to the** Science Skill Handbook.

Assessment

Performance Based on what was learned about relative ages of rock layers, have students draw a block diagram similar to those in the Lab. Tell them to include all three rock types. Have students exchange and interpret their diagrams. Use **Performance Assessment in the Science Classroom,** p. 127.

Communicating Your Data

Oldest to youngest: A: shale, sandstone, limestone, disconformity, sandstone, limestone, intrusion, fault; B: sandstone, shale, limestone, sandstone, disconformity, limestone, shale, intrusion (left), fault, disconformity, sandstone, intrusion (right)

section 3

Absolute Ages of Rocks

Absolute Ages

As you sort through your stack of magazines looking for that article about the car you saw, you decide that you need to restack them into a neat pile. By now, they're in a jumble and no longer in order of their relative age, as shown in **Figure 17.** How can you stack them so the oldest are on the bottom and the newest are on top? Fortunately, magazine dates are printed on the cover. Thus, stacking magazines in order is a simple process. Unfortunately, rocks don't have their ages stamped on them. Or do they? **Absolute age** is the age, in years, of a rock or other object. Geologists determine absolute ages by using properties of the atoms that make up materials.

Radioactive Decay

INTEGRATE Physics

Atoms consist of a dense central region called the nucleus, which is surrounded by a cloud of negatively charged particles called electrons. The nucleus is made up of protons, which have a positive charge, and neutrons, which have no electric charge. The number of protons determines the identity of the element, and the number of neutrons determines the form of the element, or isotope. For example, every atom with a single proton is a hydrogen atom. Hydrogen atoms can have no neutrons, a single neutron, or two neutrons. This means that there are three isotopes of hydrogen.

Reading Check *What particles make up an atom's nucleus?*

Some isotopes are unstable and break down into other isotopes and particles. Sometimes a lot of energy is given off during this process. The process of breaking down is called **radioactive decay.** In the case of hydrogen, atoms with one proton and two neutrons are unstable and tend to break down. Many other elements have stable and unstable isotopes.

as you read

What You'll Learn

- **Identify** how absolute age differs from relative age.
- **Describe** how the half-lives of isotopes are used to determine a rock's age.

Why It's Important

Events in Earth's history can be better understood if their absolute ages are known.

Review Vocabulary

isotopes: atoms of the same element that have different numbers of neutrons

New Vocabulary

- absolute age
- radioactive decay
- half-life
- radiometric dating
- uniformitarianism

Figure 17 The magazines that have been shuffled through no longer illustrate the principle of superposition.

1 Motivate

INTERACTIVE CHALKBOARD PowerPoint® Presentations

Bellringer

Section Focus Transparencies also are available on the Interactive Chalkboard CD-ROM. L2 ELL

Tie to Prior Knowledge

Atom Parts and Positions Have students recall the main parts of the atom and their positions within an atom. Protons and neutrons—in nucleus; electrons—around nucleus

Reading Check

Answer protons and neutrons

Section 3 Resource Manager

Chapter *Fast File* Resources

- Transparency Activity, p. 46
- Directed Reading for Content Mastery, pp. 21, 22
- MiniLAB, p. 4
- Enrichment, p. 32
- Reinforcement, p. 29
- Lab Worksheets, pp. 7–8

Mathematics Skill Activities, p. 3

Cultural Diversity, p. 69

Lab Management and Safety, p. 39

2 Teach

Mini LAB

Purpose Students will model the principle of carbon-14 dating.

Materials red and green jelly beans

Teaching Strategy Advise students not to eat the jelly beans and to wash hands thoroughly after the activity.

Analysis

1. As in carbon-14, each half-life resulted in half of the remaining jelly beans being removed.
2. 4
3. 22, 920 years old

Assessment

Performance Have students determine the relative and absolute ages of students in the classroom. Use **Performance Assessment in the Science Classroom**, p. 99. P

Discussion

Dating Errors What would happen to your estimate of age if some of the daughter isotope escaped? Since this would make it appear that less daughter has formed, the rock will seem younger.

Mini LAB

Modeling Carbon-14 Dating

Procedure

1. Count out 80 **red jelly beans.**
2. Remove half the red jelly beans and replace them with **green jelly beans.**
3. Continue replacing half the red jelly beans with green jelly beans until only 5 red jelly beans remain. Count the number of times you replace half the red jelly beans.

Analysis

1. How did this activity model the decay of carbon-14 atoms?
2. How many half lives of carbon-14 did you model during this activity?
3. If the atoms in a bone experienced the same number of half lives as your jelly beans, how old would the bone be?

Alpha and Beta Decay In some isotopes, a neutron breaks down into a proton and an electron. This type of radioactive decay is called beta decay because the electron leaves the atom as a beta particle. The nucleus loses a neutron but gains a proton. When the number of protons in an atom is changed, a new element forms. Other isotopes give off two protons and two neutrons in the form of an alpha particle. Alpha and beta decay are shown in **Figure 18.**

Half-Life In radioactive decay reactions, the parent isotope undergoes radioactive decay. The daughter product is produced by radioactive decay. Each radioactive parent isotope decays to its daughter product at a certain rate. Based on this decay rate, it takes a certain period of time for one half of the parent isotope to decay to its daughter product. The **half-life** of an isotope is the time it takes for half of the atoms in the isotope to decay. For example, the half-life of carbon-14 is 5,730 years. So it will take 5,730 years for half of the carbon-14 atoms in an object to change into nitrogen-14 atoms. You might guess that in another 5,730 years, all of the remaining carbon-14 atoms will decay to nitrogen-14. However, this is not the case. Only half of the atoms of carbon-14 remaining after the first 5,730 years will decay during the second 5,730 years. So, after two half-lives, one fourth of the original carbon-14 atoms still remain. Half of them will decay during another 5,730 years. After three half-lives, one eighth of the original carbon-14 atoms still remain. After many half-lives, such a small amount of the parent isotope remains that it might not be measurable.

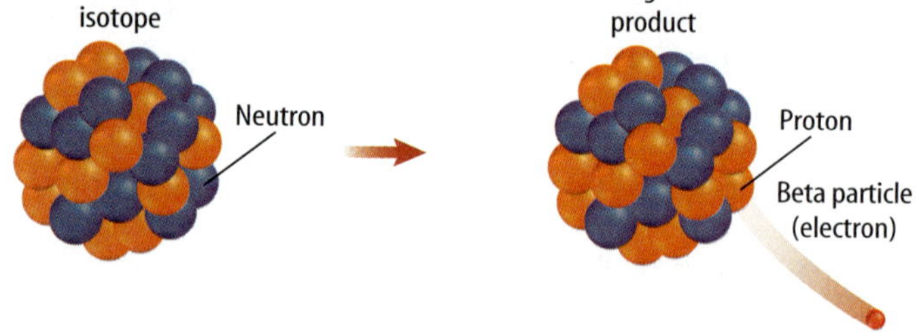

Figure 18 In beta decay, a neutron changes into a proton by giving off an electron. This electron has a lot of energy and is called a beta particle.

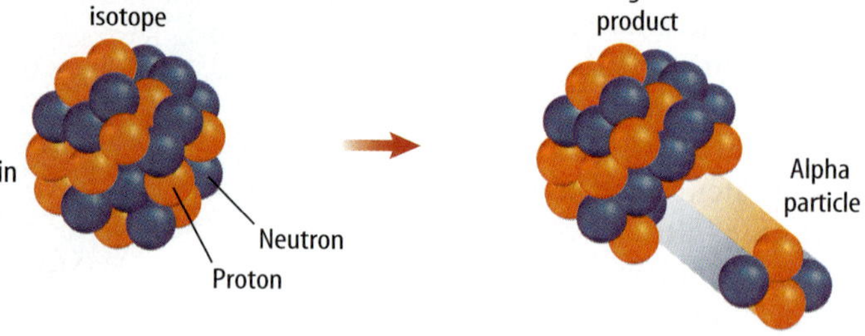

In the process of alpha decay, an unstable parent isotope nucleus gives off an alpha particle and changes into a new daughter product. Alpha particles contain two neutrons and two protons.

LAB DEMONSTRATION

Purpose to demonstrate how half-life works

Materials 100 pennies, 100 paper clips, box with lid

Preparation Place the pennies in the box, heads up.

Procedure Close the lid and shake the box twice (half-life). Open the lid and count and record the flipped (decayed) pennies. Remove the flipped pennies and replace them with paper clips. Repeat until all pennies have been replaced.

Expected Outcome About half of the remaining pennies decay during each half-life.

Assessment

What do the pennies represent? atoms of the parent isotope What do the paper clips represent? daughter products What is each two shakes? a half-life Why do fewer isotopes decay in each successive half-life? There are fewer atoms left to decay.

Radiometric Ages

Decay of radioactive isotopes is like a clock keeping track of time that has passed since rocks have formed. As time passes, the amount of parent isotope in a rock decreases as the amount of daughter product increases, as in **Figure 19.** By measuring the ratio of parent isotope to daughter product in a mineral and by knowing the half-life of the parent, in many cases you can calculate the absolute age of a rock. This process is called **radiometric dating**.

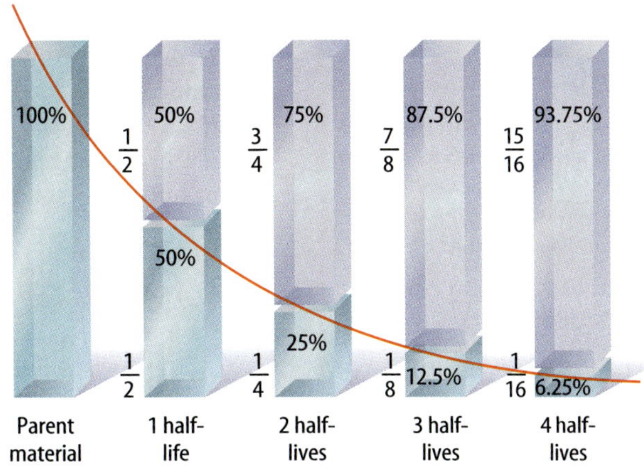

Figure 19 During each half-life, one half of the parent material decays to the daughter product. **Explain** *how one uses both parent and daughter material to estimate age.*

A scientist must decide which parent isotope to use when measuring the age of a rock. If the object to be dated seems old, then the geologist will use an isotope with a long half-life. The half-life for the decay of potassium-40 to argon-40 is 1.25 billion years. As a result, this isotope can be used to date rocks that are many millions of years old. To avoid error, conditions must be met for the ratios to give a correct indication of age. For example, the rock being studied must still retain all of the argon-40 that was produced by the decay of potassium-40. Also, it cannot contain any contamination of daughter product from other sources. Potassium-argon dating is good for rocks containing potassium, but what about other things?

Radiocarbon Dating Carbon-14 is useful for dating bones, wood, and charcoal up to 75,000 years old. Living things take in carbon from the environment to build their bodies. Most of that carbon is carbon-12, but some is carbon-14, and the ratio of these two isotopes in the environment is always the same. After the organism dies, the carbon-14 slowly decays. By determining the amounts of the isotopes in a sample, scientists can evaluate how much the isotope ratio in the sample differs from that in the environment. For example, during much of human history, people built campfires. The wood from these fires often is preserved as charcoal. Scientists can determine the amount of carbon-14 remaining in a sample of charcoal by measuring the amount of radiation emitted by the carbon-14 isotope in labs like the one in **Figure 20.** Once they know the amount of carbon-14 in a charcoal sample, scientists can determine the age of the wood used to make the fire.

Figure 20 Radiometric ages are determined in labs like this one.

Activity

Calculate Absolute Age The half-life of potassium-40 is 1.3 billion years. An igneous rock is found with 25% potassium-40 and 75% daughter product. Calculate the absolute age of this rock. 2.6 billion years (two half-lives)

L2 LS **Logical-Mathematical**

Quick Demo

Paper Half-Lives

Materials piece of paper

Estimated time 3 minutes

Procedure Have students imagine that a sheet of paper is analogous to parent isotope in a rock. The half-life of this isotope is 10 seconds. After 10 seconds, tear the paper in half and set one of the halves aside as the daughter isotope. Repeat the process every 10 seconds with the remaining paper.

Visual Learning

Figure 19 After students have studied the graph, discuss whether the overall amount of material changes after each half-life. Reinforce that the overall amount of material stays nearly the same; it is the makeup of the material that changes.

Caption Answer

Figure 19 The ratio of parent to daughter materials allows scientists to estimate how many half-lives have passed.

Fun Fact

All organisms contain both stable and radioactive carbon isotopes. The radioactive isotope decays after death, changing the carbon isotope ratio.

Teacher FYI

Half-Lives The half-life of carbon-14 is about 5,730 years. The half-life of uranium-238 in its decay to lead-206 is 4.5 billion years (nearly the age of Earth). By contrast, the half-life of uranium-235 in its decay to lead-207 is 713 million years.

Reading Check

Answer Sedimentary rocks are made of particles eroded from older rocks. Dating these particles gives the age of the preexisting rock, not the newer sedimentary rock.

Applying Science

Teaching Strategy

Explain that the age given by this method is an estimate. Radiometric dating does not give exact answers.

Answers to Questions

1. Since the Iceman is about 5,300 years old, students should estimate that about 55% of the original carbon-14 was still present.
2. Using the chart, you can see that after 17, 190 years, 12.5 percent of the carbon-14 would remain; $10 \times 0.125 = 1.25$ grams

IDENTIFYING Misconceptions

Uniformitarianism Students may think that the present is not the key to the past. Refer to page F at the beginning of this chapter for teaching strategies that address this misconception.

Science Online

Topic: Isotopes in Ice Cores

Visit bookg.msscience.com for Web links to information about ice cores and how isotopes in ice are used to learn about Earth's past.

Activity Prepare a report that shows how isotopes in ice cores can tell us about past Earth environments. Include how these findings can help us understand today's climate.

Age Determinations Aside from carbon-14 dating, rocks that can be radiometrically dated are mostly igneous and metamorphic rocks. Most sedimentary rocks cannot be dated by this method. This is because many sedimentary rocks are made up of particles eroded from older rocks. Dating these pieces only gives the age of the preexisting rock from which it came.

The Oldest Known Rocks Radiometric dating has been used to date the oldest rocks on Earth. These rocks are about 3.96 billion years old. By determining the age of meteorites, and using other evidence, scientists have estimated the age of Earth to be about 4.5 billion years. Earth rocks greater than 3.96 billion years old probably were eroded or changed by heat and pressure.

Reading Check *Why can't most sedimentary rocks be dated radiometrically?*

Applying Science

When did the Iceman die?

Carbon-14 dating has been used to date charcoal, wood, bones, mummies from Egypt and Peru, the Dead Sea Scrolls, and the Italian Iceman. The Iceman was found in 1991 in the Italian Alps, near the Austrian border. Based on carbon-14 analysis, scientists determined that the Iceman is 5,300 years old. Determine approximately in what year the Iceman died.

Half-Life of Carbon-14

Percent Carbon-14	Years Passed
100	0
50	5,730
25	11,460
12.5	17,190
6.25	22,920
3.125	

Reconstruction of Iceman

Identifying the Problem

The half-life chart shows the decay of carbon-14 over time. Half-life is the time it takes for half of a sample to decay. Fill in the years passed when only 3.125 percent of carbon-14 remain. Is there a point at which no carbon-14 would be present? Explain.

Solving the Problem

1. Estimate, using the data table, how much carbon-14 still was present in the Iceman's body that allowed scientists to determine his age.
2. If you had an artifact that originally contained 10.0 g of carbon-14, how many grams would remain after 17,190 years?

Science Journal

Uniformitarianism Have students research and write a brief report in their Science Journals about uniformitarianism and the contributions of James Hutton, John Playfair, and Sir Charles Lyell. James Hutton thought Earth processes of today were similar to those of the past. John Playfair wrote about Hutton's ideas, but Sir Charles Lyell's written works were easier to read. L2 LS **Linguistic** P

Differentiated Instruction

English-Language Learners Have students prepare 100 squares of paper, one side red and the other side blue. Start with all of one color facing up and have a student keep time. Every 30 seconds, have someone turn over half of the squares. Explain half-lives and show that the total number of atoms remains constant.

142 ◆ G CHAPTER 5 Clues to Earth's Past

Uniformitarianism

Can you imagine trying to determine the age of Earth without some of the information you know today? Before the discovery of radiometric dating, many people estimated that Earth is only a few thousand years old. But in the 1700s, Scottish scientist James Hutton estimated that Earth is much older. He used the principle of **uniformitarianism.** This principle states that Earth processes occurring today are similar to those that occurred in the past. Hutton's principle is often paraphrased as "the present is the key to the past."

Figure 21 The rugged highlands of Scotland were shaped by erosion and uplift.

Hutton observed that the processes that changed the landscape around him were slow, and he inferred that they were just as slow throughout Earth's history. Hutton hypothesized that it took much longer than a few thousand years to form the layers of rock around him and to erode mountains that once stood kilometers high. **Figure 21** shows Hutton's native Scotland, a region shaped by millions of years of geologic processes.

Today, scientists recognize that Earth has been shaped by two types of change: slow, everyday processes that take place over millions of years, and violent, unusual events such as the collision of a comet or asteroid about 65 million years ago that might have caused the extinction of the dinosaurs.

section 3 review

Summary

Absolute Ages

- The absolute age is the actual age of an object.

Radioactive Decay

- Some isotopes are unstable and decay into other isotopes and particles.
- Decay is measured in half-lives, the time it takes for half of a given isotope to decay.

Radiometric Ages

- By measuring the ratio of parent isotope to daughter product, one can determine the absolute age of a rock.
- Living organisms less than 75,000 years old can be dated using carbon-14.

Uniformitarianism

- Processes observable today are the same as the processes that took place in the past.

Self Check

1. **Evaluate** the age of rocks. You find three undisturbed rock layers. The middle layer is 120 million years old. What can you say about the ages of the layers above and below it?
2. **Determine** the age of a fossil if it had only one eighth of its original carbon-14 content remaining.
3. **Explain** the concept of uniformitarianism.
4. **Describe** how radioactive isotopes decay.
5. **Think Critically** Why can't scientists use carbon-14 to determine the age of an igneous rock?

Applying Math

6. **Make and use a table** that shows the amount of parent material of a radioactive element that is left after four half-lives if the original parent material had a mass of 100 g.

Visual Learning

Figure 22 Have students list some of the processes that helped shape this landscape.

3 Assess

DAILY INTERVENTION

Check for Understanding

Visual-Spatial Give students red and blue marbles, and have them illustrate the decay of blue marbles to red marbles, one half-life at a time.

Reteach

Radiometric Dates Have students determine the age of a rock containing 1/32 of the radium-226 originally in the rock. The half-life of radium-226 is 1,600 years. After five half-lives, or 8,000 years, 1/32 of the original isotope remains. L2 LS **Logical-Mathematical**

Assessment

Performance Have students plot the data from the tables they make in Question 6. Half-lives should appear on the horizontal axis, and the percentage of parent material should appear on the vertical axis. Use **Performance Assessment in the Science Classroom,** p. 111.

section 3 review

1. The top layer is less than 120 million years old; the bottom layer is more than 120 million years old.
2. 17,190 years old
3. Earth processes occurring today are like those that have occurred in the past.
4. In some isotopes, a neutron breaks down into a proton and an electron (beta decay). In others, an isotope gives off two protons and two neutrons (alpha decay).
5. Carbon-14 dating works only for materials that were once alive and contain carbon. Igneous rocks are nonliving materials.
6. See chart in side column.

Half-Life	Parent
0	100 g
1	50 g
2	25 g
3	12.5 g
4	6.25 g

SECTION 3 Absolute Ages of Rocks G ◆ 143

BENCH TESTED

Real-World Question

Purpose Students will research information and create a model of trace fossils that shows some behaviors of animals.

Process Skills make a model, use reference materials, observe and infer

Time Required one or two class periods, depending on the amount of research required

Make a Model

Teaching Strategies Have references available for students in the classroom where they can look up information.

Troubleshooting Tell students to limit the number of specific behaviors that they show so that the model does not get too confusing.

LAB Model and Invent

Trace Fossils

Goals

- **Construct** a model of trace fossils.
- **Describe** the information that you can learn from looking at your model.

Possible Materials

construction paper
wire
plastic (a fairly rigid type)
scissors
plaster of paris
toothpicks
sturdy cardboard
clay
pipe cleaners
glue

Safety Precautions

Real-World Question

Trace fossils can tell you a lot about the activities of organisms that left them. They can tell you how an organism fed or what kind of home it had. How can you model trace fossils that can provide information about the behavior of organisms? What materials can you use to model trace fossils? What types of behavior could you show with your trace fossil model?

Make a Model

1. **Decide** how you are going to make your model. What materials will you need?
2. **Decide** what types of activities you will demonstrate with your model. Were the organisms feeding? Resting? Traveling? Were they predators? Prey? How will your model indicate the activities you chose?
3. What is the setting of your model? Are you modeling the organism's home? Feeding areas? Is your model on land or water? How can the setting affect the way you build your model?
4. Will you only show trace fossils from a single species or multiple species? If you include more than one species, how will you provide evidence of any interaction between the species?

Check the Model Plans

1. Compare your plans with those of others in your class. Did other groups mention details that you had forgotten to think about? Are there any changes you would like to make to your plan before you continue?
2. Make sure your teacher approves your plan before you continue.

Test Your Model

1. Following your plan, construct your model of trace fossils.
2. Have you included evidence of all the behaviors you intended to model?

Analyze Your Data

1. **Evaluate** Now that your model is complete, do you think that it adequately shows the behaviors you planned to demonstrate? Is there anything that you think you might want to do differently if you were going to make the model again?
2. **Describe** how using different kinds of materials might have affected your model. Can you think of other materials that would have allowed you to show more detail than you did?

Conclude and Apply

1. **Compare and contrast** your model of trace fossils with trace fossils left by real organisms. Is one more easily interpreted than the other? Explain.
2. **List** behaviors that might not leave any trace fossils. Explain.

Communicating Your Data

Ask other students in your class or another class to look at your model and describe what information they can learn from the trace fossils. Did their interpretations agree with what you intended to show?

Test Your Model

Expected Outcome Students will create a model that includes trace fossils that indicate certain behaviors of one or more species.

Analyze Your Data

1. Answers will be individualized and based on the students' opinion of their research. Look for depth and quality of research performed.
2. Answers will vary depending on the materials students used. Clay would show three-dimensional fossils whereas paper or fabric would only be two-dimensional.

Conclude and Apply

1. The models may be easier to interpret because they will not have any extraneous features and were not subject to weather or other factors that could damage them.
2. Answers will vary. Look for behaviors like mating rituals or communication that wouldn't necessarily leave any marks on sediments for later fossilization.

Assessment

Process Have students use reference materials, such as textbooks or magazines, to find photographs of trace fossils. Students should write a brief description of the type of animals and the behavior indicated. Then, groups of students can trade photographs and try to interpret those found by others. Use **Performance Assessment in the Science Classroom,** p. 159.

Communicating Your Data

Students should discuss why their interpretations did or did not agree. They can cite their references to support their arguments.

Content Background

The coelacanths are an ancient group, appearing in the fossil record about 395 million years ago. They are characterized by lobelike, fleshy fins, and live at great depths, where they are difficult to find. The limblike skeletal structure of the fleshy fins of coelacanths is thought to be an ancestral condition of all animals with four limbs.

Discussion

Survival Hypothesize how a fish could survive when many of the animals that evolved from it on land no longer exist. Conditions in the sea, such as temperature, salinity, and food sources are slow to change and are relatively constant, allowing marine species long periods to adjust to environmental changes.

Analyze the Event

Serendipity Ask students to brainstorm the characteristics or circumstances that helped Courtenay-Latimer to recognize the significance of her find. Possible answers: She was observant; she had friends who fished; she was curious; she knew to whom to send her sketch. Point out that she was not a scientist but had characteristics that are important for scientists, such as curiosity.

Oops! Accidents in SCIENCE

SOMETIMES GREAT DISCOVERIES HAPPEN BY ACCIDENT!

The World's Oldest Fish Story

A catch-of-the-day set science on its ears

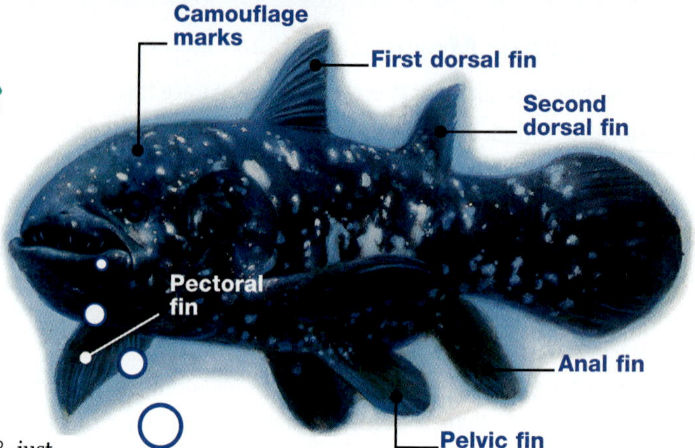

Some scientists call the coelacanth "Old Four Legs." It got its nickname because the fish has paired fins that look something like legs.

On a December day in 1938, just before Christmas, Marjorie Courtenay-Latimer went to say hello to her friends on board a fishing boat that had just returned to port in South Africa. Courtenay-Latimer, who worked at a museum, often went aboard her friends' ship to check out the catch. On this visit, she received a surprise Christmas present—an odd-looking fish. As soon as the woman spotted its strange blue fins among the piles of sharks and rays, she knew it was special.

Courtenay-Latimer took the fish back to her museum to study it. "It was the most beautiful fish I had ever seen, five feet long, and a pale mauve blue with iridescent silver markings," she later wrote. Courtenay-Latimer sketched it and sent the drawing to a friend of hers, J. L. B. Smith.

Smith was a chemistry teacher who was passionate about fish. After a time, he realized it was a coelacanth (SEE luh kanth). Fish experts knew that coelacanths had first appeared on Earth 400 million years ago. But the experts thought the fish were extinct. People had found fossils of coelacanths, but no one had seen one alive. It was assumed that the last coelacanth species had died out 65 million years ago. They were wrong. The ship's crew had caught one by accident.

Smith figured there might be more living coelacanths. So he decided to offer a reward for anyone who could find a living specimen. After 14 years of silence, a report came in that a coelacanth had been caught off the east coast of Africa.

Today, scientists know that there are at least several hundred coelacanths living in the Indian Ocean, just east of central Africa. Many of these fish live near the Comoros Islands. The coelacanths live in underwater caves during the day but move out at night to feed. The rare fish are now a protected species. With any luck, they will survive for another hundred million years.

Write a short essay describing the discovery of the coelacanths and describe the reaction of scientists to this discovery.

Science Online
For more information, visit bookg.msscience.com/oops

Write People around the world read about the discovery in newspapers and were fascinated by it. Courtenay-Latimer became famous. Thousands of people gathered to view the mounted fish. A prehistoric animal, the fish appealed to people the way dinosaurs appeal to many people today.

Resources for Teachers and Students

"Dispute Over a Legendary Fish," by Constance Holden, *Science*, April 2, 1999

A Fish Caught in Time, by Samantha Weinberg, HarperCollins Publishers, 2000

Reviewing Main Ideas

Section 1 Fossils

1. Fossils are more likely to form if hard parts of the dead organisms are buried quickly.
2. Some fossils form when original materials that made up the organisms are replaced with minerals. Other fossils form when remains are subjected to heat and pressure, leaving only a carbon film behind. Some fossils are the tracks or traces left by ancient organisms.

Section 2 Relative Ages of Rocks

1. The principle of superposition states that, in undisturbed layers, older rocks lie underneath younger rocks.
2. Unconformities, or gaps in the rock record, are due to erosion or periods of time during which no deposition occurred.
3. Rock layers can be correlated using rock types and fossils.

Section 3 Absolute Ages of Rocks

1. Absolute dating provides an age in years for the rocks.
2. The half-life of a radioactive isotope is the time it takes for half of the atoms of the isotope to decay into another isotope.

Visualizing Main Ideas

Copy and complete the following concept map on fossils.

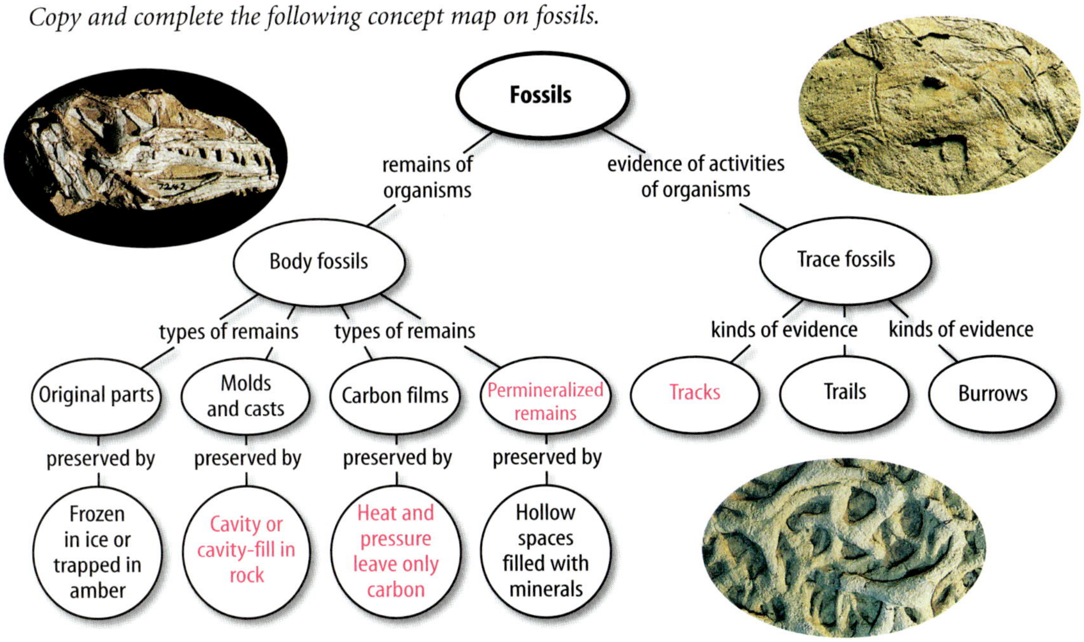

Identifying Misconceptions

Assess

Use the assessment as follow-up to page 122F after students have completed the chapter.

Procedure Set up a stream table with a meandering stream in it that flows into a lake. Have students observe the water flowing through the stream into the lake. After several minutes, ask the students to make a drawing of the feature created where the river flows into the lake. Show the students a photo of the Mississippi River delta. Ask them how they think the Mississippi River delta was formed.

Expected Outcome Students will recognize that the process of delta formation is a natural Earth process and that the process has been the same throughout Earth's history.

chapter 5 Study Guide

Reviewing Main Ideas

Summary statements can be used by students to review the major concepts of the chapter.

Visualizing Main Ideas

See student page.

Science Online

Visit bookg.msscience.com
/self_check_quiz
/interactive_tutor
/vocabulary_puzzlemaker
/chapter_review
/standardized_test

Assessment Transparency

For additional assessment questions, use the *Assessment Transparency* located in the transparency book.

Assessment

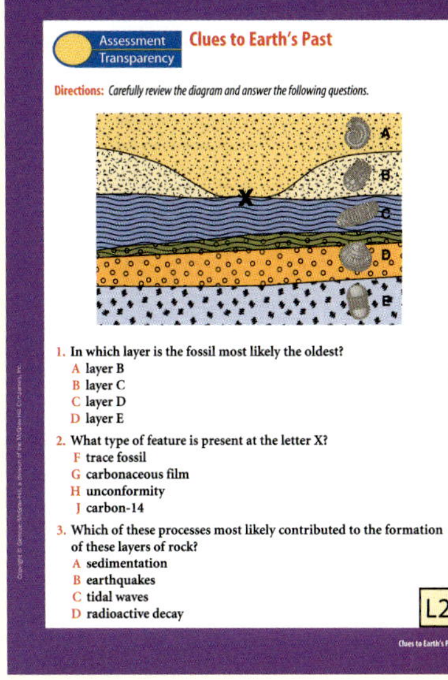

chapter 5 Review

Using Vocabulary

1. A <u>carbon film</u> is a thin film of carbon preserved as a fossil.
2. The <u>principle of superposition</u> states that older rocks lie under younger rocks in an undisturbed area.
3. According to <u>uniformitarianism</u>, processes occur today as they did in the past.
4. An <u>unconformity</u> is a gap in the rock record.
5. The time needed for half of the atoms to decay is called <u>half-life</u>.
6. An <u>index fossil</u> is the fossil of a group that lived for a short time but was widespread.
7. <u>Absolute ages</u> give the ages of rocks in years.
8. In <u>permineralized remains</u>, tiny spaces inside a fossil are filled with minerals.
9. A <u>cast</u> is a copy of a fossil produced inside a mold.

Checking Concepts

10. B
11. C
12. B
13. D
14. D
15. C
16. D
17. D

Thinking Critically

18. Possible answers: Decay can destroy organic remains if they are not buried quickly; erosion destroys layers of rock; deeply buried materials are frequently destroyed by metamorphism.
19. Since the lava contains radioactive elements, an absolute age for the flow can be determined. This age can then be used to determine relative ages for the rocks above and below the flow.

chapter 5 Review

Using Vocabulary

absolute age p. 139	principle of superposition p. 132
carbon film p. 126	radioactive decay p. 139
cast p. 127	radiometric dating p. 141
fossil p. 125	relative age p. 133
half-life p. 140	unconformity p. 134
index fossil p. 129	uniformitarianism p. 143
mold p. 127	
permineralized remains p. 126	

Write an original sentence using the vocabulary word to which each phrase refers.

1. thin film of carbon preserved as a fossil
2. older rocks lie under younger rocks
3. processes occur today as they did in the past
4. gap in the rock record
5. time needed for half the atoms to decay
6. fossil organism that lived for a short time
7. gives the age of rocks in years
8. minerals fill spaces inside fossil
9. a copy of a fossil produced by filling a mold with sediment or crystals

Checking Concepts

Choose the word or phrase that best answers the question.

10. What is any evidence of ancient life called?
    A) half-life **C)** unconformity
    B) fossil **D)** disconformity
11. Which of the following conditions makes fossil formation more likely?
    A) buried slowly
    B) attacked by scavengers
    C) made of hard parts
    D) composed of soft parts
12. What are cavities left in rocks when a shell or bone dissolves called?
    A) casts **C)** original remains
    B) molds **D)** carbon films
13. To say "the present is the key to the past" is a way to describe which of the following principles?
    A) superposition **C)** radioactivity
    B) succession **D)** uniformitarianism
14. A fault can be useful in determining which of the following for a group of rocks?
    A) absolute age **C)** radiometric age
    B) index age **D)** relative age
15. Which of the following is an unconformity between parallel rock layers?
    A) angular unconformity
    B) fault
    C) disconformity
    D) nonconformity

Use the illustration below to answer question 16.

16. Which of the following puts the layers in order from oldest to youngest?
    A) 5-4-3-2-1 **C)** 2-3-4-5-1
    B) 1-2-3-4-5 **D)** 4-3-2-5-1
17. Which process forms new elements?
    A) superposition
    B) uniformitarianism
    C) permineralization
    D) radioactive decay

Science Online bookg.msscience.com/vocabulary_puzzlemaker

Use the ExamView® Pro Testmaker CD-ROM to:

- create multiple versions of tests
- create modified tests with one mouse click for inclusion students
- edit existing questions and add your own questions
- build tests aligned with state standards using built-in State Curriculum Tags
- change English tests to Spanish with one mouse click and vice versa

Thinking Critically

18. Explain why the fossil record of life on Earth is incomplete. Give some reasons why.

19. Infer Suppose a lava flow was found between two sedimentary rock layers. How could you use the lava flow to learn about the ages of the sedimentary rock layers? *(Hint: Most lava contains radioactive isotopes.)*

20. Infer Suppose you're correlating rock layers in the western United States. You find a layer of volcanic ash deposits. How can this layer help you in your correlation over a large area?

21. Recognize Cause and Effect Explain how some woolly mammoths could have been preserved intact in frozen ground. What conditions must have persisted since the deaths of these animals?

22. Classify each of the following fossils in the correct category in the table below: *dinosaur footprint, worm burrow, dinosaur skull, insect in amber, fossil woodpecker hole,* and *fish tooth.*

Types of Fossils

Trace Fossils	Body Fossils
dinosaur footprint	dinosaur skull
worm burrow	insect in amber
fossil woodpecker hole	fish tooth

23. Compare and contrast the three different kinds of unconformities. Draw sketches of each that illustrate the features that identify them.

24. Describe how relative and absolute ages differ. How might both be used to establish ages in a series of rock layers?

25. Discuss uniformitarianism in the following scenario. You find a shell on the beach, and a friend remembers seeing a similar fossil while hiking in the mountains. What does this suggest about the past environment of the mountain?

Performance Activities

26. Illustrate Create a model that allows you to explain how to establish the relative ages of rock layers.

27. Use a Classification System Start your own fossil collection. Label each find as to type, approximate age, and the place where it was found. Most state geological surveys can provide you with reference materials on local fossils.

Applying Math

28. Calculate how many half-lives have passed in a rock containing one-eighth the original radioactive material and seven-eighths of the daughter product.

Use the graphs below to answer question 29.

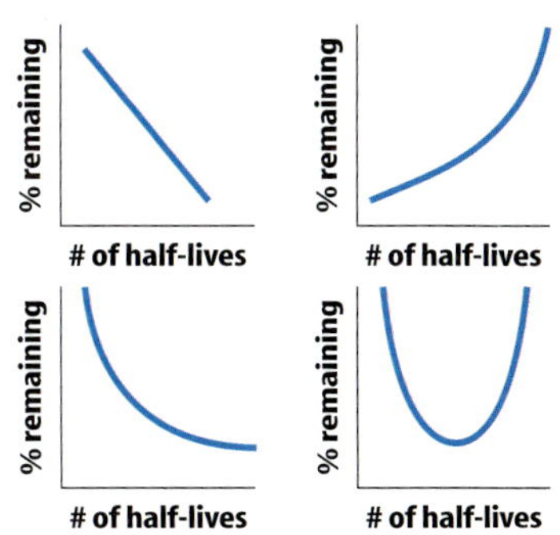

29. Interpret Data Which of the above curves best illustrates radioactive decay?

Assessment Resources

Reproducible Masters

Chapter *Fast File* Resources
Chapter Review, pp. 37–38
Chapter Tests, pp. 39–42
Assessment Transparency Activity, p. 49

Glencoe Science Web site
Chapter Review Test
Standardized Test Practice

Glencoe Technology
Assessment Transparency
Exam*View*® Pro Testmaker
MindJogger Videoquiz
Interactive Chalkboard

chapter Review 5

20. The layer of volcanic ash would have been deposited quickly during a volcanic eruption. Therefore, you have a marker rock layer that you know is the same age everywhere and can be dated by using radioactive isotopes.

21. The cold temperatures preserved the entire animal, much like freezing helps preserve foods.

22. See student page.

23. Sketches should show angular unconformities with parallel layers over tilted layers, disconformities with parallel layers over an eroded surface, and nonconformities with sedimentary over igneous or metamorphic rock.

24. Relative age is in relation to other rocks while absolute age gives an actual estimate. Any rocks aged absolutely can be used to determine relative ages of adjacent layers.

25. Answers will vary; students should recognize that the material that formed the mountain was once under water.

Performance Activities

26. Model should show layers, preferably disturbed by fault or unconformity. Student should then tell oldest to youngest and explain how it is possible to do so.

27. Fossil collections should indicate type of fossil formation, how old each is, and where each was found. Use **Performance Assessment in the Science Classroom**, p. 121.

Applying Math

National Math Standards
1, 2, 5, 6, 10

28. three

29. the lower left graph

Answer Sheet A practice answer sheet can be found at bookg.msscience.com/answer_sheet.

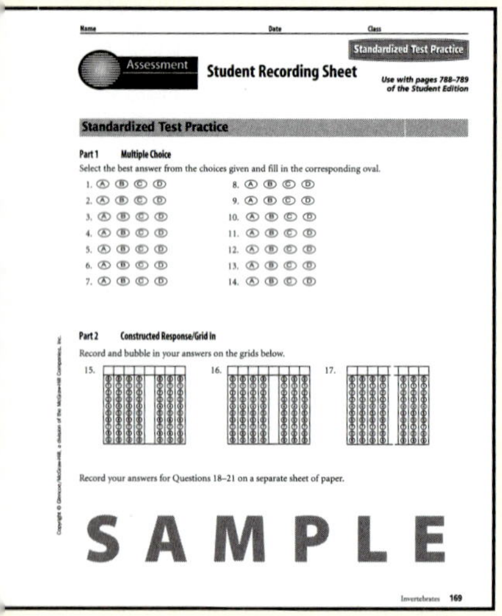
Name Date Class
Assessment Student Recording Sheet
Standardized Test Practice
Use with pages 788–789 of the Student Edition
Standardized Test Practice
Part 1 Multiple Choice
Select the best answer from the choices given and fill in the corresponding oval.
Part 2 Constructed Response/Grid In
Record and bubble in your answers on the grids below.
Record your answers for Questions 18–21 on a separate sheet of paper.
SAMPLE

Part 1 Multiple Choice

1. A 5. B 9. B
2. C 6. D 10. A
3. D 7. B 11. A
4. B 8. A

Part 2 Short Response

12. A fossil is the remains, imprints, or traces of prehistoric organisms.
13. A mold is an opening in rock that has the shape of a fossil. Molds form when a fossil, such as a shell, dissolves. A cast forms when that space fills with mineral cement or sediment.
14. The principle of uniformitarianism states that the processes that occur on Earth today are similar to those that occurred in the past.
15. The insect gets trapped in tree sap. The sap hardens with the insect inside and is incorporated into a sediment layer.

chapter 5 Standardized Test Practice

Part 1 Multiple Choice

Record your answers on the answer sheet provided by your teacher or on a sheet of paper.

Use the photo below to answer question 1.

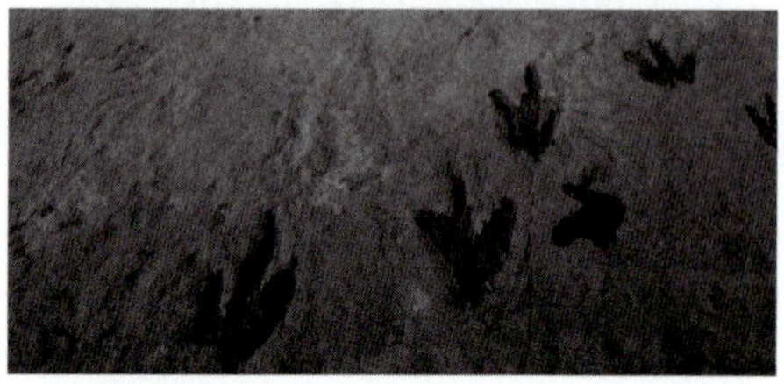

1. Which type of fossil preservation is shown above?
   A. trace fossil **C.** carbon film
   B. original remains **D.** permineralized remains

2. Which principle states that the oldest rock layer is found at the bottom in an undisturbed stack of rock layers?
   A. half-life **C.** superposition
   B. absolute dating **D.** uniformitarianism

3. Which type of scientist studies fossils?
   A. meteorologist **C.** astronomer
   B. chemist **D.** paleontologist

4. Which are the remains of species that existed on Earth for relatively short periods of time, were abundant, and were widespread geographically?
   A. trace fossils **C.** carbon films
   B. index fossils **D.** body fossils

5. Which term means matching up rock layers in different places?
   A. superposition **C.** uniformitarianism
   B. correlation **D.** absolute dating

6. Which of the following is least likely to be found as a fossil?
   A. clam shell **C.** snail shell
   B. shark tooth **D.** jellyfish imprint

7. Which type of fossil preservation is a thin carbon silhouette of the original organism?
   A. cast **C.** mold
   B. carbon film **D.** permineralized remains

8. Which isotope is useful for dating wood and charcoal that is less than about 75,000 years old?
   A. carbon-14 **C.** uranium-238
   B. potassium-40 **D.** argon-40

Use the diagram below to answer questions 9–11.

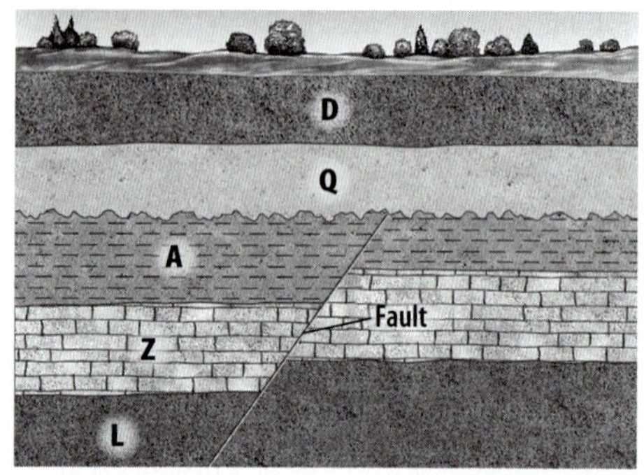

9. Which sequence of letters describes the rock layers in the diagram from oldest to youngest?
   A. D, Q, A, Z, L **C.** Z, L, A, D, Q
   B. L, Z, A, Q, D **D.** Q, D, L, Z, A

10. What does the wavy line between layers A and Q represent?
    A. a disconformity **C.** a nonconformity
    B. a fault **D.** an angular unconformity

11. Which of the following correctly describes the relative age of the fault?
    A. younger than A, but older than Q
    B. younger than Z, but older than L
    C. younger than Q, but older than A
    D. younger than D, but older than Q

Part 2 Short Response/Grid In

16. The evidence is mostly from radioactive dating. Some meteorites are thought to represent the age when the solar system formed. These have ages of about 4.5 billion years. Additional evidence that supports this age comes from the ratio of various lead isotopes on Earth.
17. The nuclei of some isotopes are unstable and break down into other isotopes and particles in a process called radioactive decay. The time it takes for half the atoms of an isotope to decay is called its half-life.
18. 50
19. 6.25 [note: because these are theoretical, they have infinite significant figures.]
20. The amount of parent isotope is reduced by half for each elapsed half-life.

Part 2 Short Response/Grid In

Record your answers on the answer sheet provided by your teacher or on a sheet of paper.

12. What is a fossil?

13. How is a fossil cast different from a fossil mold?

14. Describe the principle of uniformitarianism.

15. Explain how the original remains of an insect can be preserved as a fossil in amber.

16. Why do scientists hypothesize that Earth is about 4.5 billion years old?

17. Describe the process of radioactive decay. Use the terms *isotope, nucleus,* and *half-life* in your answer.

Use the table below to answer questions 18–20.

Number of Half-lives	Parent Isotope Remaining (%)
1	100
2	X
3	25
4	12.5
5	Y

18. What value should replace the letter X in the table above?

19. What value should replace the letter Y in the table above?

20. Explain the relationship between the number of half-lives that have elapsed and the amount of parent isotope remaining.

21. Compare and contrast the three types of unconformities.

22. Why are index fossils useful for estimating the age of rock layers?

Part 3 Open Ended

Record your answers on a sheet of paper.

23. Why are fossils important? What information do they provide?

24. List three different types of trace fossils. Explain how each type forms.

Examine the graph below and answer questions 25–27.

Relationship Between Sediment Burial Rate and Potential for Remains to Become Fossils

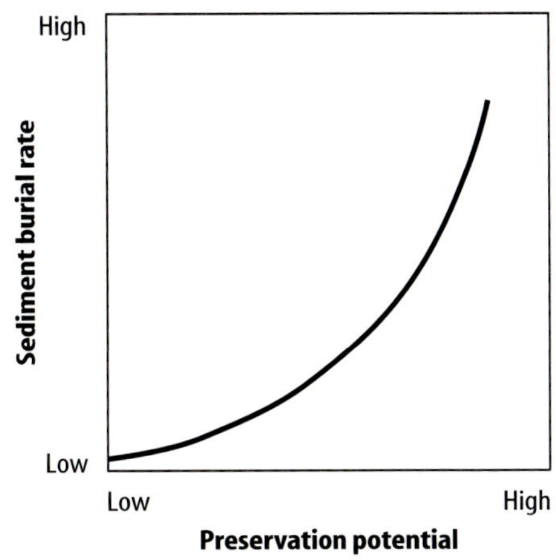

25. How does the potential for remains to be preserved change as the rate of burial by sediment increases?

26. Why do you think this relationship exists?

27. What other factors affect the potential for the remains of organisms to become fossils?

28. How could a fossil of an organism that lived in ocean water millions of years ago be found in the middle of North America?

Test-Taking Tip

Check It Again Double check your answers before turning in the test.

Science Online bookg.msscience.com/standardized_test

24. Students might pick one of the following: burrow, forms when an animal burrows into sediment; track, forms when an animal steps over a sediment substrate; resting trace, forms when an organism rests on a sediment substrate; trail, forms as an organism slides over a sediment substrate; other answers also might be correct.

25. The preservation potential increases as the burial rate increases.

26. Burial by sediment slows decay and protects remains from scavengers, wave action, and other destructive processes.

27. Students should answer that the presence of hard parts increases the likelihood of fossilization. Low oxygen, marine environments also favor fossilization because aerobic decay cannot occur.

28. In the geologic past, shallow oceans covered much of North America.

Rubrics

For more help evaluating open-ended assessment questions, see the rubric on p. 10T.

21. All form because of erosion or nondeposition. Angular unconformities separate rock layers with different orientations. Disconformities separate parallel rock layers. Nonconformities separate igneous or metamorphic rock from sedimentary layers.

22. The species only lived for a short time. Therefore, any rock containing a fossil of that species formed during that short time range.

Part 3 Open Ended

23. Fossils are important because they provide information about past life on Earth. Fossils also provide information about the evolution of organisms, past environments, past climates, among others.

chapter 6 Organizer

Section/Objectives	Standards		Labs/Features
	National	State/Local	
Chapter Opener	See pp. 9T–10T for a Key to Standards.		**Launch Lab:** Survival Through Time, p. 153 **Foldables,** p. 153
Section 1 Life and Geologic Time 2 sessions 1 block 1. **Explain** how geologic time can be divided into units. 2. **Relate** changes of Earth's organisms to divisions on the geologic time scale. 3. **Describe** how plate tectonics affects species.	National Content Standards: UCP.1, UCP.2, UCP.3, UCP.4, UCP.5, A.1, A.2, D.1		
Section 2 Early Earth History 3 sessions 1 block 4. **Identify** characteristic Precambrian and Paleozoic life-forms. 5. **Draw** conclusions about how species adapted to changing environments in Precambrian time and the Paleozoic Era. 6. **Describe** changes in Earth and its life-forms at the end of the Paleozoic Era.	National Content Standards: UCP.1, UCP.2, UCP.3, UCP.4, UCP.5, A.1, A.2, D.1		**Integrate Chemistry,** p. 163 **MiniLAB:** Dating Rock Layers with Fossils, p. 164 **Visualizing Unusual Life Forms,** p. 165 **Science Online,** p. 166 **Lab:** Changing Species, p. 169
Section 3 Middle and Recent Earth History 3 sessions 2 blocks 7. **Compare** and contrast characteristic life-forms in the Mesozoic and Cenozoic Eras. 8. **Explain** how changes caused by plate tectonics affected organisms during the Mesozoic Era. 9. **Identify** when humans first appeared on Earth.	National Content Standards: UCP.1, UCP.2, UCP.3, UCP.5, A.1, A.2, D.1		**Science Online,** p. 171 **Applying Math:** Calculating Extinction Using Percentages, p. 173 **MiniLAB:** Calculating the Age of the Atlantic Ocean, p. 174 **Lab:** Discovering the Past, p. 176 **Science Stats:** Extinct!, p. 178

Glencoe Exclusive! TeacherWorks™ All-In-One Planner and Resource Center

Lab Materials	Reproducible Resources	Section Assessment	Technology
Launch Lab: green, orange, and blue yarn; scissors; green construction paper; tweezers; scissors; stopwatch	**Chapter *Fast File* Resources** Foldables Worksheet, p. 15 Directed Reading Overview, p. 17 Note-taking Worksheets, pp. 31–33	Glencoe's Assessment Advantage	**TeacherWorks** includes: • Interactive Teacher Edition • Lesson Planner with calendar • Access to all program blacklines • Correlations to standards • Web links
Need materials? Contact Science Kit at 1-800-828-7777 or www.sciencekit.com on the Internet.	**Chapter *Fast File* Resources** Transparency Activity, p. 42 Enrichment, p. 28 Reinforcement, p. 25 Directed Reading, p. 18 Lab Activity, pp. 9–10 Transparency Activity, pp. 45–46	**Portfolio** Science Journal, p. 157 **Performance** Applying Skills, p. 161 **Content** Section Review, p. 161	Section Focus Transparency Teaching Transparency Virtual Labs CD-ROM Guided Reading Audio Program Interactive Chalkboard CD-ROM
MiniLAB: paper, pencil **Lab:** deck of playing cards	**Chapter *Fast File* Resources** Transparency Activity, p. 43 MiniLAB, p. 3 Enrichment, p. 29 Reinforcement, p. 26 Directed Reading, p. 18 Lab Worksheets, pp. 5–6 **Reading and Writing Skill Activities,** p. 27 **Science Inquiry Labs,** pp. 21–22 **Cultural Diversity,** p. 31	**Portfolio** Integrate Chemistry, p. 163 **Performance** MiniLAB, p. 164 Applying Skills, p. 168 **Content** Section Review, p. 168	Section Focus Transparency Virtual Labs CD-ROM Guided Reading Audio Program Interactive Chalkboard CD-ROM Video Lab
MiniLAB: world map or globe, metric ruler **Activity:** Internet and other resources on Earth's history	**Chapter *Fast File* Resources** Transparency Activity, p. 44 MiniLAB, p. 4 Enrichment, p. 30 Reinforcement, p. 27 Directed Reading, pp. 19, 20 Lab Worksheets, pp. 7–8 Lab Activity, pp. 11–14 **Mathematics Skill Activities,** p. 5 **Lab Management and Safety,** p. 73	**Performance** Applying Math, p. 173 MiniLAB, p. 174 Applying Math, p. 175 **Content** Section Review, p. 175	Section Focus Transparency Virtual Labs CD-ROM Guided Reading Audio Program Interactive Chalkboard CD-ROM

End of Chapter Assessment

Blackline Masters	Technology	Professional Series
Chapter *Fast File* Resources Chapter Review, pp. 35–36 Chapter Tests, pp. 37–40 **Standardized Test Practice,** pp. 27–30	MindJogger Videoquiz Virtual Labs CD-ROM Exam*View*® Pro Testmaker TeacherWorks CD-ROM Interactive Chalkboard CD-ROM	**Performance Assessment in the Science Classroom (PASC)**

Transparencies

Section Focus

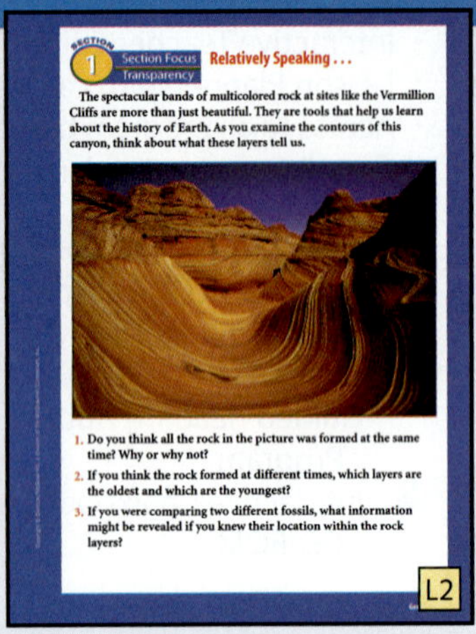
Section 1 Section Focus Transparency — Relatively Speaking . . .

The spectacular bands of multicolored rock at sites like the Vermillion Cliffs are more than just beautiful. They are tools that help us learn about the history of Earth. As you examine the contours of this canyon, think about what these layers tell us.

1. Do you think all the rock in the picture was formed at the same time? Why or why not?
2. If you think the rock formed at different times, which layers are the oldest and which are the youngest?
3. If you were comparing two different fossils, what information might be revealed if you knew their location within the rock layers?

L2

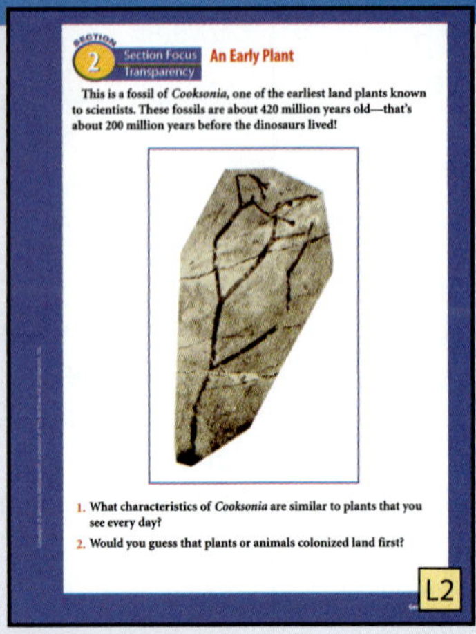
Section 2 Section Focus Transparency — An Early Plant

This is a fossil of *Cooksonia*, one of the earliest land plants known to scientists. These fossils are about 420 million years old—that's about 200 million years before the dinosaurs lived!

1. What characteristics of *Cooksonia* are similar to plants that you see every day?
2. Would you guess that plants or animals colonized land first?

L2

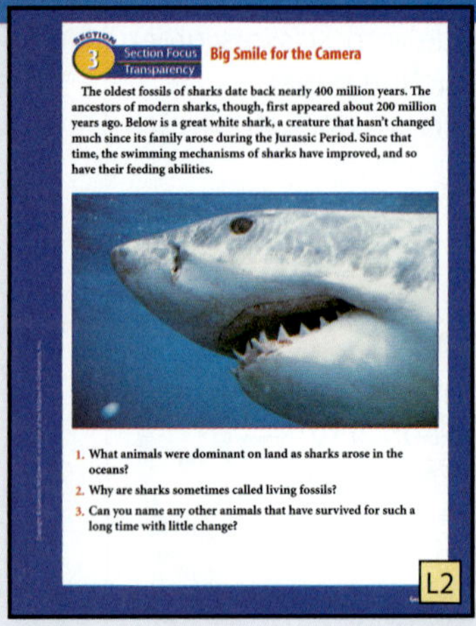
Section 3 Section Focus Transparency — Big Smile for the Camera

The oldest fossils of sharks date back nearly 400 million years. The ancestors of modern sharks, though, first appeared about 200 million years ago. Below is a great white shark, a creature that hasn't changed much since its family arose during the Jurassic Period. Since that time, the swimming mechanisms of sharks have improved, and so have their feeding abilities.

1. What animals were dominant on land as sharks arose in the oceans?
2. Why are sharks sometimes called living fossils?
3. Can you name any other animals that have survived for such a long time with little change?

L2

This is a representation of key blackline masters available in the Teacher Classroom Resources. See Resource Manager boxes within the chapter for additional information.

Assessment

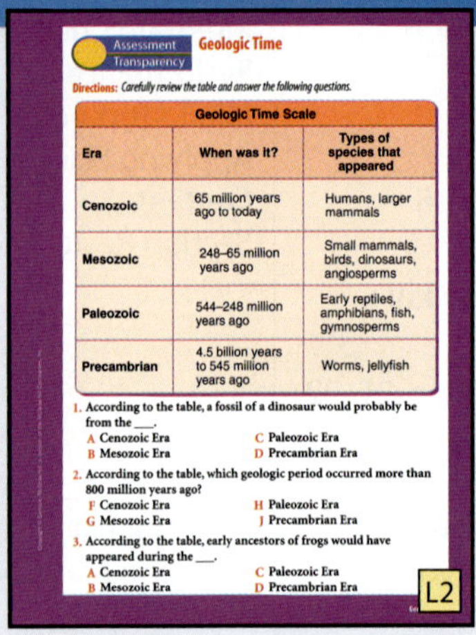
Assessment Transparency — Geologic Time

Directions: Carefully review the table and answer the following questions.

Geologic Time Scale

Era	When was it?	Types of species that appeared
Cenozoic	65 million years ago to today	Humans, larger mammals
Mesozoic	248–65 million years ago	Small mammals, birds, dinosaurs, angiosperms
Paleozoic	544–248 million years ago	Early reptiles, amphibians, fish, gymnosperms
Precambrian	4.5 billion years to 545 million years ago	Worms, jellyfish

1. According to the table, a fossil of a dinosaur would probably be from the ___.
A Cenozoic Era
B Mesozoic Era
C Paleozoic Era
D Precambrian Era
2. According to the table, which geologic period occurred more than 800 million years ago?
F Cenozoic Era
G Mesozoic Era
H Paleozoic Era
J Precambrian Era
3. According to the table, early ancestors of frogs would have appeared during the ___.
A Cenozoic Era
B Mesozoic Era
C Paleozoic Era
D Precambrian Era

L2

Teaching

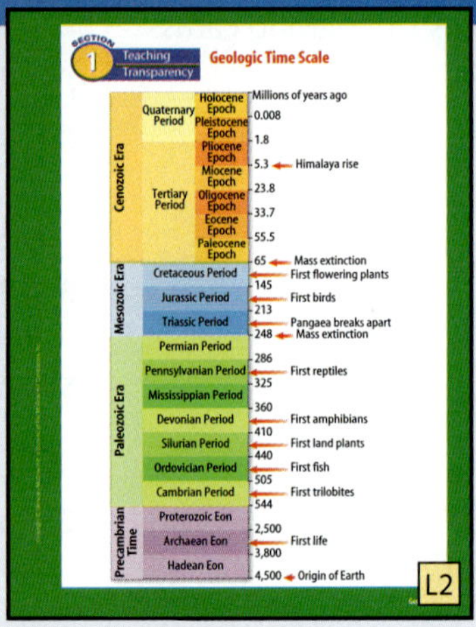
Section 1 Teaching Transparency — Geologic Time Scale

L2

Key to Teaching Strategies

The following designations will help you decide which activities are appropriate for your students.

L1 Level 1 activities should be appropriate for students with learning difficulties.

L2 Level 2 activities should be within the ability range of all students.

L3 Level 3 activities are designed for above-average students.

ELL ELL activities should be within the ability range of English Language Learners.

COOP LEARN Cooperative Learning activities are designed for small group work.

LS Multiple Learning Styles logos, as described on page 6T, are used throughout to indicate strategies that address different learning styles.

P These strategies represent student products that can be placed into a best-work portfolio.

PBL Problem-Based Learning activities apply real-world situations to learning.

Hands-on Activities

Student Text Lab Worksheet

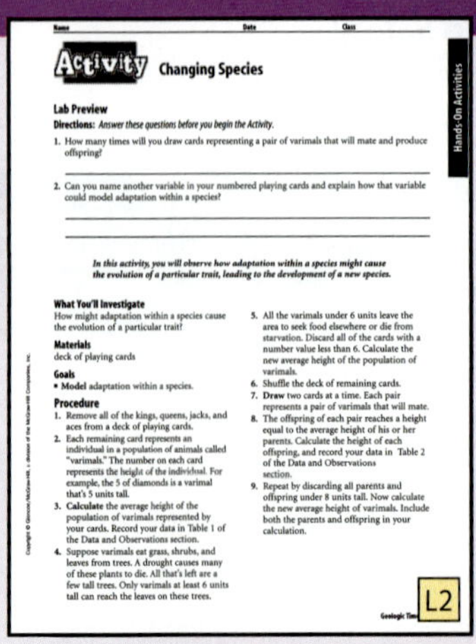
Activity — Changing Species

L2

Laboratory Activities

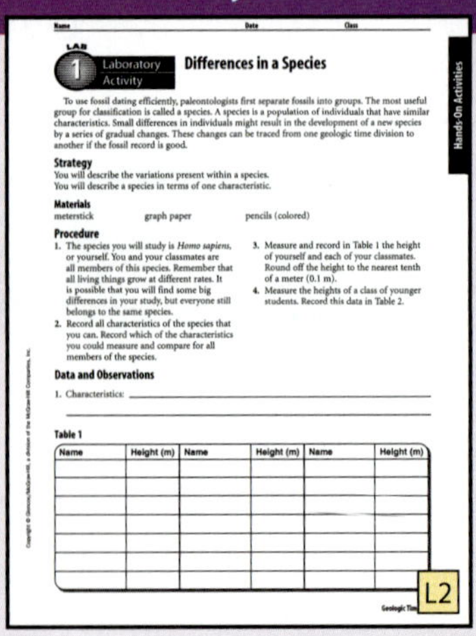
Laboratory Activity 1 — Differences in a Species

L2

Resource Manager

Meeting Different Ability Levels

Content Outline

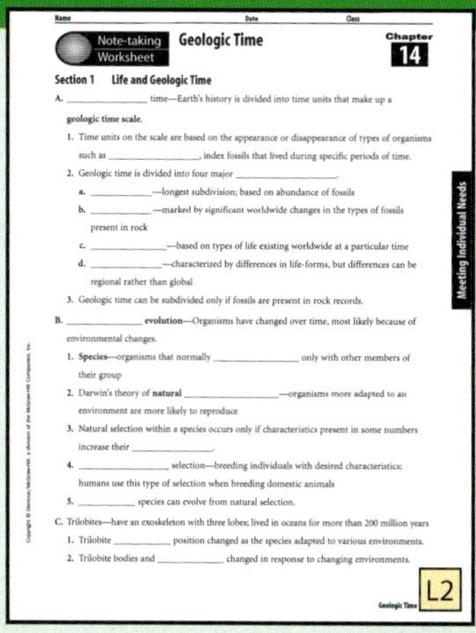

Reinforcement

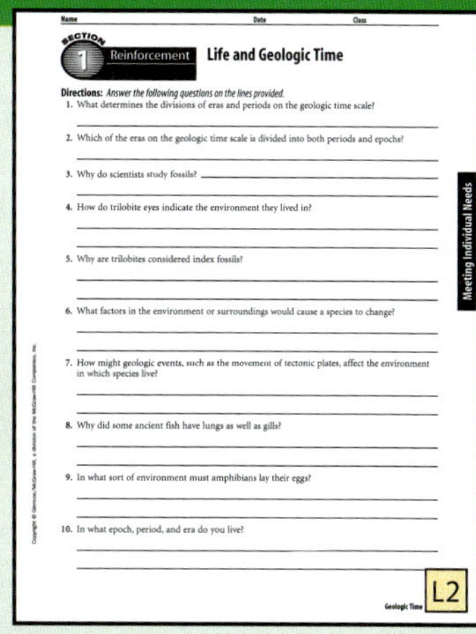

Enrichment

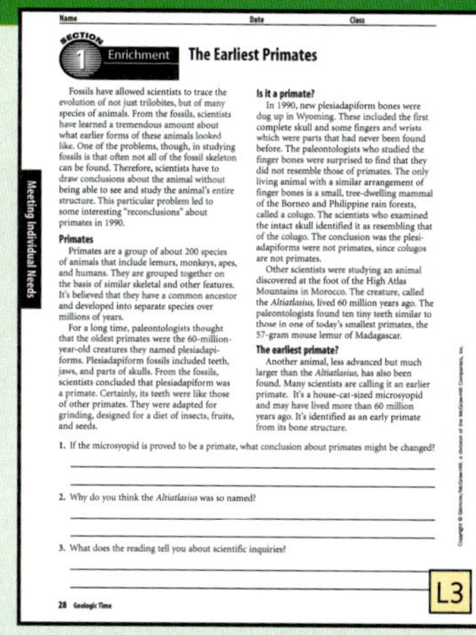

Directed Reading (English/Spanish)

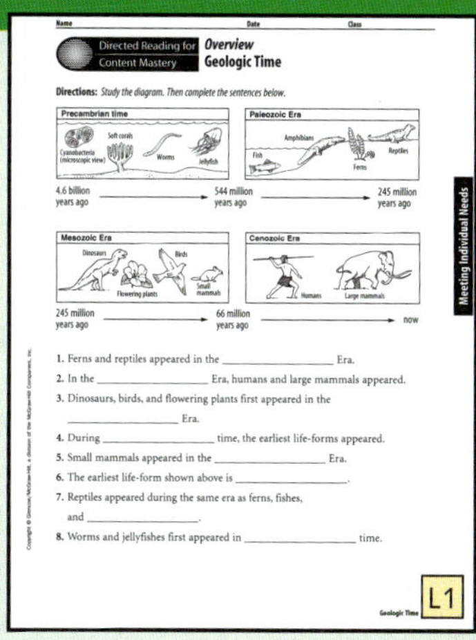

Study Guide

Reading Essentials

Assessment

Test Practice Workbook

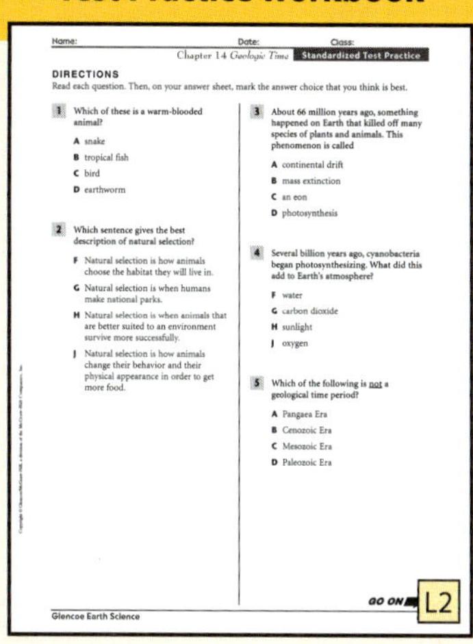

Chapter Review

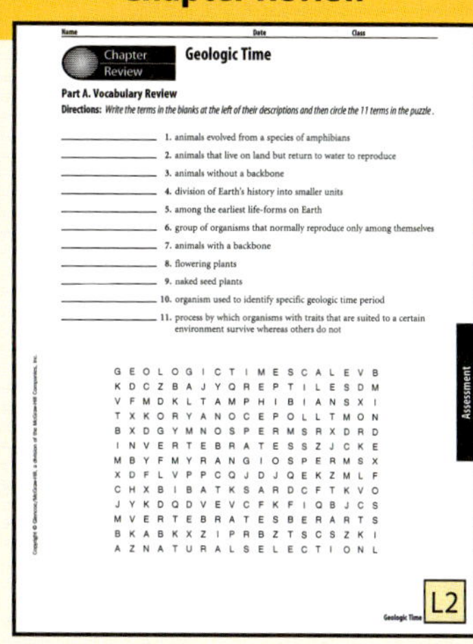

Chapter Tests

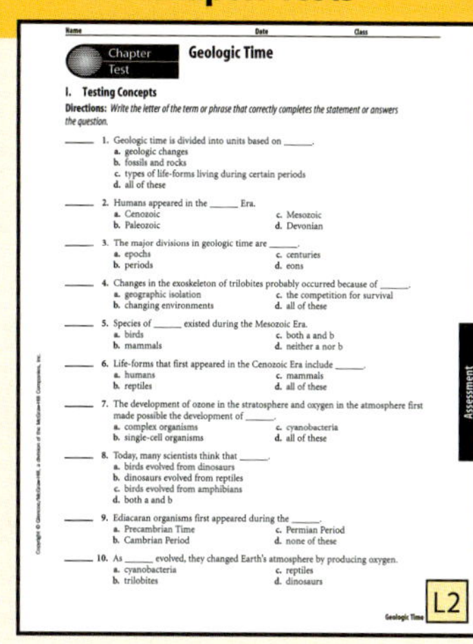

Science Content Background

section 1 Life and Geologic Time

Understanding Earth's Past

Although the geologic time scale was originally based on relative dating, absolute methods such as radiometric dating have enabled geologists to determine absolute dates of many rocks and some geologic events. Some scientists believe that supercontinents such as Pangaea have existed multiple times during Earth's history.

section 2 Early Earth History

The End of the Paleozoic Era

The largest of all mass extinctions occurred at the end of the Paleozoic Era. Some evidence suggests that a huge volcanic eruption occurred in what is now Siberia. The eruption would have released large concentrations of poisonous gases and dust into the atmosphere. The release of these gases and dust, along with global changes in climates associated with the eruption, could have been the cause of the mass extinction. Other evidence suggests the cause may have been a global turnover of ocean water. The upwelling of deep ocean water would have brought large amounts of carbon dioxide to the surface, some of which would have been released into the atmosphere. The excess carbon dioxide can lead to the death of many organisms. Carbon dioxide is also a greenhouse gas. Added to the atmosphere in sufficiently high quantities, the gas could have led to a global rise in temperatures and subsequent mass extinctions.

section 3 Middle and Recent Earth History

Separating Middle and Recent Earth History

The K-T (Cretaceous/Tertiary) boundary is the point in the geologic time scale that separates the Mesozoic Era from the Cenozoic Era. This boundary contains evidence of the factors that may have caused the extinction of the dinosaurs. An iridium-rich clay layer found in rocks of this age led to the hypothesis that a collision with an asteroid was the final blow to the already dying dinosaur population.

The Development of Humans

Humans and their direct ancestors make up a group of animals called hominids. Some early hominids were of the genus *Australopithecus*. These hominids were bipedal and had rounded jaws. They lived in the open grasslands of what is now eastern and southern Africa. Australopithecines were smaller than modern humans. Standing less than 1.5 m tall, they had a mass of about 20 kg. Their brains were half the size of present-day humans' brains. Fossil evidence indicates that these early human ancestors may have used simple tools made of bone.

chapter content resources

Internet Resources

For additional content background, visit bookg.msscience.com to:

- access your book online
- find references to related articles in popular science magazines
- access Web links with related content background
- access current events with science journal topics

Print Resources

Dinosaur Encyclopedia: From Dinosaurs to Dawn of Man, edited by Kitty Blount and Maggie Crowley, Darling Kindersley Limited, 2001

Life: A Natural History of the First Four Billion Years of Life on Earth, Richard Fortey, Random House, 1999

IDENTIFYING Misconceptions

Find Out What Students Think

Students may think that . . .

Time is measured only with a watch in minutes and seconds.

Students probably have had few experiences with the concept of geologic time. The vast amount of time represented by the geologic time scale is beyond the grasp of many students and adults alike. To make the geologic time scale more manageable, it has been divided into eras based on the fossil record and major events in Earth's history.

Activity

Access the *Geological Time Scale Metaphor Calculator* at bookg.msscience.com. Have small groups of students work together to develop their own metaphors for understanding geologic time. For example, one group of students might choose their average age to represent all of geologic time. They then could determine how old they would have been when dinosaurs lived on Earth or when human history started. Have students discuss with the class the different metaphors they develop.

Promote Understanding

Activity

Provide each group of three or four students with a meterstick, a calculator, several small self-stick notes, and a copy of the geologic time scale. If copies of time scales are not available, refer students to **Figure 2** in Section 1.

- Explain that each group will be making a geologic time scale on the meterstick. They should imagine that the meterstick represents all of geologic time.
- Review how to do percentages. Help students determine what percent of the meterstick should represent Precambrian time using this formula: 4,056 million years ÷ 4,500 million years $= x \times 1{,}000$ mm; $x = 901$ mm. Direct students to place a self-stick note at this point on the meterstick and label it "Precambrian Time Ends."
- Have them repeat this process for each era.
- Direct students to estimate where major animal groups developed and when major geologic events, such as mass extinctions and changes in continental position or shape, occurred. Encourage each group to label at least three of these events on their metersticks.
- When groups have completed their tasks, have a representative from each group present their time scale to the class. As each group finishes its presentation, combine its data with those of others to make a class geologic time scale. Display this scale throughout the study of the chapter so that students have a concrete reference for understanding the vast length of Precambrian time compared with the other eras.

Assess

After completing the chapter, see *Identifying Misconceptions* in the Study Guide at the end of the chapter.

chapter 6

Chapter Vocabulary

geologic time scale, p. 154
eon, p. 155
era, p. 155
period, p. 155
epoch, p. 155
organic evolution, p. 156
species, p. 156
natural selection, p. 157
trilobite, p. 159
Pangaea, p. 161
Precambrian time, p. 162
cyanobacteria, p. 163
Paleozoic Era, p. 164
Mesozoic Era, p. 170
Cenozoic Era, p. 174

Science Journal Answers will vary. Animals and plants living far north will probably shift south. Animals with limited mobility may become extinct. Accept answers that show careful thought.

INTERACTIVE CHALKBOARD with Image Bank

PowerPoint® Presentations

This CD-ROM is an editable Microsoft® PowerPoint® presentation that includes:

- a pre-made presentation for every chapter
- interactive graphics
- animations
- audio clips
- image bank
- all new section and chapter questions
- Standardized Test Practice
- transparencies
- pre-lab questions for all labs
- Foldables directions
- links to bookg.msscience.com

chapter 6

Geologic Time

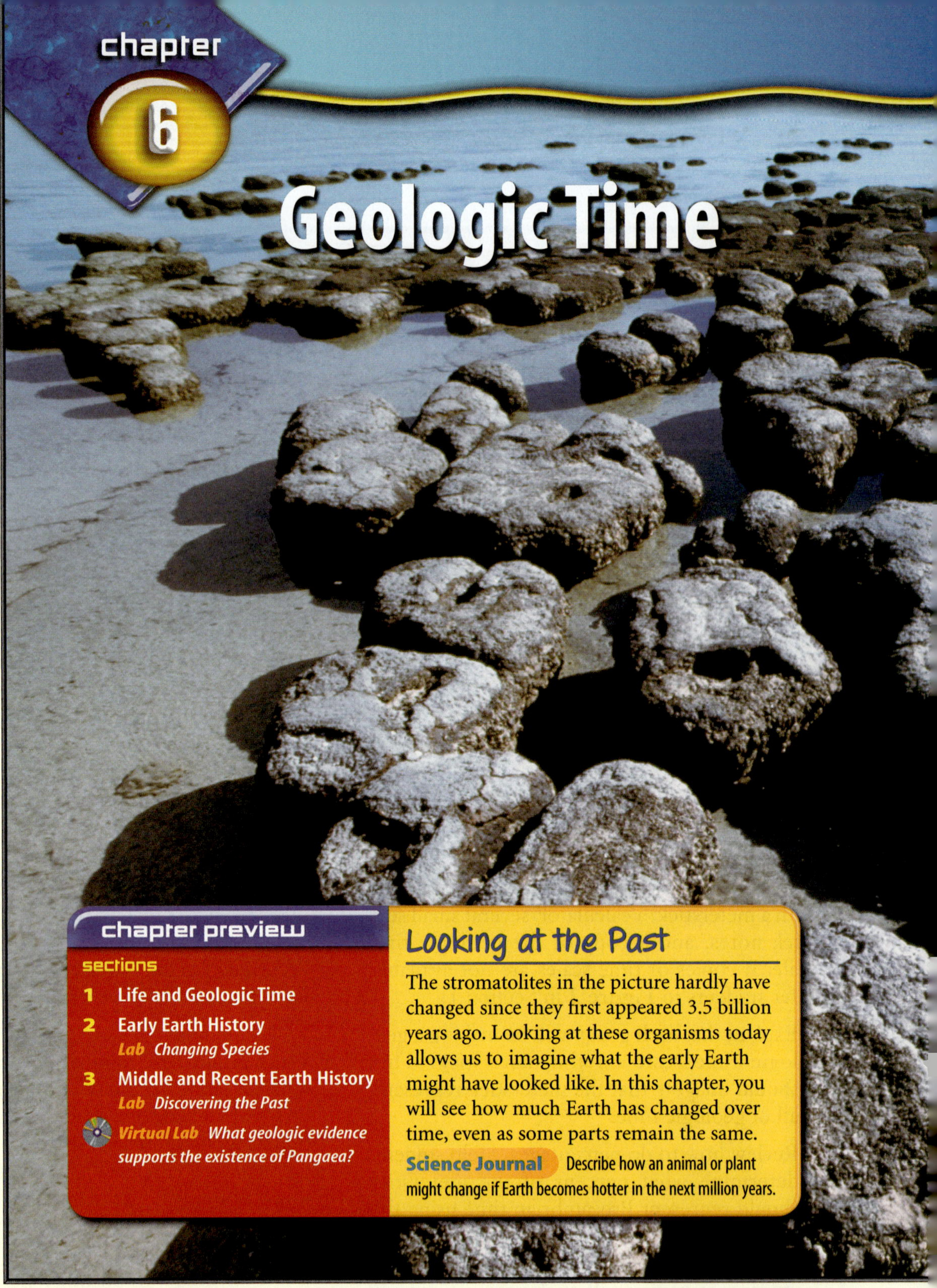

chapter preview

sections

1 Life and Geologic Time
2 Early Earth History
Lab *Changing Species*
3 Middle and Recent Earth History
Lab *Discovering the Past*
Virtual Lab *What geologic evidence supports the existence of Pangaea?*

Looking at the Past

The stromatolites in the picture hardly have changed since they first appeared 3.5 billion years ago. Looking at these organisms today allows us to imagine what the early Earth might have looked like. In this chapter, you will see how much Earth has changed over time, even as some parts remain the same.

Science Journal Describe how an animal or plant might change if Earth becomes hotter in the next million years.

Theme Connection

Stability and Change Overall, Earth's surface and the organisms that live on it may appear unchanging during the short period of time that makes up a human lifetime. However, over the millions or even billions of years that make up the divisions in the geologic time scale, major changes occur in the landscape and in species.

About the Photo

Stromatolites Some of the earliest evidence of life on Earth is fossil stromatolites. They were mats of cyanobacteria that used photosynthesis to convert sunlight into food and generated oxygen as a byproduct. After the appearance of life, oxygen may be the greatest change in Earth's history. Earth has changed dramatically since those times, but stromatolites can still be found.

Start-Up Activities

Launch LAB

Survival Through Time

Environments include the living and nonliving things that surround and affect organisms. Whether or not an organism survives in its environment depends upon its characteristics. Only if an organism survives until adulthood can it reproduce and pass on its characteristics to its offspring. In this lab, you will use a model to find out how one characteristic can determine whether individuals can survive in an environment.

1. Cut 15 pieces each of green, orange, and blue yarn into 3-cm lengths.
2. Scatter them on a sheet of green construction paper.
3. Have your partner use a pair of tweezers to pick up as many pieces as possible in 15 s.
4. **Think Critically** In your Science Journal, discuss which colors your partner selected. Which color was least selected? Suppose that the construction paper represents grass, the yarn pieces represent insects, and the tweezers represent an insect-eating bird. Which color of insect do you predict would survive to adulthood?

FOLDABLES Study Organizer

Geological Time Make the following Foldable to help you identify the major events in each era of geologic time.

STEP 1 **Fold** the top of a vertical piece of paper down and the bottom up to divide the paper into thirds.

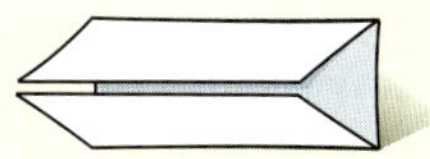

STEP 2 **Turn** the paper horizontally; **unfold and label** the three columns as shown.

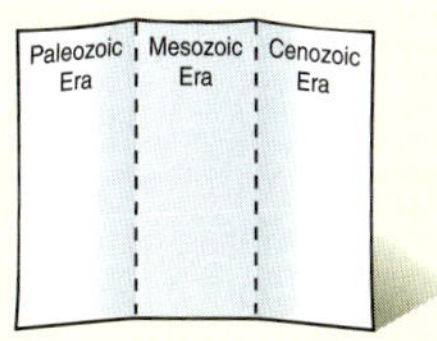

Read for Main Ideas As you read the chapter, list at least three major events that occurred in each era. Keep the events in chronological order. For each event, note the period in which it took place.

Science Online — Preview this chapter's content and activities at bookg.msscience.com

Launch LAB

Purpose Use the Launch Lab to introduce students to the concept of natural selection. L1
ELL COOP LEARN LS
Kinesthetic

Preparation To save time, cut the yarn into 3-cm lengths before class.

Materials 15 pieces each of green, orange, and blue yarn cut into 3-cm lengths; sheet of green construction paper; stopwatch; scissors; tweezers

Teaching Strategy For the lab to work well, match the color of the green yarn to the color of the green construction paper.

Safety Precautions Caution students to handle the scissors and tweezers with care.

Think Critically

Students will probably pick up more orange and blue pieces because they are more easily seen. Green was probably least selected. A green insect would blend in better with the background and have a better chance of going unnoticed by predators.

Assessment

Process Have students find photos or illustrations of animals whose coloration provides camouflage that helps them survive in their environment. Use **Performance Assessment in the Science Classroom,** p. 89.

FOLDABLES Study Organizer — Dinah Zike Study Fold

Student preparation materials for this Foldable are available in the Chapter *Fast File* Resources.

1 Motivate

INTERACTIVE CHALKBOARD
PowerPoint® Presentations

Bellringer

Section Focus Transparencies also are available on the Interactive Chalkboard CD-ROM. L2 ELL

SECTION 1 Section Focus Transparency **Relatively Speaking . . .**

The spectacular bands of multicolored rock at sites like the Vermillion Cliffs are more than just beautiful. They are tools that help us learn about the history of Earth. As you examine the contours of this canyon, think about what these layers tell us.

1. Do you think all the rock in the picture was formed at the same time? Why or why not?
2. If you think the rock formed at different times, which layers are the oldest and which are the youngest?
3. If you were comparing two different fossils, what information might be revealed if you knew their location within the rock layers?

L2

Geologic Time

Tie to Prior Knowledge

Earth's History Are any dinosaurs living today? How long ago do you think dinosaurs disappeared? Have students brainstorm answers, then explain that they disappeared about 65 million years ago, long before there were any people on Earth. Note that although this is a long period of time, it is only a fraction of Earth's history. Tell students they will learn more about the major events in Earth's history in this section.

Text Question Answer

yes

section 1

Life and Geologic Time

as you read

What You'll Learn

- **Explain** how geologic time can be divided into units.
- **Relate** changes of Earth's organisms to divisions on the geologic time scale.
- **Describe** how plate tectonics affects species.

Why It's Important

The life and landscape around you are the products of change through geologic time.

Review Vocabulary
fossils: remains, traces, or imprints of prehistoric organisms

New Vocabulary
- geologic time scale
- eon
- era
- period
- epoch
- organic evolution
- species
- natural selection
- trilobite
- Pangaea

Geologic Time

A group of students is searching for fossils. By looking in rocks that are hundreds of millions of years old, they hope to find many examples of trilobites (TRI loh bites) so that they can help piece together a puzzle. That puzzle is to find out what caused the extinction of these organisms. **Figure 1** shows some examples of what they are finding. The fossils are small, and their bodies are divided into segments. Some of them seem to have eyes. Could these interesting fossils be trilobites?

Trilobites are small, hard-shelled organisms that crawled on the seafloor and sometimes swam through the water. Most ranged in size from 2 cm to 7 cm in length and from 1 cm to 3 cm in width. They are considered to be index fossils because they lived over vast regions of the world during specific periods of geologic time.

The Geologic Time Scale The appearance or disappearance of types of organisms throughout Earth's history marks important occurrences in geologic time. Paleontologists have been able to divide Earth's history into time units based on the life-forms that lived only during certain periods. This division of Earth's history makes up the **geologic time scale**. However, sometimes fossils are not present, so certain divisions of the geologic time scale are based on other criteria.

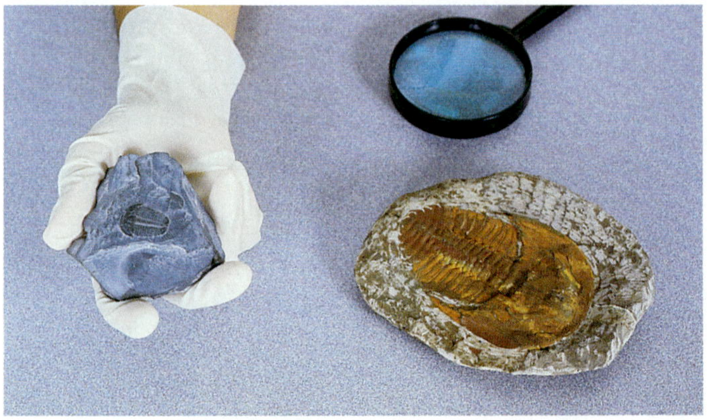

Figure 1 Many sedimentary rocks in the United States are rich in invertebrate fossils such as these trilobites.

Section 1 Resource Manager

Chapter *FAST FILE* Resources
- Transparency Activity, pp. 42, 45–46
- Directed Reading for Content Mastery, pp. 17, 18
- Note-taking Worksheets, pp. 31–33
- Enrichment, p. 28
- Lab Activity, pp. 9–10
- Reinforcement, p. 25

Earth Science Critical Thinking/Problem Solving, pp. 1, 14, 18

Major Subdivisions of Geologic Time The oldest rocks on Earth contain no fossils. Then, for many millions of years after the first appearance of fossils, the fossil record remained sparse. Later in Earth's history came an explosion in the abundance and diversity of organisms. These organisms left a rich fossil record. As shown in **Figure 2,** four major subdivisions of geologic time are used—eons, eras, periods, and epochs. The longest subdivisions—**eons**—are based upon the abundance of certain fossils.

Reading Check *What are the major subdivisions of geologic time?*

Next to eons, the longest subdivisions are the **eras,** which are marked by major, striking, and worldwide changes in the types of fossils present. For example, at the end of the Mesozoic Era, many kinds of invertebrates, birds, mammals, and reptiles became extinct.

Eras are subdivided into periods. **Periods** are units of geologic time characterized by the types of life existing worldwide at the time. Periods can be divided into smaller units of time called **epochs.** Epochs also are characterized by differences in life-forms, but some of these differences can vary from continent to continent. Epochs of periods in the Cenozoic Era have been given specific names. Epochs of other periods usually are referred to simply as early, middle, or late. Epochs are further subdivided into units of shorter duration.

Dividing Geologic Time There is a limit to how finely geologic time can be subdivided. It depends upon the kind of rock record that is being studied. Sometimes it is possible to distinguish layers of rock that formed during a single year or season. In other cases, thick stacks of rock that have no fossils provide little information that could help in subdividing geologic time.

Figure 2 Scientists have divided the geologic time scale into subunits based upon the appearance and disappearance of types of organisms.
Explain *how the even blocks in this chart can be misleading.*

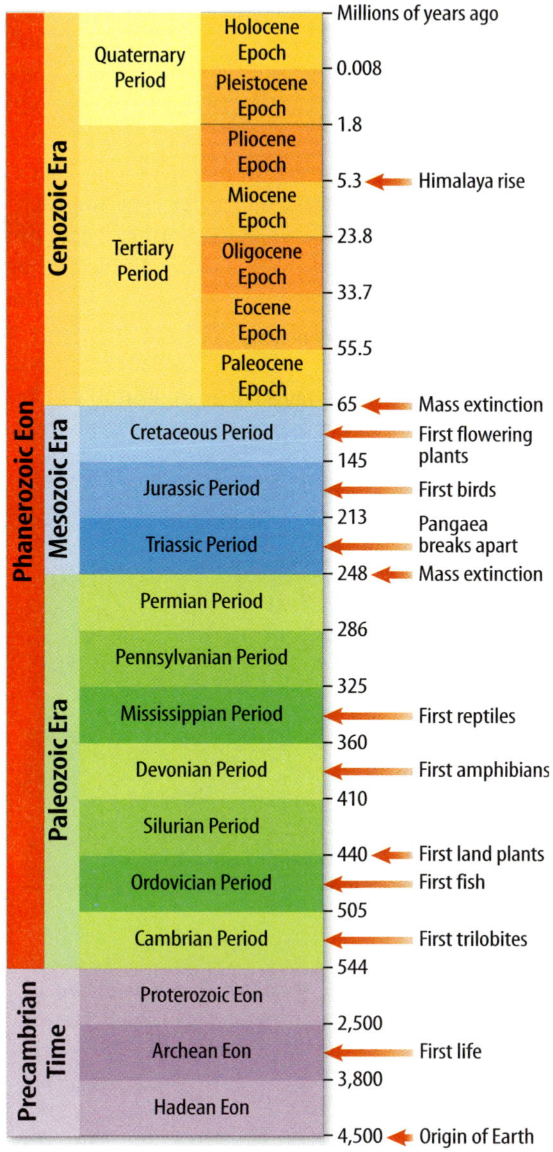

2 Teach

Visual Learning

Figure 2 What major event in Earth's history happened in the era in which we now live? What is the name of that era? Himalaya rise; Cenozoic Era

Caption Answer

Figure 2 The length of time for each box is not constant.

Reading Check

Answer eons, eras, periods, epochs

Use an Analogy

Eras, Periods, and Epochs Tell students that just as there are different periods in their lives, there are different periods in Earth's history. For example, eras in geologic time are marked by major worldwide changes in the kinds or abundance of organisms. Students might mark eras of their lives according to major events such as the birth of a sibling or changing schools. Have students discuss how they would divide their own lives into eras, periods, and epochs. L2

IDENTIFYING Misconceptions

Time Students may think that time is measured only with a watch in minutes and seconds. Refer to page F at the beginning of this chapter for teaching strategies that address this misconception.

Curriculum Connection

Geography Periods on the geologic time scale are often named for regions where the rocks of that age were first studied or are widely exposed. Have students use inference to answer these questions. Where were rocks of the Pennsylvanian Period first studied or widely exposed? Pennsylvania Which period represents rocks first studied in the Jura Mountains between France and Switzerland? Jurassic Period L2

IDENTIFYING Misconceptions

Incomplete History Students may think that scientists have constructed a complete record of Earth's past from fossils contained in rocks. Explain that very little is known about a large part of Earth's history—the first 4 billion years—because much of the fossil record has been destroyed or was not preserved. Because of erosion or nondeposition, gaps also exist in the most recent part of the fossil record as well.

Use Science Words

Word Usage Reinforce the differences between the terms *hypothesis* and *theory*. Explain that a theory is more than a hypothesis; it is an accepted (by some) explanation based on a large number of tests. Have students use each word in a sentence. Possible answer: After a hypothesis is successfully tested over many years, it may become a theory. L2

Caption Answer

Figure 3 If the lizards cannot interbreed and produce fertile offspring, they are separate species.

Organic Evolution

The fossil record shows that species have changed over geologic time. This change through time is known as **organic evolution.** According to most theories about organic evolution, environmental changes can affect an organism's survival. Those organisms that are not adapted to changes are less likely to survive or reproduce. Over time, the elimination of individuals that are not adapted can cause changes to species of organisms.

INTEGRATE Life Science **Species** Many ways of defining the term species (SPEE sheez) have been proposed. Life scientists often define a **species** as a group of organisms that normally reproduces only with other members of their group. For example, dogs are a species because dogs mate and reproduce only with other dogs. In some rare cases, members of two different species, such as lions and tigers, can mate and produce offspring. These offspring, however, are usually sterile and cannot produce offspring of their own. Even though two organisms look nearly alike, if the populations they each come from do not interbreed naturally and produce offspring that can reproduce, the two individuals do not belong to the same species. **Figure 3** shows an example of two species that look similar to each other but live in different areas and do not mate naturally with each other.

Figure 3 Just because two organisms look alike does not mean that they belong to the same species.
Describe *an experiment to test if these lizards are separate species.*

The coast horned lizard lives along the coast of central and southern California.

The desert horned lizard lives in arid regions of the southwestern United States.

Differentiated Instruction

Challenge Have students investigate Lamarck's ideas about acquired characteristics and describe how he explained change within a species. Then have students list the inadequacies of Lamarck's idea. Lamarck's hypothesis was that species evolve by keeping traits that their parents develop during their lives or losing traits that were not used. It's incorrect to think a trait acquired by a parent during its lifetime would be passed on. There is no identified mechanism by which an acquired trait can be passed on. There is also no evidence to back up Lamarck's assumption. For example, if a woman lifts weights, her children won't necessarily be stronger than children of women who do not lift weights. L3

156 ◆ G CHAPTER 6 Geologic Time

Natural Selection Charles Darwin was a naturalist who sailed around the world from 1831 to 1836 to study biology and geology. **Figure 4** shows a map of his journey. With some of the information about the plants and animals he observed on this trip in mind, he later published a book about the theory of evolution by natural selection.

In his book, he proposed that **natural selection** is a process by which organisms with characteristics that are suited to a certain environment have a better chance of surviving and reproducing than organisms that do not have these characteristics. Darwin knew that many organisms are capable of producing more offspring than can survive. This means that organisms compete with each other for resources necessary for life, such as food and living space. He also knew that individual organisms within the same species could be different, or show variations, and that these differences could help or hurt the individual organism's chance of surviving.

Some organisms that were well suited to their environment lived longer and had a better chance of producing offspring. Organisms that were poorly adapted to their environment produced few or no offspring. Because many characteristics are inherited, the characteristics of organisms that are better adapted to the environment get passed on to offspring more often. According to Darwin, this can cause a species to change over time.

Figure 4 Charles Darwin sailed around the world between 1831 and 1836 aboard the HMS *Beagle* as a naturalist. On his journey he saw an abundance of evidence for natural selection, especially on the Galápagos Islands off the western coast of South America.

Make a Model

Fossils Have pairs of students make a model clay impression "fossil" of a common item, using plaster of paris. Then have sets of partners exchange their "fossils." Have the students imagine they are paleontologists as they identify the object that formed the fossil. L1 COOP LEARN LS **Kinesthetic**

Discussion

Climate Survival According to Darwin's theory of natural selection, which type of animal species is more likely to survive in a polar climate: a cold-blooded animal with skin covered by scales or a warm-blooded animal with thick fur? Explain. A warm-blooded animal with thick fur. The thick fur serves to insulate, and warm-bloodedness means the internal temperature would be higher than a cold-blooded animal's, so the warm-blooded animal is better suited to withstand cold temperatures. L1

Science Journal

Theories of Organic Evolution Have students compare and contrast *gradualism* and *punctuated equilibrium* in their Science Journals. Both are theories of organic evolution; gradualism refers to a gradual, slow change in species; punctuated equilibrium describes sudden bursts of evolution separated by times of little change. L2 LS **Linguistic** P

Teacher FYI

Darwin Family Charles Darwin was the grandson of Erasmus Darwin. Erasmus was a noted physician and naturalist in his own right and had proposed his own theory of evolution in the 1790s.

SECTION 1 Life and Geologic Time G ◆ 157

Discussion

Desert Survival In a desert area where there is little moisture, what types of characteristics would be favored in animals? characteristics that would allow animals to live with little water, conserve water, or eat food, such as cacti, that contains water L2

Activity

Adaptations to Environment Show students photographs of different organisms. Describe the type of environment in which each organism lives. Have students list the adaptations that are important for each organism's survival in its environment. L2 LS **Visual-Spatial**

Reading Check

Answer Because short-necked and long-necked animals would have an equal opportunity for food, the number of short-necked members of the species might increase. Long-necked animals might also survive but would not necessarily continue in the majority.

Quick Demo

Nutshells and Trilobites

Materials cashew, pistachio, walnut (last 2 in the shell), pliers

Estimated Time 10 minutes

Procedure WARNING: *Avoid exposing students to potential food allergies.* Break each of the nuts using a pair of pliers. Show that the thicker the shell, the harder the nut is to crack. Compare this to the evolution in trilobites. L2

Figure 5 Giraffes can eat leaves off the branches of tall trees because of their long necks.

Natural Selection Within a Species Suppose that an animal species exists in which a few of the individuals have long necks, but most have short necks. The main food for the animal is the leafy foliage on trees in the area. What happens if the climate changes and the area becomes dry? The lower branches of the trees might not have any leaves. Now which of the animals will be better suited to survive? Clearly, the long-necked animals have a better chance of surviving and reproducing. Their offspring will have a greater chance of inheriting the important characteristic. Gradually, as the number of long-necked animals becomes greater, the number of short-necked animals decreases. The species might change so that nearly all of its members have long necks, as the giraffe in **Figure 5** has.

Reading Check *What might happen to the population of animals if the climate became wet again?*

It is important to notice that individual, short-necked animals didn't change into long-necked animals. A new characteristic becomes common in a species only if some members already possess that characteristic and if the trait increases the animal's chance of survival. If no animal in the species possessed a long neck in the first place, a long-necked species could not have evolved by means of natural selection.

Artificial Selection Humans have long used the principle of artificial selection when breeding domestic animals. By carefully choosing individuals with desired characteristics, animal breeders have created many breeds of cats, dogs, cattle, and chickens. **Figure 6** shows the great variety of cats produced by artificial selection.

Figure 6 Cat breeders have succeeded in producing a great variety of cats by using the principle of artificial selection.

The Evolution of New Species Natural selection explains how characteristics change and how new species arise. For example, if the short-necked animals migrated to a different location, they might have survived. They could have continued to reproduce in the new location, eventually developing enough different characteristics from the long-necked animals that they might not be able to breed with each other. At this point, at least one new species would have evolved.

Cultural Diversity

Domestication Domestication is the process of selectively breeding wild animals and plants into forms that better serve the needs of people. The first attempts at domestication were made as early as 9000 B.C. by tribes that engaged in hunting and in gathering wild edible plants. Dogs, goats, and possibly sheep were the first animals to be domesticated. Plants with tubers probably were domesticated before seed plants—cereals, legumes, and other vegetables. Some plants were domesticated for their fibrous stalks, which were used for making fishing nets. Suggest that interested students research some of the earliest domesticated plants and animals, such as wheat, dogs, sheep, cattle, horses, and chickens. Have them explain the different characteristics people were looking for as they bred these organisms. L2

Trilobites

Remember the trilobites? The term *trilobite* comes from the structure of the hard outer skeleton or exoskeleton. The exoskeleton of a **trilobite** consists of three lobes that run the length of the body. As shown in **Figure 7,** the trilobite's body also has a head (cephalon), a segmented middle section (thorax), and a tail (pygidium).

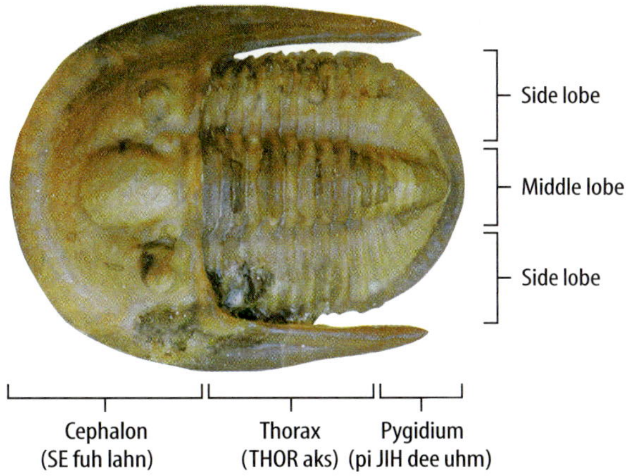

Figure 7 The trilobite's body was divided into three lobes that run the length of the body—two side lobes and one middle lobe.

Changing Characteristics of Trilobites Trilobites inhabited Earth's oceans for more than 200 million years. Throughout the Paleozoic Era, some species of trilobites became extinct and other new species evolved. Species of trilobites that lived during one period of the Paleozoic Era showed different characteristics than species from other periods of this era. As **Figure 8** shows, paleontologists can use these different characteristics to demonstrate changes in trilobites through geologic time. These changes can tell you about how different trilobites from different periods lived and responded to changes in their environments.

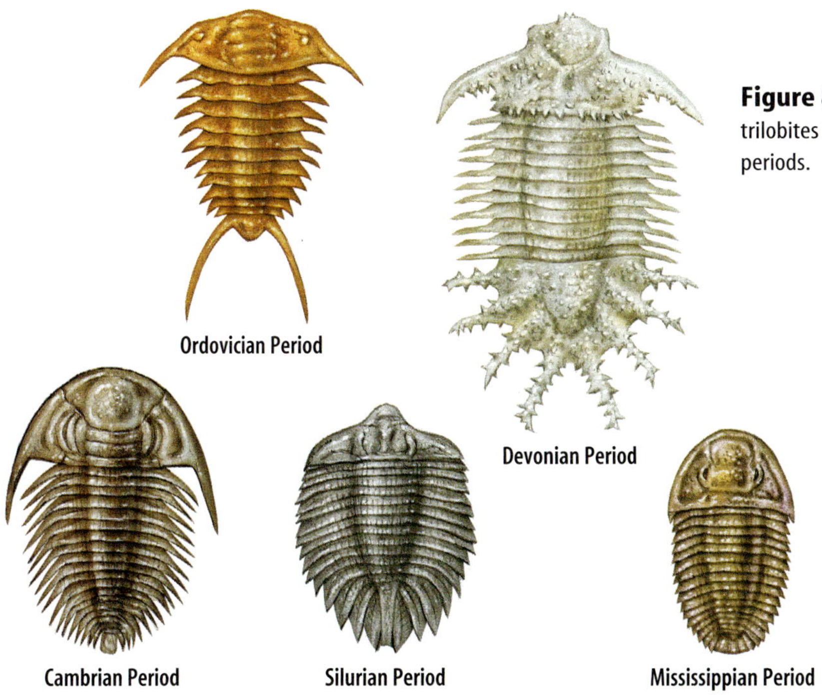

Figure 8 Different kinds of trilobites lived during different periods.

Teacher FYI

Exoskeletons The exoskeletons of trilobites did not grow, so the animals had to molt as they increased in size. These shed skeletal remains were left behind to form many of the fossils we know today.

Discussion

Environmental Changes Imagine that you find trilobites in a rock formation in a desert area. What does this tell you about how the environment of that area has changed? Because trilobites lived in oceans, it tells you that the area was once under the sea. L2

Visual Learning

Figure 8 Examine the diagram and describe how the appearance of trilobites changed through time. The trilobite's characteristics changed to suit its changing environment. Students may infer that during the Devonian Period, the environment became more hostile and the trilobite needed elaborate spines for defense. L2

Differentiated Instruction

English-Language Learners Take five minutes to illustrate deep time. Keep track of a five-minute period that will represent the 4.5 billion years of Earth's history. At about 1 minute, 7 seconds, announce that life begins. At 4 minutes, 24 seconds, announce the start of the Paleozoic Era. At 4 minutes, 43 seconds, the Mesozoic Era begins, and at 4 minutes, 55 seconds, the Cenozoic Era begins. Announce the appearance of humans just before time expires. L2

SECTION 1 Life and Geologic Time G ◆ 159

Discussion

Trilobite Preservation Why were trilobites more likely to be preserved as fossils than many other animals that lived in Precambrian Time? Trilobites had hard parts that could be preserved as body fossils. Trilobites also molted their exoskeletons. Therefore, one animal could produce many fossils. L2

Text Question Answer

Eyes on stalks would be useful to trilobites that burrowed into the sand.

Text Question Answer

Possible answers: a sudden climate change such as that caused by a major volcanic eruption that placed a lot of ash and soot into the air, decreasing the solar energy reaching Earth; results of continental and oceanic movements caused by plate tectonics; asteroid impact

Virtual Labs

Pangaea *What geologic evidence suggests the existence of Pangaea?*

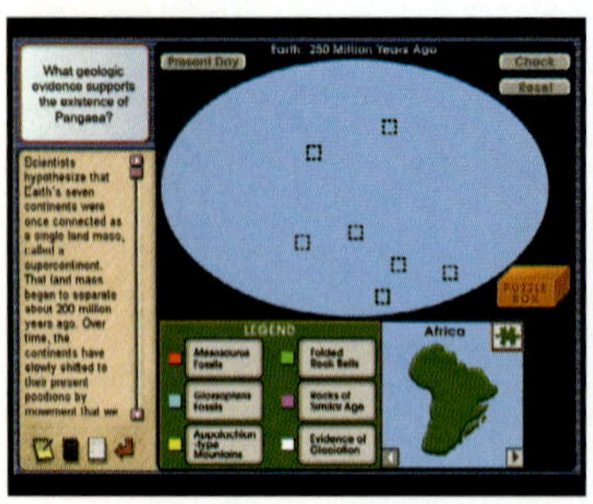

Figure 9 Trilobites had many different types of eyes. Some had eyes that contained hundreds of small circular lenses, somewhat like an insect. The blind trilobite (right) had no eyes.

Figure 10 *Olenellus* is one of the most primitive trilobite species.

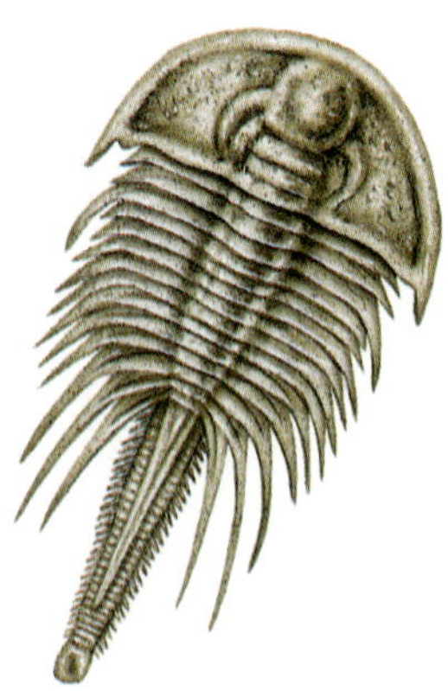

Trilobite Eyes Trilobites, shown in **Figure 9,** might have been the first organisms that could view the world with complex eyes. Trilobite eyes show the result of natural selection. The position of the eyes on an organism gives clues about where it must have lived. Eyes that are located toward the front of the head indicate an organism that was adapted for active swimming. If the eyes are located toward the back of the head, the organism could have been a bottom dweller. In most species of trilobites, the eyes were located midway on the head—a compromise for an organism that was adapted for crawling on the seafloor and swimming in the water.

Over time, the eyes in trilobites changed. In many trilobite species, the eyes became progressively smaller until they completely disappeared. Blind trilobites, such as the one on the right in **Figure 9,** might have burrowed into sediments on the seafloor or lived deeper than light could penetrate. In other species, however, the eyes became more complex. One kind of trilobite, *Aeglina,* developed large compound eyes that had numerous individual lenses. Some trilobites developed stalks that held the eyes upward. Where would this be useful?

Trilobite Bodies The trilobite body and tail also underwent significant changes in form through time, as you can see in **Figure 8** on the previous page. A special case is *Olenellus,* shown in **Figure 10.** This trilobite, which lived during the Early Cambrian Period, had an extremely segmented body—perhaps more so than any other known species of trilobite. It is thought that *Olenellus,* and other species that have so many body segments, are primitive trilobites.

Fossils Show Changes Trilobite exoskeletons changed as trilobites adapted to changing environments. Species that could not adapt became extinct. What processes on Earth caused environments to change so drastically that species adapted or became extinct?

Curriculum Connection

Math Have students construct a circle graph showing the 4.5 billion years of Earth's history divided into four major units of geologic time. The Precambrian represents about 88.3%; the Paleozoic Era about 6.4%; the Mesozoic Era about 3.9%; and the Cenozoic Era about 1.4% of Earth's history. L2 LS **Logical-Mathematical**

Active Reading

Flow Chart A flow chart helps students logically sequence events. Have students write the geologic eras in a series of large rectangles. As students read the chapter, have them write major events that occurred in each era in smaller rectangles under the large rectangles. L2

160 ◆ G CHAPTER 6 Geologic Time

Plate Tectonics and Earth History

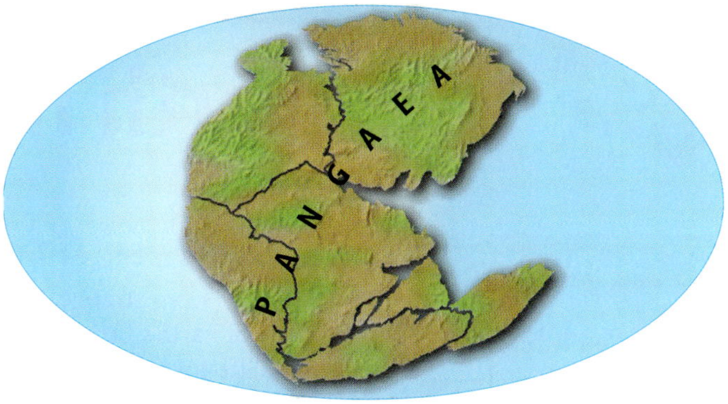

Plate tectonics is one possible answer to the riddle of trilobite extinction. Earth's moving plates caused continents to collide and separate many times. Continental collisions formed mountains and closed seas caught between continents. Continental separations created wider, deeper seas between continents. By the end of the Paleozoic Era, sea levels had dropped and the continents had come together to form one giant landmass, the supercontinent **Pangaea** (pan JEE uh). Because trilobites lived in the oceans, their environment was changed or destroyed. **Figure 11** shows the arrangement of continents at the end of the Paleozoic Era. What effect might these changes have had on the trilobite populations?

Not all scientists accept the above explanation for the extinctions at the end of the Paleozoic Era, and other possibilities—such as climate change—have been proposed. As in all scientific debates, you must consider the evidence carefully and come to conclusions based on the evidence.

Figure 11 The amount of shallow water environment was reduced when Pangaea formed.
Describe *how this change affected organisms that lived along the coasts of continents.*

section 1 review

Summary

Geologic Time

- Earth's history is divided into eons, eras, periods, and epochs, based on fossils.

Organic Evolution

- The fossil record indicates that species have changed over time.
- Charles Darwin proposed natural selection to explain change in species.
- In natural selection, organisms best suited to their environments survive and produce the most offspring.

Trilobites

- Trilobites were abundant in the Paleozoic fossil record and can be used as index fossils.

Plate Tectonics and Earth History

- Continents moving through time have influenced the environments of past organisms.

Self Check

1. **Discuss** how fossils relate to the geologic time scale.
2. **Infer** how plate tectonics might lead to extinction.
3. **Infer** how the eyes of a trilobite show how it lived.
4. **Explain** how paleontologists use trilobite fossils as index fossils for various geologic time periods.
5. **Think Critically** Aside from moving plates, what other factors could cause an organism's environment to change? How would this affect species?

Applying Skills

6. **Recognize Cause and Effect** Answer the questions below.
   a. How does natural selection cause evolutionary change to take place?
   b. How could the evolution of a characteristic within one species affect the evolution of a characteristic within another species? Give an example.

Science Online bookg.msscience.com/self_check_quiz

Caption Answer

Figure 11 The habitats of such organisms would have changed greatly, and these organisms would have become extinct if they could not adapt.

3 Assess

DAILY INTERVENTION

Check for Understanding

Visual-Spatial Make separate cutouts of the continents and have students arrange them as Pangaea. Then move the continents into their modern locations.

Reteach

Change Through Time Reinforce how life forms change through geologic time by showing students illustrations of trilobites from the Cambrian Period and then from later periods in the Paleozoic Era. L1 LS **Visual-Spatial**

Assessment

Process Assess students' abilities to observe and infer by having them look at photographs of anteaters and infer why these animals have long snouts and tongues. Use **Performance Assessment in the Science Classroom,** p. 89. L2

section 1 review

1. Appearance, disappearance, or change in abundance and diversity of fossil organisms can be used to subdivide geologic time.
2. Changes in Earth's surface that altered habitats could have caused organisms with traits less suited to the new habitat to become extinct.
3. no eyes: deep water, burrower; stalked eyes: burrower; eyes toward front of head: swimmer; eyes toward back of head: bottom-dweller
4. Trilobites lived over a wide area and during different short periods of geologic time.
5. Possible answer: Changes in climate; these changes could have caused organisms to evolve different structures and eventually result in a new species.
6. **a.** Organisms best suited to survive pass traits to young. Over time, the species evolves.
   b. Possible answer: If a species evolves so that it eats one type of food instead of another, the species that it preys on would be affected. The prey species would evolve so that traits that help it avoid capture would be emphasized.

1 Motivate

INTERACTIVE CHALKBOARD
PowerPoint® Presentations

Bellringer

Section Focus Transparencies also are available on the Interactive Chalkboard CD-ROM. L2 ELL

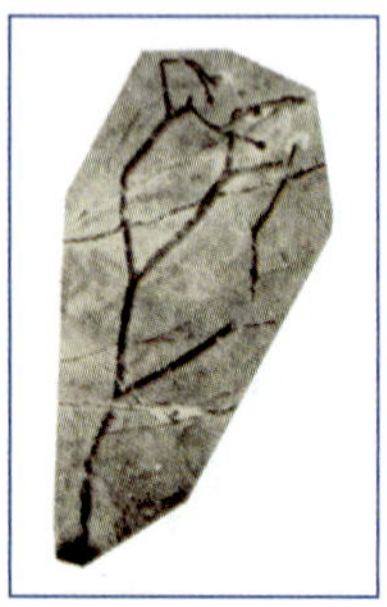

Section 2 Section Focus Transparency **An Early Plant**

This is a fossil of *Cooksonia*, one of the earliest land plants known to scientists. These fossils are about 420 million years old—that's about 200 million years before the dinosaurs lived!

1. What characteristics of *Cooksonia* are similar to plants that you see every day?
2. Would you guess that plants or animals colonized land first?

L2

Geologic Time

Tie to Prior Knowledge

Barren Land Ask students to think of a place they have been or have seen in a photo that is barren rock with no plant or animal life. You may want to show some photographs. Explain that Earth may have looked like this after it formed more than 4 billion years ago.

section 2

Early Earth History

as you read

***What* You'll Learn**

- **Identify** characteristic Precambrian and Paleozoic life-forms.
- **Draw** conclusions about how species adapted to changing environments in Precambrian time and the Paleozoic Era.
- **Describe** changes in Earth and its life-forms at the end of the Paleozoic Era.

***Why* It's Important**

The Precambrian includes most of Earth's history.

Review Vocabulary

life: state of being in which one grows, reproduces, and maintains a constant internal environment

New Vocabulary

- Precambrian time
- cyanobacteria
- Paleozoic Era

Precambrian Time

It may seem strange, but **Figure 12** is probably an accurate picture of Earth's first billion years. Over the next 3 billion years, simple life-forms began to colonize the oceans.

Look again at the geologic time scale shown in **Figure 2. Precambrian** (pree KAM bree un) **time** is the longest part of Earth's history and includes the Hadean, Archean, and Proterozoic Eons. Precambrian time lasted from about 4.5 billion years ago to about 544 million years ago. The oldest rocks that have been found on Earth are about 4 billion years old. However, rocks older than about 3.5 billion years are rare. This probably is due to remelting and erosion.

Although the Precambrian was the longest interval of geologic time, relatively little is known about the organisms that lived during this time. One reason is that many Precambrian rocks have been so deeply buried that they have been changed by heat and pressure. Many fossils can't withstand these conditions. In addition, most Precambrian organisms didn't have hard parts that otherwise would have increased their chances to be preserved as fossils.

Figure 12 During the early Precambrian, Earth was a lifeless planet with many volcanoes.

Section 2 Resource Manager

Chapter *FAST FILE* Resources
Transparency Activity, p. 43
Directed Reading for Content Mastery, p. 18
MiniLAB, p. 3
Enrichment, p. 29
Reinforcement, p. 26
Lab Worksheets, pp. 5–6

Reading and Writing Skill Activities, p. 27
Science Inquiry Labs, pp. 21–22
Cultural Diversity, p. 31
Earth Science Critical Thinking/Problem Solving, p. 11

Stromatolites have changed little throughout geologic time. These modern ones in Australia look much like ancient stromatolites.

This is a microscopic view of cyanobacteria.

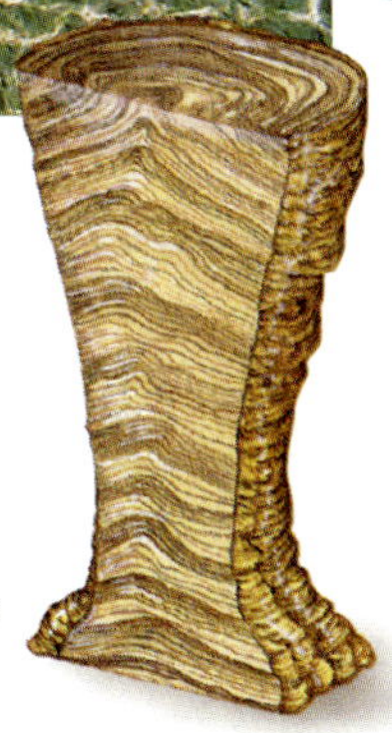

This cross section of a stromatolite head shows its layered appearance.

Figure 13 Cyanobacteria can produce layered mounds of sediment called stromatolites. The sticky surface of the cyanobacteria colony traps grains of sediment. The surface of the sediment then becomes colonized with cyanobacteria again, and the cycle repeats, producing the layers inside the stromatolite.

Early Life Many studies of the early history of life involve ancient stromatolites (stroh MA tuh lites). **Figure 13** shows stromatolites, which are layered mats formed by cyanobacteria colonies. **Cyanobacteria** are blue-green algae thought to be one of the earliest forms of life on Earth. Cyanobacteria first appeared about 3.5 billion years ago. They contained chlorophyll and used photosynthesis. This is important because during photosynthesis, they produced oxygen, which helped change Earth's atmosphere. Following the appearance of cyanobacteria, oxygen became a major atmospheric gas. Also of importance was that the ozone layer in the atmosphere began to develop, shielding Earth from ultraviolet rays. It is hypothesized that these changes allowed species of single-celled organisms to evolve into more complex organisms.

Reading Check *What atmospheric gas is produced by photosynthesis?*

Animals without backbones, called invertebrates (ihn VUR tuh brayts), appeared toward the end of Precambrian time. Imprints of invertebrates have been found in late Precambrian rocks, but because these early invertebrates were soft bodied, they weren't often preserved as fossils. Because of this, many Precambrian fossils are trace fossils.

INTEGRATE Chemistry

Earth's First Air
Cyanobacteria are thought to have been one of the mechanisms by which Earth's early atmosphere became richer in oxygen. Research the composition of Earth's early atmosphere and where these gases probably came from. Record your findings in your Science Journal.

2 Teach

Discussion

Cyanobacteria How would life on Earth be different today if cyanobacteria had not produced oxygen for the planet's atmosphere? Animals as we know them would not have evolved, since they depend on oxygen to live. In addition, without the protection of the ozone layer it is likely that organisms as we know them would not have evolved. L2

Reading Check

Answer oxygen

INTEGRATE Chemistry

Earth's First Air Before oxygen began to accumulate, Earth's atmosphere was probably composed of water vapor, carbon dioxide, and nitrogen. These gases might have been in the original material that formed Earth and been given off into the atmosphere during volcanic eruptions. They might also have been brought to Earth by comets impacting. P

Research Investigate the Miller-Urey experiment. What atmosphere did they assume for early Earth, and what were they able to show? Miller and Urey used a mix of methane, ammonia, and water plus a spark to produce amino acids. What advances have occurred since that experiment? Similar experiments with a more accurate mix of gases for early Earth have yielded similar results. L2

LAB DEMONSTRATION

Purpose to show the importance of hard parts to fossil preservation

Materials flat pan, gelatin dessert, small seashells, board that fits inside the pan, clay

Procedure Place a layer of clay in the bottom of a flat pan. Sprinkle pieces of the gelatin dessert and the shells on top of the clay. Cover with a second clay layer. Apply pressure to the top layer of clay with the board. Peel away the top clay layer to expose the shell layer.

Expected Outcome Students observe that hard parts (shells) retain their shape and leave an imprint, but the soft gelatin is smashed and leaves no recognizable trace.

Assessment

Compare what you observed in this activity with the formation of fossils. Many fossils form from hard parts of living organisms. Most soft parts do not become fossilized. L2

Reading Check

Answer Ediacaran organisms had shapes similar to modern jellyfish, worms, and soft corals.

Mini LAB

Purpose Students use fossils to determine the age of rock layers. L2 LS **Visual-Spatial**

Teaching Strategy Suggest that students make a chart that lists the periods during which each rock layer could have formed.

Analysis

1. Layer 2; it contains fossils A and B. These fossil organisms lived together during the Ordovician Period.
2. There is only one fossil in each layer, and that fossil existed during two different periods.

Assessment

Content Which rock layer is the oldest? Which is the youngest? Explain. Layer 1 is oldest; layer 3 is youngest; unless they have been disturbed, the oldest layer is always at the bottom and the youngest at the top. Use **Performance Assessment in the Science Classroom,** p. 89.

Mini LAB

Dating Rock Layers with Fossils

Procedure

1. Draw three rock layers.
2. Number the layers 1 to 3, bottom to top.
3. Layer 1 contains fossil A. Layer 2 contains fossils A and B. Layer 3 contains fossil C.
4. Fossil A lived from the Cambrian through the Ordovician. Fossil B lived from the Ordovician through the Silurian. Fossil C lived in the Silurian and Devonian.

Analysis

1. Which layers were you able to date to a specific period?
2. Why isn't it possible to determine during which specific period the other layers formed?

Figure 14 This giant predatory fish lived in seas that were present in North America during the Devonian Period. It grew to about 6 m in length.

Unusual Life-Forms A group of animals with shapes similar to modern jellyfish, worms, and soft corals was living late in Precambrian time. Fossils of these organisms were first found in the Ediacara Hills in southern Australia. This group of organisms has become known as the Ediacaran (ee dee uh KAR un) fauna. **Figure 15** shows some of these fossils.

Reading Check *What modern organisms do some Ediacaran organisms resemble?*

Ediacaran animals were bottom dwellers and might have had tough outer coverings like air mattresses. Trilobites and other invertebrates might have outcompeted the Ediacarans and caused their extinction, but nobody knows for sure why these creatures disappeared.

The Paleozoic Era

As you have learned, fossils are unlikely to form if organisms have only soft parts. An abundance of organisms with hard parts, such as shells, marks the beginning of the Paleozoic (pay lee uh ZOH ihk) Era. The **Paleozoic Era,** or era of ancient life, began about 544 million years ago and ended about 248 million years ago. Traces of life are much easier to find in Paleozoic rocks than in Precambrian rocks.

Paleozoic Life Because warm, shallow seas covered large parts of the continents during much of the Paleozoic Era, many of the life-forms scientists know about were marine, meaning they lived in the ocean. Trilobites were common, especially early in the Paleozoic. Other organisms developed shells that were easily preserved as fossils. Therefore, the fossil record of this era contains abundant shells. However, invertebrates were not the only animals to live in the shallow, Paleozoic seas.

Vertebrates, or animals with backbones, also evolved during this era. The first vertebrates were fishlike creatures without jaws. Armoured fish with jaws such as the one shown in **Figure 14** lived during the Devonian Period. Some of these fish were so huge that they could eat large sharks with their powerful jaws. By the Devonian Period, forests had appeared and vertebrates began to adapt to land environments, as well.

Differentiated Instruction

Learning Disabled and Visually Impaired Provide plastic, plaster, or clay models of important fossils of animals and plants that lived during the Paleozoic Era. Handling the fossils will help students picture these organisms' shapes, sizes, and external features. L1

Teacher FYI

Ediacaran Fossils Fossils of Ediacaran organisms are so different from later organisms that the time when these organisms lived can be thought of as a distinct period, the Vendian. This period is the latest part of the Proterozoic Eon and occurred before the Cambrian Period of the Paleozoic Era.

NATIONAL GEOGRAPHIC

VISUALIZING UNUSUAL LIFE-FORMS

Figure 15

A variety of 600-million-year-old fossils—known as Ediacaran (eed ee uh KAR un) fauna—have been found on every continent except Antarctica. These unusual organisms were originally thought to be descendants of early animals such as jellyfish, worms, and coral. Today, paleontologists debate whether these organisms were part of the animal kingdom or belonged to an entirely new kingdom whose members became extinct about 545 million years ago.

DICKENSONIA (dihk un suh NEE uh) Impressions of *Dickensonia,* a bottom-dwelling wormlike creature, have been discovered. Some are nearly one meter long.

RANGEA (rayn JEE uh) As it lay rooted in sea-bottom sediments, *Rangea* may have snagged tiny bits of food by filtering water through its body.

SPRIGGINA (sprih GIHN uh) Some scientists hypothesize that the four-centimeter-long *Spriggina* was a type of crawling, segmented organism. Others suggest that it sat upright while attached to the seafloor.

CYCLOMEDUSA (si kloh muh DEW suh) Although it looks a lot like a jellyfish, *Cyclomedusa* may have had more in common with modern sea anemones. Some paleontologists, however, hypothesize that it is unrelated to any living organism.

NATIONAL GEOGRAPHIC

Visualizing Unusual Life Forms

Have students examine the pictures and read the captions. Then ask the following question.

If Ediacaran fauna were soft-bodied and did not have hard skeletons or shells, how might they have been fossilized? Their body shapes were probably preserved by quick burial by sand or other sediments on the sea floor. L2

Activity

Ediacaran Fossils Have small groups of students research the locations around the world where Ediacaran fossils have been found and mark these locations on a world map. Ask each group to select one location and make an oral presentation about the fossils found there. L2

IDENTIFYING Misconceptions

Shark Evolution Some students may think of sharks as having evolved quite recently because they are living today. Explain that, like many other animals living today, fossil evidence indicates that sharks evolved long ago. Sharks evolved during the Devonian Period, over 360 million years ago.

Reading Check

Answer lungs and leglike fins

Inquiry Lab

Eggs Out of Water

Question What materials could be used to design an egg that will survive out of water?

Possible Materials clay, paper, cardboard, aluminum foil, cloth, wax paper

Estimated Time 1 class period

Teaching Strategies

- An egg must retain water and allow gas exchange. Students will need to solve both problems, so encourage them to use a mix of materials.
- Ask students how this activity relates to the evolution of eggs.

For additional inquiry activities, see *Science Inquiry Labs*. L2

Figure 16 Amphibians probably evolved from fish like *Panderichthys* (pan dur IHK theez), which had leglike fins and lungs.

Life on Land Based on their structure, paleontologists know that many ancient fish had lungs as well as gills. Lungs enabled these fish to live in water with low oxygen levels—when needed they could swim to the surface and breathe air. Today's lungfish also can get oxygen from the water through gills and from the air through lungs.

One kind of ancient fish had lungs and leglike fins, which were used to swim and crawl around on the ocean bottom. Paleontologists hypothesize that amphibians might have evolved from this kind of fish, shown in **Figure 16.** The characteristics that helped animals survive in oxygen-poor waters also made living on land possible. Today, amphibians live in a variety of habitats in water and on land. They all have at least one thing in common, though. They must lay their eggs in water or moist places.

Reading Check *What are some characteristics of the fish from which amphibians might have evolved?*

By the Pennsylvanian Period, some amphibians evolved an egg with a membrane that protected it from drying out. Because of this, these animals, called reptiles, no longer needed to lay eggs in water. Reptiles also have skin with hard scales that prevent loss of body fluids. This adaptation enables them to survive farther from water and in relatively dry climates, as shown in **Figure 17,** where many amphibians cannot live.

Science Online

Topic: Paleozoic Life

Visit bookg.msscience.com for Web links to information about Paleozoic life.

Activity Prepare a presentation on the organisms of one period of the Paleozoic Era. Describe a few animals from different groups, including how and where they lived. Are any of these creatures alive today, and if not, when did they become extinct?

Figure 17 Reptiles have scaly skins that allow them to live in dry places.

Visual Learning

Figure 17 Have students list another characteristic of reptiles that allows them to live on land. Students should cite eggs with a membrane that keeps it from drying out. This membrane allows eggs to be laid on land instead of in the water. L2 LS **Linguistic**

Science Journal

Extinctions Have students write reports on species that became extinct and causes of the extinctions that occurred at the ends of the Ordovician and Permian periods. Possible answers: Ordovician: species of graptolites, trilobites, brachiopods due to global cooling; Permian: >90% of marine species and 70% of land species due to several factors. L2

166 ◆ G CHAPTER 6 Geologic Time

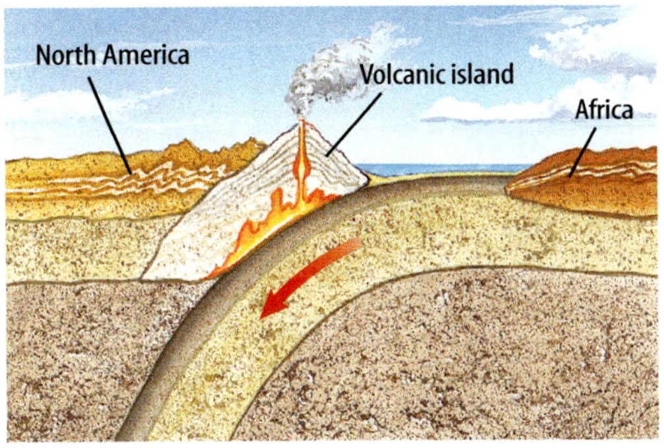

More than 375 million years ago, volcanic island chains formed in the ocean and were pushed against the coast as Africa moved toward North America.

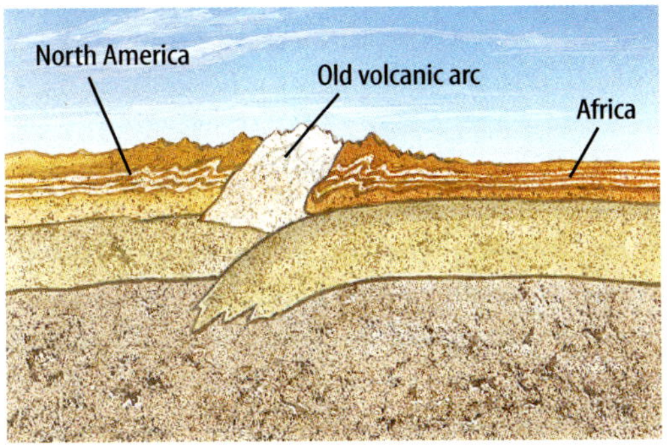

About 375 million years ago, the African plate collided with the North American plate, forming mountains on both continents.

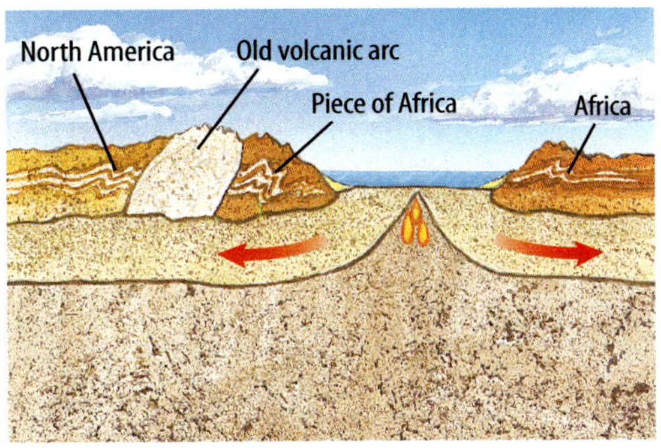

About 200 million years ago, the Atlantic Ocean opened up, separating the two continents.

Mountain Building Several mountain-building episodes occurred during the Paleozoic Era. The Appalachian Mountains, for example, formed during this time. This happened in several stages, as shown in **Figure 18.** The first mountain-building episode occurred as the ocean separating North America from Europe and Africa closed. Several volcanic island chains that had formed in the ocean collided with the North American Plate, as shown in the top picture of **Figure 18.** The collision of the island chains generated high mountains.

The next mountain-building episode was a result of the African Plate colliding with the North American Plate, as shown in the left picture of **Figure 18.** When Africa and North America collided, rock layers were folded and faulted. Some rocks originally deposited near the eastern coast of the North American Plate were pushed along faults as much as 65 km westward by the collision. Sediments were uplifted to form an immense mountain belt, part of which still remains today.

Figure 18 The Appalachian Mountains formed in several stages.
Infer *how these movements affected species in the Appalachians.*

Discussion

Mountain Formation Show a physical-features map of the U.S. to students. Remind students that the Appalachians formed along the east coast of North America. Why aren't they on the coast today? Possible answers: erosion and deposition have extended land between the mountains and the coast; a drop in sea level has uncovered land that was once sea bottom. L2

Activity

Fossil Field Trip Take students on a field trip to an exposure of Paleozoic Era rocks near your school. Show them how and where to look for fossils. Have students take photos or make sketches of the fossils. Back in the classroom, have students try to identify the fossils they found. If no exposure exists, use fossils of Paleozoic age ordered from a science supply house. Display the labeled fossils in class. L2 LS
Visual-Spatial

Caption Answer

Figure 18 Local climate would change, causing the extinction or disposal of species that were not adapted.

Quick Demo

Soft-bodied Fossils

Materials gummi worms, soft clay, plaster of paris

Estimated Time 10 minutes

Procedure Place a layer of soft clay in a shallow pan. Press gummi worms into clay to leave an impression. Cover with a layer of plaster of paris. When plaster dries, it should leave behind impressions of the worm. Compare this to the fossils of the Ediacaran fauna. Preparing a second tray with the plaster beforehand will allow the demo to proceed without waiting for it to dry.

Differentiated Instruction

Challenge Have students research the Pennsylvanian Period's swamps, where much of the coal used today was formed. Have students describe the process that must occur for plant remains to change into coal. Plant material must collect under water or be buried quickly in the absence of air so that the material's carbon content is not lost. Over time, with pressure and heat, the material increases in hardness, and carbon is concentrated. Plant material changes from peat to lignite to bituminous coal and finally to anthracite coal. L2 L3

SECTION 2 Early Earth History G ◆ 167

Discussion

Mass Extinctions What would be the result of mass extinctions today? Possible answers: Depending on the plant or animal species, there could be widespread starvation or destruction of whole ecosystems. If the event that causes the mass extinction is catastrophic and of worldwide impact, the human population on Earth could be reduced or eliminated.

3 Assess

DAILY INTERVENTION

Check for Understanding

Visual-Spatial Instruct pairs of students to search discarded magazines for photographs that show the characteristics that make amphibians different from fish and reptiles different from amphibians. Have partners quiz each other on the differences. LS L2

Reteach

Early Earth History Have small groups of students work together to outline the section, focusing on major topics contained in heads, subheads, captions, and Reading Check answers. L1 LS **Interpersonal**

Assessment

Oral Have students write questions on the material in the section, and then take turns quizzing each other. Use **Performance Assessment in the Science Classroom**, p. 91. L2

Figure 19 Hyoliths were organisms that became extinct at the end of the Paleozoic Era.

End of an Era At the end of the Paleozoic Era, more than 90 percent of all marine species and 70 percent of all land species died off. **Figure 19** shows one such animal. The cause of these extinctions might have been changes in climate and a lowering of sea level.

Near the end of the Permian Period, the continental plates came together and formed the supercontinent Pangaea. Glaciers formed over most of its southern part. The slow, gradual collision of continental plates caused mountain building. Mountain-building processes caused seas to close and deserts to spread over North America and Europe. Many species, especially marine organisms, couldn't adapt to these changes, and became extinct.

Other Hypotheses Other explanations also have been proposed for this mass extinction. During the late Paleozoic Era, volcanoes were extremely active. If the volcanic activity was great enough, it could have affected the entire globe. Another recent theory is similar to the one proposed to explain the extinction of dinosaurs. Perhaps a large asteroid or comet collided with Earth some 248 million years ago. This event could have caused widespread extinctions just as many paleontologists suggest happened at the end of the Mesozoic Era, 65 million years ago. Perhaps the extinction at the end of the Paleozoic Era was caused by several or all of these events happening at about the same time.

section 2 review

Summary

Precambrian Time

- Precambrian time covers almost 4 billion years of Earth history, but little is known about the organisms of this time.
- Cyanobacteria were among the earliest life-forms.

The Paleozoic Era

- Invertebrates developed shells and other hard parts, leaving a rich fossil record.
- Vertebrates—animals with backbones—appeared during this era.
- Plants and amphibians first moved to land during the Paleozoic Era.
- Adaptations in reptiles allow them to move away from water for reproduction.
- Geologic events at the end of the Paleozoic Era led to a mass extinction.

Self Check

1. **List** the geologic events that ended the Paleozoic Era.
2. **Infer** how geologic events at the end of the Paleozoic Era might have caused extinctions.
3. **Discuss** the advance that allowed reptiles to reproduce away from water. Why was this an advantage?
4. **Identify** the major change in life-forms that occurred at the end of Precambrian time.
5. **Think Critically** How did cyanobacteria aid the evolution of complex life on land? Do you think cyanobacteria are as significant to this process today as they were during Precambrian time?

Applying Skills

6. **Use a Database** Research trilobites and describe these organisms and their habitats in your Science Journal. Include hand-drawn illustrations and compare them with the illustrations in your references.

Science Online bookg.msscience.com/self_check_quiz

section 2 review

1. mountain building, continental movement, closure of seas, spread of deserts
2. Mountain building caused seas to close and deserts to spread over wide areas. Many species, especially marine organisms, could not adapt to these changes.
3. The development of a membrane-encased egg allowed reptiles to reproduce away from water. This allowed these animals to live in a greater variety of habitats than their amphibian ancestors.
4. The first invertebrates appeared in Earth's oceans.
5. They produced oxygen, which helped make Earth's atmosphere suitable for most animal life and helped develop Earth's protective ozone shield; no.
6. Trilobites are marine invertebrates. Descriptions should provide basic information about the appearance, habitat, and habits of these animals.

LAB

Changing Species

In this lab, you will observe how adaptation within a species might cause the evolution of a particular trait, leading to the development of a new species.

Real-World Question

How might adaptation within a species cause the evolution of a particular trait?

Goals

- **Model** adaptation within a species.

Materials

deck of playing cards

Procedure

1. **Remove** all of the kings, queens, jacks, and aces from a deck of playing cards.
2. Each remaining card represents an individual in a population of animals called "varimals." The number on each card represents the height of the individual.
3. **Calculate** the average height of the population of varimals represented by your cards.
4. Suppose varimals eat grass, shrubs, and leaves from trees. A drought causes many of these plants to die. All that's left are a few tall trees. Only varimals at least 6 units tall can reach the leaves on these trees.
5. All the varimals under 6 units leave the area or die from starvation. Discard all of the cards with a number less than 6. Calculate the new average height of the varimals.
6. Shuffle the deck of remaining cards.
7. **Draw** two cards at a time. Each pair represents a pair of varimals that will mate.

8. The offspring of each pair reaches the average height of its parents. Calculate and record the height of each offspring.
9. Discard all parents and offspring under 8 units tall and repeat steps 6–8. Now calculate the new average height of varimals. Include both the parents and offspring in your calculation.

Conclude and Apply

1. **Describe** how the height of the population changed.
2. **Explain** If you hadn't discarded the shortest varimals, would the average height of the population have changed as much?
3. Suppose the offspring grew to the height of one of its parents. How would the results change in each of the following scenarios?
   a. The height value for the offspring is chosen by coin toss.
   b. The height value for the offspring is whichever parent is tallest.
4. **Explain** If there had been no variation in height before the droughts occurred, would the species have been able to evolve?

BENCH TESTED

Real-World Question

Purpose Students will formulate a model showing the role of natural selection in creating variations within a species. L2 COOP LEARN IS **Visual-Spatial and Interpersonal**

Process Skills formulate models, use numbers, interpret data, form operational definitions, recognize cause and effect, communicate

Time Required 30 minutes

Procedure

Teaching Strategy To save time, before class remove the cards that will not be used in this activity. This will eliminate step 1 of the procedure.

Conclude and Apply

1. It increased.
2. No; the average would remain at about 6. The shortest varimals would still have been part of the calculation for the average height.
3. a. Since the choices are made at random, the average height should be no different than that in the original activity. Individual variation will be greater because no individuals will exist at in-between heights.
   b. With the offspring always growing to the taller parents' height, the average height of the population should increase over time until, given enough generations, the population is entirely composed of tall individuals. For a challenge, you could have the students repeat the exercise until all of the individuals are 10 units tall. How many steps does that require?
4. Probably not; if all the individuals would have been the same height, no group of individuals would have had an advantage in reaching high food. In that case, any evolution of the varimal population would likely have occurred from some variation of a trait other than height.

Assessment

Oral To further assess students' understanding of evolution, ask them whether any individual varimal increased in height because of natural selection. Encourage students to explain their answers. No; individuals do not evolve inheritable adaptations. L2

1 Motivate

INTERACTIVE CHALKBOARD PowerPoint® Presentations

Bellringer

Section Focus Transparencies also are available on the Interactive Chalkboard CD-ROM. L2 ELL

SECTION 3 Section Focus Transparency **Big Smile for the Camera**

The oldest fossils of sharks date back nearly 400 million years. The ancestors of modern sharks, though, first appeared about 200 million years ago. Below is a great white shark, a creature that hasn't changed much since its family arose during the Jurassic Period. Since that time, the swimming mechanisms of sharks have improved, and so have their feeding abilities.

1. What animals were dominant on land as sharks arose in the oceans?
2. Why are sharks sometimes called living fossils?
3. Can you name any other animals that have survived for such a long time with little change?

L2

Geologic Time

Tie to Prior Knowledge

Dinosaur Extinction Students are probably aware of the extinction of the dinosaurs, which is a major event of the Mesozoic Era. Have them list reasons they have heard for the extinction. Then tell students they will find out about this event and others in the most recent two eras of Earth's history in this section. L2

section 3

Middle and Recent Earth History

as you read

What You'll Learn

- **Compare and contrast** characteristic life-forms in the Mesozoic and Cenozoic Eras.
- **Explain** how changes caused by plate tectonics affected organisms during the Mesozoic Era.
- **Identify** when humans first appeared on Earth.

Why It's Important

Many important groups of animals, like birds and mammals, appeared during the Mesozoic Era.

Review Vocabulary
dinosaur: a reptile from one of two orders that dominated the Mesozoic Era

New Vocabulary
- Mesozoic Era
- Cenozoic Era

The Mesozoic Era

Dinosaurs have captured people's imaginations since their bones first were unearthed more than 150 years ago. Dinosaurs and other interesting animals lived during the Mesozoic Era, which was between 248 and 65 million years ago. The Mesozoic Era also was marked by rapid movement of Earth's plates.

The Breakup of Pangaea The **Mesozoic** (meh zuh ZOH ihk) **Era,** or era of middle life, was a time of many changes on Earth. At the beginning of the Mesozoic Era, all continents were joined as a single landmass called Pangaea, as shown in **Figure 11.**

Pangaea separated into two large landmasses during the Triassic Period, as shown in **Figure 20.** The northern mass was Laurasia (law RAY zhuh), and Gondwanaland (gahn DWAH nuh land) was the southern landmass. As the Mesozoic Era continued, Laurasia and Gondwanaland broke apart and eventually formed the present-day continents.

Species that had adapted to the new environments survived the mass extinction at the end of the Paleozoic Era. Recall that a reptile's skin helps it retain bodily fluids. This characteristic, along with their shelled eggs, enabled reptiles to adapt readily to the drier climate of the Mesozoic Era. Reptiles became the most conspicuous animals on land by the Triassic Period.

Figure 20 At the end of the Triassic Period, Pangaea began to break up into the northern supercontinent, Laurasia, and the southern supercontinent, Gondwanaland.

Section 3 Resource Manager

Chapter *FAST FILE* Resources
- Transparency Activity, p. 44
- Directed Reading for Content Mastery, pp. 19, 20
- Enrichment, p. 30
- Reinforcement, p. 27
- MiniLAB, p. 4
- Lab Worksheets, pp. 7–8
- Lab Activity, pp. 11–14

Earth Science Critical Thinking/Problem Solving, p. 12
Cultural Diversity, p. 13
Mathematics Skill Activities, p. 5
Home and Community Involvement, p. 27
Lab Management and Safety, p. 73

Dinosaurs What were the dinosaurs like? Dinosaurs ranged in height from less than 1 m to enormous creatures like *Apatosaurus* and *Tyrannosaurus.* The first small dinosaurs appeared during the Triassic Period. Larger species appeared during the Jurassic and Cretaceous Periods. Throughout the Mesozoic Era, new species of dinosaurs evolved and other species became extinct.

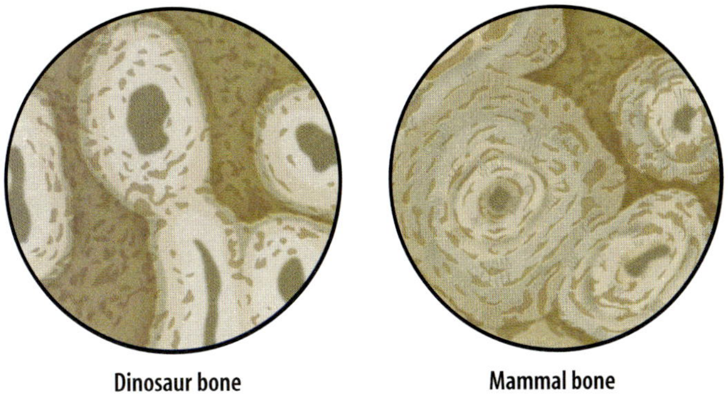

Figure 21 Some dinosaur bones show structural features that are like mammal bones, leading some paleontologists to think that dinosaurs were warm blooded like mammals.

Dinosaurs Were Active Studying fossil footprints sometimes allows paleontologists to calculate how fast animals walked or ran. Some dinosaur tracks indicate that these animals were much faster runners than you might think. *Gallimimus* was 4 m long and could reach speeds of 65 km/h—as fast as a modern racehorse.

Some studies also indicate that dinosaurs might have been warm blooded, not cold blooded like present-day reptiles. The evidence that leads to this conclusion has to do with their bone structure. Slices through some cold-blooded animal bones show rings similar to growth rings in trees. The bones of some dinosaurs don't show this ring structure. Instead, they are similar to bones found in modern mammals, as you can see in **Figure 21.**

Reading Check ***Why do some paleontologists think that dinosaurs were warm blooded?***

These observations indicate that some dinosaurs might have been warm-blooded, fast-moving animals somewhat like present-day mammals and birds. They might have been quite different from present-day reptiles.

Good Mother Dinosaurs The fossil record also indicates that some dinosaurs nurtured their young and traveled in herds in which the adults surrounded their young.

One such dinosaur is *Maiasaura.* This dinosaur built nests in which it laid its eggs and raised its offspring. Nests have been found in relatively close clusters, indicating that more than one family of dinosaurs built in the same area. Some fossils of hatchlings have been found near adult animals, leading paleontologists to think that some dinosaurs nurtured their young. In fact, *Maiasaura* hatchlings might have stayed in the nest while they grew in length from about 35 cm to more than 1 m.

Science Online

Topic: Warm Versus Cold

Visit bookg.msscience.com for Web links to information about dinosaurs.

Activity Work with a partner to research the debate on warm-blooded versus cold-blooded dinosaurs. Present your finding to the class in the form of a debate. Be sure to cover the main points of disagreement between the two sides.

2 Teach

Visual Learning

Figure 20 How might the separation of Pangaea have affected organisms living there? Environments would have changed as the new oceans formed, (for example, the climates may have changed) so species would have had to adapt or become extinct. In addition, populations of the same species may have been separated by the growing oceans, forming subgroups from which different species could arise. L2

Reading Check

Answer Their bone structure is similar to that of bones found in modern mammals.

Discussion

Dinosaurs and Birds Some scientists believe that some dinosaurs evolved into birds. What evidence do you see that might support this idea? Possible answers include dinosaurs being warm-blooded, laying eggs, caring for young. L2

Activity

Measuring Speed Have students investigate how tracks will be different for different running speeds. Have students run at different speeds and mark where their feet land. Measure the distance between steps for each pace.

Differentiated Instruction

Challenge Have students research the debate on warm or cold-blooded dinosaurs. Students should find evidence from both positions and prepare a poster outlining their findings. For an additional challenge, have the students summarize the poster with their opinion, supported by evidence. L3

SECTION 3 Middle and Recent Earth History G ◆ 171

Fun Fact

One of the longest dinosaurs was *Seismosaurus,* which might have reached lengths of 40 m.

Visual Learning

Figure 22 What about this dinosaur tells you that it was probably a meat eater? It has sharp teeth. L2

Use Science Words

Word Meaning The Triassic Period, the earliest period in the Mesozoic Era, is named for three related units of rock first studied in Germany. The Jurassic Period is named after the Jura Mountains between France and Switzerland, where rock from this period was first studied. Have students research the derivation of the most recent period of the Mesozoic Era, the Cretaceous Period. It is named for the white chalk cliffs of Dover, England. Cretaceous is derived from the Latin word for chalk, *creta.* L2

Quick Demo

Running and Walking

Materials flour

Estimated Time 15 minutes

Procedure Take the students outside. Have 2 volunteers step in flour. Have one walk and the other run on the sidewalk. Point out the differences in the mark left by the two kinds of gaits and compare to evidence from dinosaur tracks. L2

Figure 22 Birds might have evolved from dinosaurs.

A *Bambiraptor feinberger*, above, is a 75-million-year-old member of a family of meat-eating dinosaurs thought by some paleontologists to be closely related to birds.

B Considered one of the world's most priceless fossils, *Archaeopteryx*, above, was first found in a limestone quarry in Germany in 1861.

Birds Birds appeared during the Jurassic Period. Some paleontologists think that birds evolved from small, meat-eating dinosaurs much like *Bambiraptor feinberger* in **Figure 22A.** The earliest bird, *Archaeopteryx,* shown in **Figure 22B,** had wings and feathers. However, because *Archaeopteryx* had features not shared with modern birds, scientists know it was not a direct ancestor of today's birds.

Mammals Mammals first appeared in the Triassic Period. The earliest mammals were small, mouselike creatures, as shown in **Figure 23.** Mammals are warm-blooded vertebrates that have hair covering their bodies. The females produce milk to feed their young. These two characteristics have enabled mammals to survive in many changing environments.

Gymnosperms During most of the Mesozoic Era, gymnosperms (JIHM nuh spurmz), which first appeared in the Paleozoic Era, dominated the land. Gymnosperms are plants that produce seeds but not flowers. Many gymnosperms are still around today. These include pines and ginkgo trees.

Figure 23 The earliest mammals were small creatures that resembled today's mice and shrews.

Angiosperms Angiosperms (AN jee uh spurmz), or flowering plants, first evolved during the Cretaceous Period. Angiosperms produce seeds with hard outer coverings.

Because their seeds are enclosed and protected, angiosperms can live in many environments. Angiosperms are the most diverse and abundant land plants today. Present-day angiosperms that evolved during the Mesozoic Era include magnolia and oak trees.

Differentiated Instruction

Learning Disabled Supply students with drawings or models of six different dinosaurs. Have them determine whether the dinosaurs were meat or plant eaters. Have students describe what characteristics were useful for their analysis. Possible answers: Different types of teeth will suggest the type of food they ate. Predators' eyes were in the front of the head for depth perception; plant-eaters' eyes were on the side of the head for a greater field of vision. L2

Science Journal

Dinosaur Questions Have students take a few minutes to think of a question about dinosaurs for which they still do not have an answer. Put the questions in a box, and have each student take a question, making sure it is not his or her own. Have students research the answers to the questions, writing the answers in their Science Journals. L2

End of an Era The Mesozoic Era ended about 65 million years ago with a major extinction of land and marine species. Many groups of animals, including the dinosaurs, disappeared suddenly at this time. Many paleontologists hypothesize that a comet or asteroid collided with Earth, causing a huge cloud of dust and smoke to rise into the atmosphere, blocking out the Sun. Without sunlight the plants died, and all the animals that depended on these plants also died. Not everything died, however. All the organisms that you see around you today are descendants of the survivors of the great extinction at the end of the Mesozoic Era.

Applying Math — Calculate Percentages

CALCULATING EXTINCTION BY USING PERCENTAGES At the end of the Cretaceous Period, large numbers of plants and animals became extinct. Scientists still are trying to understand why some types of plants and animals survived while others died out. Looking at data about amphibians, reptiles, and mammals that lived during the Cretaceous Period, can you determine what percentage of amphibians survived this mass extinction?

Solution

1 *This is what you know:*

Animal Extinctions

Animal Type	Groups Living Before Extinction Event (n)	Groups Left After Extinction Event (t)
Amphibians	12	4
Reptiles	63	30
Mammals	24	8

2 *This is what you need to find out:* p = the percentage of amphibian groups that survived the Cretaceous extinction

3 *This is the equation you need to use:*

- $p = t / n \times 100$
- Both t and n are shown on the above chart.

4 *Substitute the known values:* $p = 4 / 12 \times 100 = 33.3\%$

Practice Problems

1. Using the same equation as demonstrated above, calculate the percentage of reptiles and then the percentage of mammals that survived. Which type of animal was least affected by the extinction?
2. What percentage of all groups survived?

Science Online — For more practice, visit bookg.msscience.com/math_practice

Teacher FYI

Asteroid Angle One explanation of how the asteroid that may have killed off the dinosaurs caused such widespread damage concerns the angle at which it struck Earth. It is thought to have collided with Earth at a shallow angle on what is now the Yucatan peninsula. This would have sent large amounts of hot debris over much of what is now North America.

Applying Math

National Math Standards

Correlation to Mathematics Objectives

1

Answers to Practice Problems

1. **Reptiles:**
   p = percent of reptile groups that survived the Cretaceous extinction
   $p = t/n$; $p = 5\ 30/63 = 0.476$ or 48%
   Mammals:
   p = percent of mammal groups that survived the Cretaceous extinction
   $p = t/n$; $p = 8/24 = 0.333$ or 33%
   Reptiles were less affected.
2. p = percent of all groups that survived the Cretaceous extinction
   $p = t/n$; $p = 4 + 30 + 8 / 12 + 63 + 24 = 42/99 = 0.424$ or 42%

Teacher FYI

The K-T Boundary The K-T boundary separates the Cretaceous Period (Mesozoic Era) from the Tertiary Period (Cenozoic Era). An iridium-rich clay layer found at this boundary led to the theory that a collision with an asteroid resulted in the demise of the dinosaurs.

Reading Check

Answer It became much cooler, and ice ages occurred.

IDENTIFYING Misconceptions

Humans and Dinosaurs Some students may think that humans were alive when dinosaurs lived on Earth. Explain that human ancestors *Australopithicines* appeared about 4.2 million years ago, long after dinosaurs became extinct about 65 million years ago.

Mini LAB

Purpose Students use the rate of seafloor spreading to calculate the age of the Atlantic Ocean. LS **Logical-Mathematical**

Materials world map, metric ruler

Teaching Strategy The rate given is an average for several points along the Mid-Atlantic Ridge.

Analysis

1. Student answers will depend on the points they choose on each continent and their measurements. Separations of 6,000 to 7,000 km would take from 160 to 200 million years to achieve.
2. Triassic values probably slightly older than estimates.

Assessment

Portfolio Have students make sketches of something that may have happened on land, in the ocean, or to organisms during the time the Atlantic was forming. Use **Performance Assessment in the Science Classroom**, p. 127. L2

Try at Home

Mini LAB

Calculating the Age of the Atlantic Ocean

Procedure

1. On a **world map** or **globe**, measure the distance in kilometers between a point on the east coast of South America and a point on the west coast of Africa.
2. Measure in SI several times and take the average of your results.
3. Assuming that Africa has been moving away from South America at a rate of 3.5 cm per year, calculate how many years it took to create the Atlantic Ocean.

Analysis

1. Did the values used to obtain your average value vary much?
2. How close did your age come to the accepted estimate for the beginning of the breakup of Pangaea in the Triassic Period?

Try at Home

The Cenozoic Era

The **Cenozoic (se nuh ZOH ihk) Era,** or era of recent life, began about 65 million years ago and continues today. Many mountain ranges in North and South America and Europe began to form in the Cenozoic Era. In the late Cenozoic, the climate became much cooler and ice ages occurred. The Cenozoic Era is subdivided into two periods. The first of these is the Tertiary Period. The present-day period is the Quaternary Period. It began about 1.8 million years ago.

Reading Check *What happened to the climate during the late Cenozoic Era?*

Times of Mountain Building Many mountain ranges formed during the Cenozoic Era. These include the Alps in Europe and the Andes in South America. The Himalaya, shown in **Figure 24,** formed as India moved northward and collided with Asia. The collision crumpled and thickened Earth's crust, raising the highest mountains presently on Earth. Many people think the growth of these mountains has helped create cooler climates worldwide.

Figure 24 The Himalaya extend along the India-Tibet border and contain some of the world's tallest mountains. India drifted north and finally collided with Asia, forming the Himalaya.

Further Evolution of Mammals

Throughout much of the Cenozoic Era, expanding grasslands favored grazing plant eaters like horses, camels, deer, and some elephants. Many kinds of mammals became larger. Horses evolved from small, multi-toed animals into the large, hoofed animals of today. However, not all mammals remained on land. Ancestors of the present-day whales and dolphins evolved to live in the sea.

As Australia and South America separated from Antarctica during the continuing breakup of the continents, many species became isolated. They evolved separately from life-forms in other parts of the world. Evidence of this can be seen today in Australia's marsupials. Marsupials are mammals such as kangaroos, koalas, and wombats (shown in **Figure 25**) that carry their young in a pouch.

Your species, *Homo sapiens,* probably appeared about 140,000 years ago. Some people suggest that the appearance of humans could have led to the extinction of many other mammals. As their numbers grew, humans competed for food that other animals relied upon. Also, fossil bones and other evidence indicate that early humans were hunters.

Figure 25 The wombat is one of many Australian marsupials. As a result of human activities, the number and range of wombats have diminished.

section 3 review

Summary

The Mesozoic Era

- During the Triassic Period, Pangaea split into two continents.
- Dinosaurs were the dominant land animals of the Mesozoic Era.
- Birds, mammals, and flowering plants all appeared during this era.
- The Mesozoic Era ended 65 million years ago with a mass extinction.

The Cenozoic Era

- The Cenozoic Era has been a mountain-building period with cooler climates.
- Mammals became dominant with many new life-forms appearing after the dinosaurs disappeared.
- Humans also appeared in the Cenozoic Era, probably about 140,000 years ago.

Self Check

1. **List** the era, period, and epoch in which *Homo sapiens* first appeared.
2. **Discuss** whether mammals became more or less abundant after the extinction of the dinosaurs, and explain why.
3. **Infer** how seeds with a hard outer covering enabled angiosperms to survive in a wide variety of climates.
4. **Explain** why some paleontologists hypothesize that dinosaurs were warm-blooded animals.
5. **Think Critically** How could two species that evolved on separate continents have many similarities?

Applying Math

6. **Convert Units** A fossil mosasaur, a giant marine reptile, measured 9 m in length and had a skull that measured 45 cm in length. What fraction of the mosasaur's total length did the skull account for? Compare your length with the mosasaur's length.

Science Online bookg.msscience.com/self_check_quiz

3 Assess

DAILY INTERVENTION

Check for Understanding

Interpersonal Have students compare and contrast marsupials that evolved in Australia with their placental counterparts that evolved in similar environments in Europe and America. Allow students to work in groups and make short presentations to the class. L3 COOP LEARN IN

Reteach

Time Line Have students use information from this section to make a time line that shows major Earth events during the Mesozoic and Cenozoic Eras. Some students might want to add small sketches at various places along the timeline, illustrating the events. L2 IN **Visual-Spatial**

Assessment

Process Have students predict what would happen, in general, to plant and animal species if the Atlantic Ocean were to close up again. Students should predict that the closing of the Atlantic Ocean would cause climate changes in North America and Europe (among other places). It also would cause species to adapt to new conditions or face extinction. Use **Performance Assessment in the Science Classroom,** p. 89. L2

section 3 review

1. Cenozoic Era, Quaternary Period, Pleistocene Epoch
2. More abundant; dinosaur extinction reduced the competition for food and habitat.
3. They survived better because their seeds are enclosed and protected.
4. Studies show they may have been fast moving and that some had bones similar in structure to those of modern mammals.
5. They could have evolved from a common ancestor, especially if the continents were once one land mass. (They also could have evolved in similar environments.)
6. 9 m = 900 cm; 45 cm/900 cm = 1/20; Students should use a similar formula, but substitute their height in centimeters for the length of mosasaur's skull.

BENCH TESTED

Real-World Question

Purpose

Internet Students will gather data from internet sites that can be accessed by visiting bookg.msscience.com/internet_lab. Students can post their data on the site and get data from other schools around the country. Click on the geologic time period you are interested in. This will open a database of fossil and paleogeographic data for that time period. The site provides links to other sites where fossils and paleogeographic data are available. Maps and data tables are also provided bookg.msscience.com/internet_lab. Students can print them out or use them as a guide.

Non-Internet Sources If you do not have Internet access, use historical geology references or fossil guidebooks in a library or natural history museum. Use a map of North America with an overlay to reconstruct the environment during the chosen time period, paying special attention to your local region. Record your data in a table similar to the one shown.

Process Skills determine, compare, gather information, record, make and use tables, form a hypothesis, review, describe, infer L2

LAB Use the Internet

Discovering the Past

Goals

- **Gather** information about fossils found in your area.
- **Communicate** details about fossils found in your area.
- **Synthesize** information from sources about the fossil record and the changes in your area over time.

Data Source

Science Online

Visit bookg.msscience.com/internet_lab for more information about fossils and changes over geologic time and for data collected by other students.

Real-World Question

Imagine what your state was like millions of years ago. What animals might have been roaming around the spot where you now sit? Can you picture a *Tyrannosaurus rex* roaming the area that is now your school? The animals and plants that once inhabited your region might have left some clues to their identity—fossils. Scientists use fossils to piece together what Earth looked like in the geologic past. Fossils can help determine whether an area used to be dry land or underwater. Fossils can help uncover clues about how plants and animals have evolved over the course of time. Using the resources of the Internet and by sharing data with your peers, you can start to discover how North America has changed through time. How has your area changed over geologic time? How might the area where you are now living have looked thousands or millions of years ago? Do you think that the types of animals and plants have changed much over time? Form a hypothesis concerning the change in organisms and geography from long ago to the present day in your area.

Fossils in Your Area

Fossil Name	Plant or Animal Fossil	Age of Fossils	Details About Plant or Animal Fossil	Location of Fossil	Additional Information
			Answers will vary.		

Alternative Inquiry Lab

Explore the Past Have students investigate changes in Earth through geologic time. Let them choose a geographical area to follow over time. This will encourage students to work on an area of interest and make the class presentations less repetitive. Students could show a progression of animals and plants, or changes in the environment. Advanced students could investigate the length of a day or atmospheric chemistry. L2

Using Scientific Methods

Make a Plan

1. **Determine** the age of the rocks that make up your area. Were they formed during Precambrian time, the Paleozoic Era, the Mesozoic Era, or the Cenozoic Era?
2. Gather information about the plants and animals found in your area during one of the above geologic time intervals. Find specific information on when, where, and how the fossil organisms lived. If no fossils are known from your area, find out information about the fossils found nearest your area.

Follow Your Plan

1. Make sure your teacher approves your plan before you start.
2. Go to bookg.msscience.com/internet_lab to post your data in the table. Add any additional information you think is important to understanding the fossils found in your area.

Analyze Your Data

1. What present-day relatives of prehistoric animals or plants exist in your area?
2. How have the organisms in your area changed over time? Is your hypothesis supported? Why or why not?
3. What other information did you discover about your area's climate or environment from the geologic time period you investigated?

Conclude and Apply

1. **Describe** the plant and animal fossils that have been discovered in your area. What clues did you discover about the environment in which these organisms lived? How do these compare to the environment of your area today?
2. **Infer** from the fossil organisms found in your area what the geography and climate were like during the geologic time period you chose.

Communicating Your Data

Find this lab using the link below.

Science Online

bookg.msscience.com/internet_lab

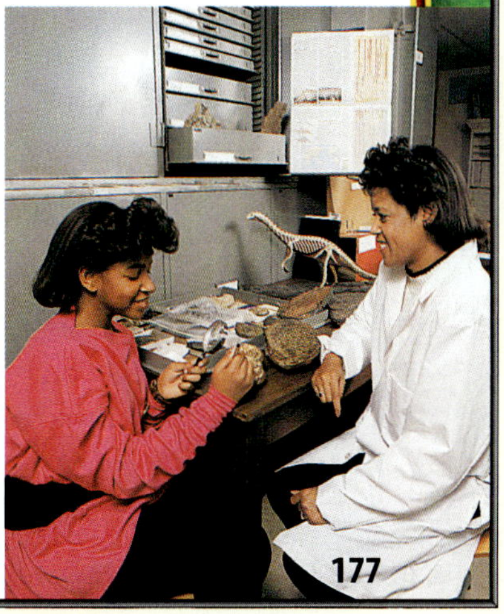

177

Time Required one week for data collection, one week to interpret the data

Make a Plan

Preparation

Internet Visit bookg.msscience.com/internet_lab to run through the steps the students will follow.

Non-Internet Sources Obtain a larger map of North America and a thin paper or clear plastic overlay. Have some reference books about fossils available. Have students visit or write to a natural history museum.

Follow Your Plan

Teaching Strategies

- This lab will allow students to discover what their area and the rest of North America looked like in the past. The main goal is to use the fossil record to determine the paleoclimate and paleogeography of North America during the four selected geologic time periods.
- Make sure students research where the fossils they find lived—whether on land or in a freshwater or ocean environment. L2

Analyze Your Data

Answers will be subjective and based on students' individual research.

Conclude and Apply

Answers will be individualized and often based on the student's opinion of his or her research. Look for depth and quality of research performed.

Communicating Your Data

Post results at bookg.msscience.com/internet_lab. Either compare your findings for a site to someone else's or look for other areas that have a similar history.

Assessment

Performance Have students mount their data tables on poster board. Display them around the room. Provide time for each group to present their findings to the class. Use **Performance Assessment in the Science Classroom,** p. 143. L2

Content Background

Current estimates of the wild tiger population range from 5,000 to 7,200 individuals. That figure is only about 5 percent of the population in 1900. Their range is steadily shrinking and the Bali, Caspian, and Javan tiger subspecies have already become extinct. Demand for tiger pelts, body parts used for medicinal purposes, and continuing loss of habitat make the extinction of tigers in the wild likely by 2010.

Discussion

Cause of Extinction What might have caused the extinction of *Smilodon?* Possible answers: competition with other predators, humans, climate change. L2

Activity

Time Line Have small groups of students use long sheets of butcher paper to make a time line, showing the mass extinctions shown in the graph in this feature. Ask them to research some representative organisms that became extinct during each mass extinction and illustrate their time lines with drawings of those organisms. Display the groups' time lines in the classroom. L2

Applying Math

Answer

About 1.6 million years

SCIENCE Stats

Extinct!

Did you know...

...The saber-toothed cat lived in the Americas from about 1.6 million to 8,000 years ago. *Smilodon*, the best-known saber-toothed cat, was among the most ferocious carnivores. It had large canine teeth, about 15 cm long, which it used to pierce the flesh of its prey.

Saber-toothed cat

Applying Math How many years did *Smilodon* live in the Americas before it became extinct?

Woolly mammoth

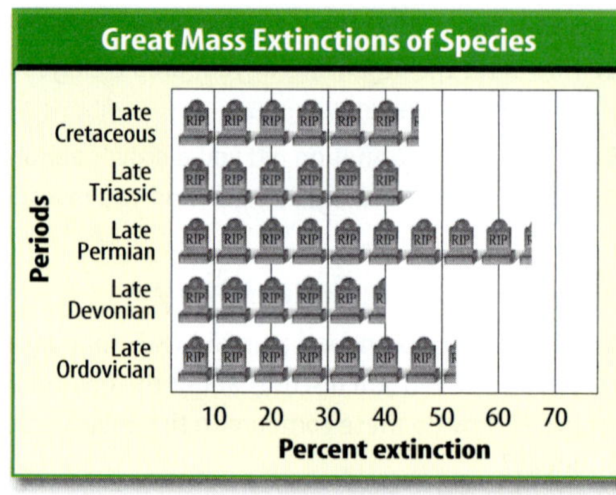

...The woolly mammoth lived in the cold tundra regions during the Ice Age. It looked rather like an elephant with long hair, had a mass between 5,300 kg and 7,300 kg, and was between 3 m and 4 m tall.

Write About It

Visit bookg.msscience.com/science_stats to research extinct animals. Trace the origins of each of the species and learn how long its kind existed on Earth.

Visual Learning

Graph of Mass Extinctions Which period had the largest percentage of extinctions? Permian What percentage of organisms became extinct at that time? almost 70 percent Which period had the smallest percentage of extinctions? Devonian What percentage of organisms became extinct at that time? about 40 percent

Write About It

Students will find that some organisms existed on Earth for hundreds of millions of years, while others existed for much shorter periods of time. Have students select several different organisms and make a bar graph that compares the length of time the organisms existed on Earth. Discuss with students how to select a scale for the graph that will allow them to graph the organisms they have chosen. L2

178 ◆ G CHAPTER 6 Geologic Time

Reviewing Main Ideas

Section 1 Life and Geologic Time

1. Geologic time is divided into eons, eras, periods, and epochs.
2. Divisions within the geologic time scale are based largely on major evolutionary changes in organisms.
3. Plate movements affect organic evolution.

Section 2 Early Earth History

1. Cyanobacteria evolved during Precambrian time. Trilobites, fish, and corals were abundant during the Paleozoic Era.
2. Plants and animals began to move onto land during the middle of the Paleozoic Era.
3. The Paleozoic Era was a time of mountain building. The Appalachian Mountains formed when several islands and finally Africa collided with North America.
4. At the end of the Paleozoic Era, many marine invertebrates became extinct.

Section 3 Middle and Recent Earth History

1. Reptiles and gymnosperms were dominant land life-forms in the Mesozoic Era. Mammals and angiosperms began to dominate the land in the Cenozoic Era.
2. Pangaea broke apart during the Mesozoic Era. Many mountain ranges formed during the Cenozoic Era.

Visualizing Main Ideas

Copy and complete the concept map on geologic time using the following choices: Cenozoic, Trilobites in oceans, Mammals common, Paleozoic, Dinosaurs roam Earth, *and* Abundant gymnosperms.

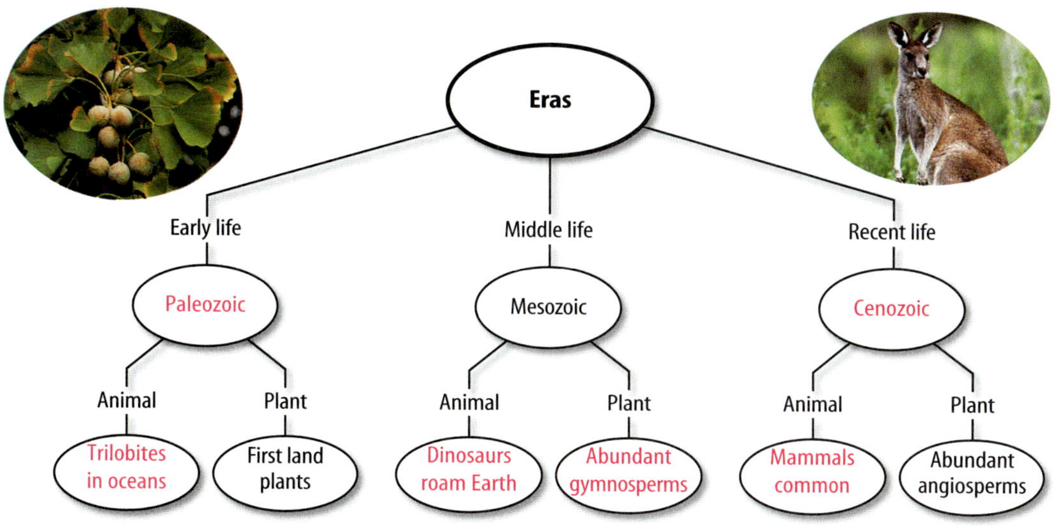

Identifying Misconceptions

Assess

Use the assessment as follow-up to page F after students have completed the chapter.

Demonstration Have students relate numerical and relative ages for the geologic time scale by accessing the *Virtual Dating Simulation* through the link at bookg.msscience.com.

Expected Outcome Students will recognize that time is measured not only by a watch in minutes and seconds, but also in huge dimensions using both relative and numerical methods.

Reinforcement Have a geologist visit the classroom to provide examples of how the geologic time scale is used in their work. L2

chapter 6 Study Guide

Reviewing Main Ideas

Summary statements can be used by students to review the major concepts of the chapter.

Visualizing Main Ideas

See student page.

Science Online
Visit bookg.msscience.com
/self_check_quiz
/interactive_tutor
/vocabulary_puzzlemaker
/chapter_review
/standardized_test

Assessment Transparency

For additional assessment questions, use the *Assessment Transparency* located in the transparency book.

Assessment

Assessment Transparency **Geologic Time**

Directions: *Carefully review the table and answer the following questions.*

Geologic Time Scale		
Era	When was it?	Types of species that appeared
Cenozoic	65 million years ago to today	Humans, larger mammals
Mesozoic	248–65 million years ago	Small mammals, birds, dinosaurs, angiosperms
Paleozoic	544–248 million years ago	Early reptiles, amphibians, fish, gymnosperms
Precambrian	4.5 billion years to 545 million years ago	Worms, jellyfish

1. According to the table, a fossil of a dinosaur would probably be from the ___.
   A Cenozoic Era
   B Mesozoic Era
   C Paleozoic Era
   D Precambrian Era
2. According to the table, which geologic period occurred more than 800 million years ago?
   F Cenozoic Era
   G Mesozoic Era
   H Paleozoic Era
   J Precambrian Era
3. According to the table, early ancestors of frogs would have appeared during the ___.
   A Cenozoic Era
   B Mesozoic Era
   C Paleozoic Era
   D Precambrian Era

L2

Geologic Time

Using Vocabulary

1. A change in the hereditary features of a species over a long period of time is organic evolution.
2. A record of events in Earth history is the geologic time scale.
3. The largest subdivision of geologic time is the eon.
4. The process by which the best-suited individuals survive in their environment is natural selection.
5. A group of individuals that normally breed only among themselves is a species.

Checking Concepts

6. D	11. A
7. D	12. A
8. D	13. C
9. A	14. B
10. B	15. A

Thinking Critically

16. Too much harmful ultraviolet radiation from the Sun reached Earth's surface.
17. Trilobites were widespread and individual species existed for a limited amount of time before becoming extinct.
18. Most organisms of Precambrian age lacked the hard parts that developed in the Paleozoic Era.
19. Evolution occurs through natural selection, which is the process by which organisms with characteristics best suited to their environment survive and reproduce.
20. Acquired traits are not hereditary.

Using Vocabulary

Cenozoic Era p. 174
cyanobacteria p. 163
eon p. 155
epoch p. 155
era p. 155
geologic time scale p. 154
Mesozoic Era p. 170
natural selection p. 157
organic evolution p. 156
Paleozoic Era p. 164
Pangaea p. 161
period p. 155
Precambrian time p. 162
species p. 156
trilobite p. 159

Fill in the blank with the correct word or words.

1. A change in the hereditary features of a species over a long period is ________.
2. A record of events in Earth history is the ________.
3. The largest subdivision of geologic time is the ________.
4. The process by which the best-suited individuals survive in their environment is ________.
5. A group of individuals that normally breed only among themselves is a(n) ________.

Checking Concepts

Choose the word or phrase that best completes the sentence.

6. How many millions of years ago did the era in which you live begin?
   A) 650 **C)** 1.6
   B) 245 **D)** 65
7. What is the process by which better-suited organisms survive and reproduce?
   A) endangerment **C)** gymnosperm
   B) extinction **D)** natural selection
8. During what period did the most recent ice age occur?
   A) Pennsylvanian **C)** Tertiary
   B) Triassic **D)** Quaternary
9. What is the next smaller division of geologic time after the era?
   A) period **C)** epoch
   B) stage **D)** eon
10. What was one of the earliest forms of life?
   A) gymnosperm **C)** angiosperm
   B) cyanobacterium **D)** dinosaur

Use the illustration below to answer question 11.

11. Consider the undisturbed rock layers in the figure above. If fossil X were a *Tyrannosaurus rex* bone, and fossil Y were a trilobite; then fossil Z could be which of the following?
   A) stromatolite **C)** angiosperm
   B) sabre-tooth cat **D)** *Homo sapiens*
12. During which era did the dinosaurs live?
   A) Mesozoic **C)** Miocene
   B) Paleozoic **D)** Cenozoic
13. Which type of plant has seeds without protective coverings?
   A) angiosperms **C)** gymnosperms
   B) apples **D)** magnolias
14. Which group of plants evolved during the Mesozoic Era and is dominant today?
   A) gymnosperms **C)** ginkgoes
   B) angiosperms **D)** algae
15. In which era did the Ediacaran fauna live?
   A) Precambrian **C)** Mesozoic
   B) Paleozoic **D)** Cenozoic

Science Online bookg.msscience.com/vocabulary_puzzlemaker

Use the Exam*View*® Pro Testmaker CD-ROM to:

- create multiple versions of tests
- create modified tests with one mouse click for inclusion students
- edit existing questions and add your own questions
- build tests aligned with state standards using built-in State Curriculum Tags
- change English tests to Spanish with one mouse click and vice versa

180 ◆ G CHAPTER REVIEW

chapter 6 Review

Thinking Critically

16. Infer why plants couldn't move onto land until an ozone layer formed.

17. Discuss why trilobites are classified as index fossils.

18. Compare and contrast the most significant difference between Precambrian life-forms and Paleozoic life-forms.

19. Describe how natural selection is related to organic evolution.

20. Explain In the early 1800s, a naturalist proposed that the giraffe species has a long neck as a result of years of stretching their necks to reach leaves in tall trees. Why isn't this true?

21. Infer Use the outlines of the present-day continents to make a sketch of Pangaea.

22. Form Hypotheses Suggest some reasons why trilobites might have become extinct at the end of the Paleozoic Era.

23. Interpret Data A student found what she thought was a piece of dinosaur bone in Pleistocene sediment. How likely is it that she is right? Explain.

24. Infer why mammals didn't become dominant until after the dinosaurs disappeared.

Performance Activities

25. Make a Model In the Section 2 Lab, you learned how a particular characteristic might evolve within a species. Modify the experimental model by using color instead of height as a characteristic. Design your activity with the understanding that varimals live in a dark-colored forest environment.

26. Make a Display Certain groups of animals have dominated the land throughout geologic time. Use your textbook and other references to discover some of the dominant species of each era. Make a display that illustrates some animals from each era. Be sure to include appropriate habitats.

Applying Math

Use the graph below to answer questions 27 and 28.

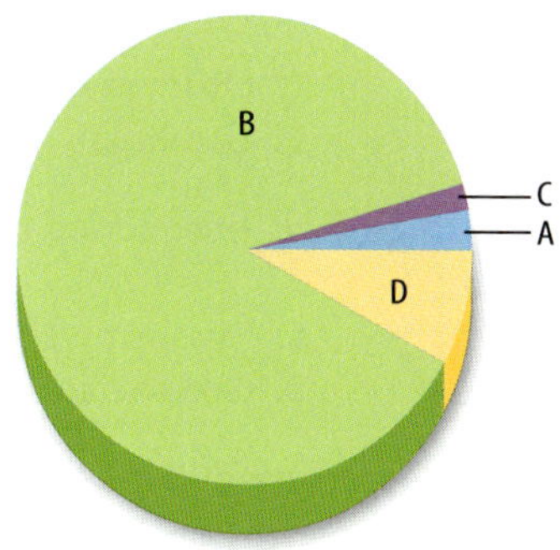

27. Modeling Geologic Time The circle graph above represents geologic time. Determine which interval of geologic time is represented by each portion of the graph. Which interval was longest? Which do we know the least about? Which of these intervals is getting larger?

28. Interpret Data The Cenozoic Era has lasted 65 million years. What percentage of Earth's 4.5-billion-year history is that?

chapter 6 Review

21. Students' illustrations should be similar to **Figure 11.**

22. Possible answers: plate tectonic processes that changed the configuration of land and seas, disappearance of shallow seas where they lived; meteorite impact

23. It is not likely that she is correct, because dinosaurs became extinct long before the Pleistocene Epoch.

24. Dinosaurs were better-adapted to the habitat until they became extinct. Mammals were then the best-adapted.

Performance Activities

25. Students should realize that black, as opposed to red, is an easier color to camouflage in a dark-colored forest environment. Use the **Performance Task Assessment List for Models** in **Performance Assessment in the Science Classroom,** p. 123.

26. Answers will vary

Applying Math

National Math Standards

5

27. A—Mesozoic Era, B—Precambrian time, C—Cenozoic Era, D—Paleozoic Era; Precambrian; Cenozoic Era

28. The Cenozoic Era represents about 1.4% of Earth's history; 65,000,000/4,500,000,000 years = 0.014.

Assessment Resources

Reproducible Masters

Chapter *Fast File* Resources
- Chapter Review, pp. 35–36
- Chapter Tests, pp. 37–40
- Assessment Transparency Activity, p. 47

Glencoe Science Web site
- Chapter Review Test
- Standardized Test Practice

Glencoe Technology

- Assessment Transparency
- Exam*View*® Pro Testmaker
- MindJogger Videoquiz
- Interactive Chalkboard

FAST FILE

Answer Sheet A practice answer sheet can be found at bookg.msscience.com/answer_sheet.

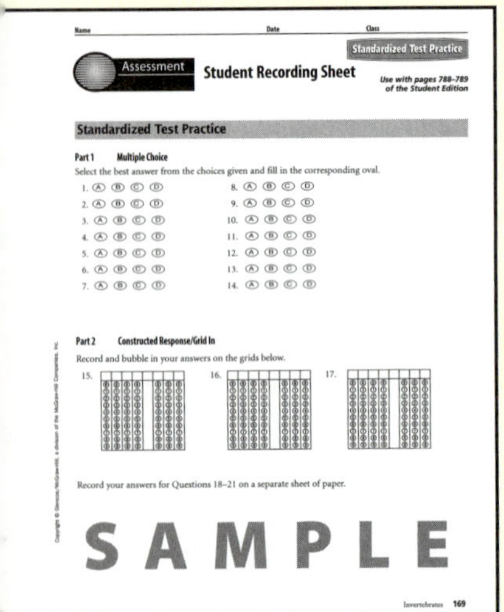

Part 1 Multiple Choice

1. A
2. C
3. B
4. C
5. D
6. A
7. C
8. D
9. B
10. B
11. C

Part 2 Short Response

12. Charles Darwin published a book describing evolution by natural selection.
13. The most favored hypothesis is that a meteorite struck Earth and caused severe environmental damage. Other hypotheses include climate change and volcanic gas emissions.

Part 1 Multiple Choice

Record your answers on the answer sheet provided by your teacher or on a sheet of paper.

Examine the diagram below. Then answer questions 1–3.

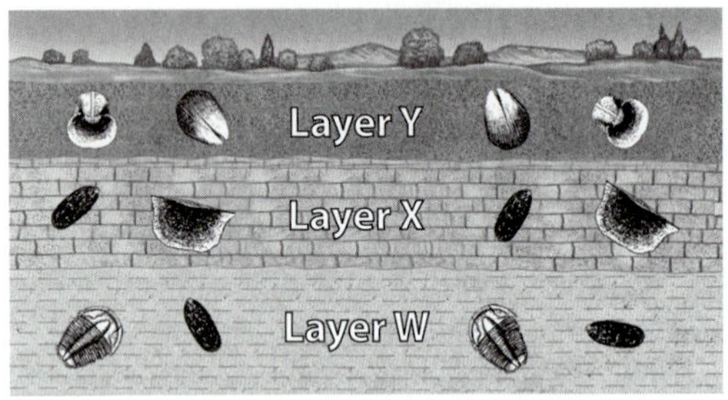

1. During which geologic time period did layer W form?
   A. Cambrian **C.** Devonian
   B. Ordovician **D.** Silurian

2. During which geologic time period did layer X form?
   A. Devonian **C.** Ordovician
   B. Silurian **D.** Cambrian

3. During which geologic time period did layer Y form?
   A. Cambrian **C.** Mississippian
   B. Silurian **D.** Ordovician

4. When did dinosaurs roam Earth?
   A. Precambrian time
   B. Paleozoic Era
   C. Mesozoic Era
   D. Cenozoic Era

5. What is the name of the supercontinent that formed at the end of the Paleozoic Era?
   A. Gondwanaland
   B. Eurasia
   C. Laurasia
   D. Pangaea

6. During which geologic period did modern humans evolve?
   A. Quaternary
   B. Triassic
   C. Ordovician
   D. Tertiary

7. How many body lobes did trilobites have?
   A. one **C.** three
   B. two **D.** four

8. Which mountain range formed because India collided with Asia?
   A. Alps **C.** Ural
   B. Andes **D.** Himalaya

Use the diagram below to answer questions 9–11.

Cenozoic Era	Quaternary Period	Holocene Epoch
		Pleistocene Epoch
	Tertiary Period	Pliocene Epoch
		Miocene Epoch
		Oligocene Epoch
		Eocene Epoch
		Paleocene Epoch

9. What is the oldest epoch in the Cenozoic Era?
   A. Pleistocene **C.** Miocene
   B. Paleocene **D.** Holocene

10. What is the youngest epoch in the Cenozoic Era?
    A. Miocene **C.** Paleocene
    B. Holocene **D.** Eocene

11. Which epoch is part of the Quaternary Period?
    A. Oligocene **C.** Pleistocene
    B. Eocene **D.** Pliocene

Part 2 Short Response/Grid In

14. Archaeopteryx is an early bird that had some dinosaur-like features. This fossil provides evidence that birds evolved from small, meat-eating dinosaurs.
15. Some dinosaur bones have a structure that is similar to modern mammal bones.
16. Stromatolites are layered, mound-shaped bodies of calcium carbonate. They form as cyanobacteria mats trap carbonate mud.
17. A species can be defined as a group of organisms that normally reproduces only with other members of its group.
18. 18.18 s
19. 100 s
20. 10.35 m/s; *T. rex* might have been slightly faster than Michael Johnson. However, if they could race, it probably would be close.

Part 2 Short Response/Grid In

Record your answers on the answer sheet provided by your teacher or on a sheet of paper.

12. Who was Charles Darwin? How did he contribute to science?

13. Explain one hypothesis about why dinosaurs might have become extinct.

14. Describe *Archaeopteryx*. Why is this an important fossil?

15. Why do many scientists think that dinosaurs were warm-blooded?

16. What are stromatolites? How do they form?

17. Define the term *species.*

Select one of the equations below to help you answer questions 18–20.

time = distance ÷ speed

or

speed = distance ÷ time

18. It recently was estimated that *T. rex* could run no faster than about 11 m/s. At this speed, how long would it take *T. rex* to run 200 m?

19. A typical ornithopod (plant-eating dinosaur that walked on two legs) probably moved at a speed of about 2 m/s. How long would it take this dinosaur to run 200 m?

20. In 1996, Michael Johnson ran 200 m in 19.32 s. What was his average speed? How does this compare with *T. rex?*

Test-Taking Tip

Show Your Work For constructed response questions, show all of your work and any calculations on your answer sheet.

Part 3 Open Ended

Record your answers on a sheet of paper.

Use the diagram below to answer questions 21 and 22. It shows the time ranges of various types of organisms on Earth. When a bar is wider, there were more species of that type of organism (higher diversity).

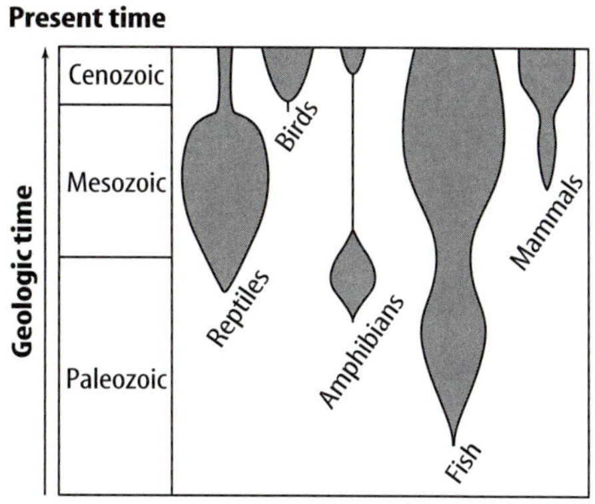

21. Describe how the diversity of reptiles changed through time.

22. How has the diversity of mammals changed through time? Do you see any relationship with how reptile diversity has changed?

23. Why might mammals in Australia be so much different than mammals on other continents?

24. Describe how natural selection might cause a species to change through time.

25. How did early photosynthetic organisms change the conditions on Earth to allow more advanced organisms to flourish?

26. What are mass extinctions? How have they affected life on Earth?

27. Write a description of what Earth was like during Precambrian time. Summarize how Earth was different than it is now.

Part 3 Open Ended

21. Reptiles were more diverse during the Mesozoic and less diverse during the Cenozoic.

22. Mammals were less diverse during the Mesozoic and became more diverse during the Cenozoic. Mammal diversity increased as reptile diversity decreased. Mammals likely filled some niches vacated by reptiles after the Cretaceous/Tertiary extinction.

23. Mammals evolved in isolation in Australia.

24. A population has natural variability, and random mutations occur. Organisms that have characteristics that are well suited to the environment survive longer and produce more offspring. The gene pool changes as a result.

25. These organisms produced oxygen. At first, the oxygen changed the composition of the seas. Then, ozone gas formed in the upper atmosphere and oxygen became an abundant atmospheric gas.

26. Mass extinctions occur when the extinction rate across a variety of groups is significantly higher than normal.

27. There was no life on land. Earth's interior was hotter, and volcanism was more common. During the earliest Precambrian, the seas also were lifeless. Later, microscopic life evolved. The Ediacaran fauna and some invertebrates lived in the seas during the latest Precambrian.

Rubrics

For more help evaluating open-ended assessment questions, see the rubric on p. 4T.

Student Resources

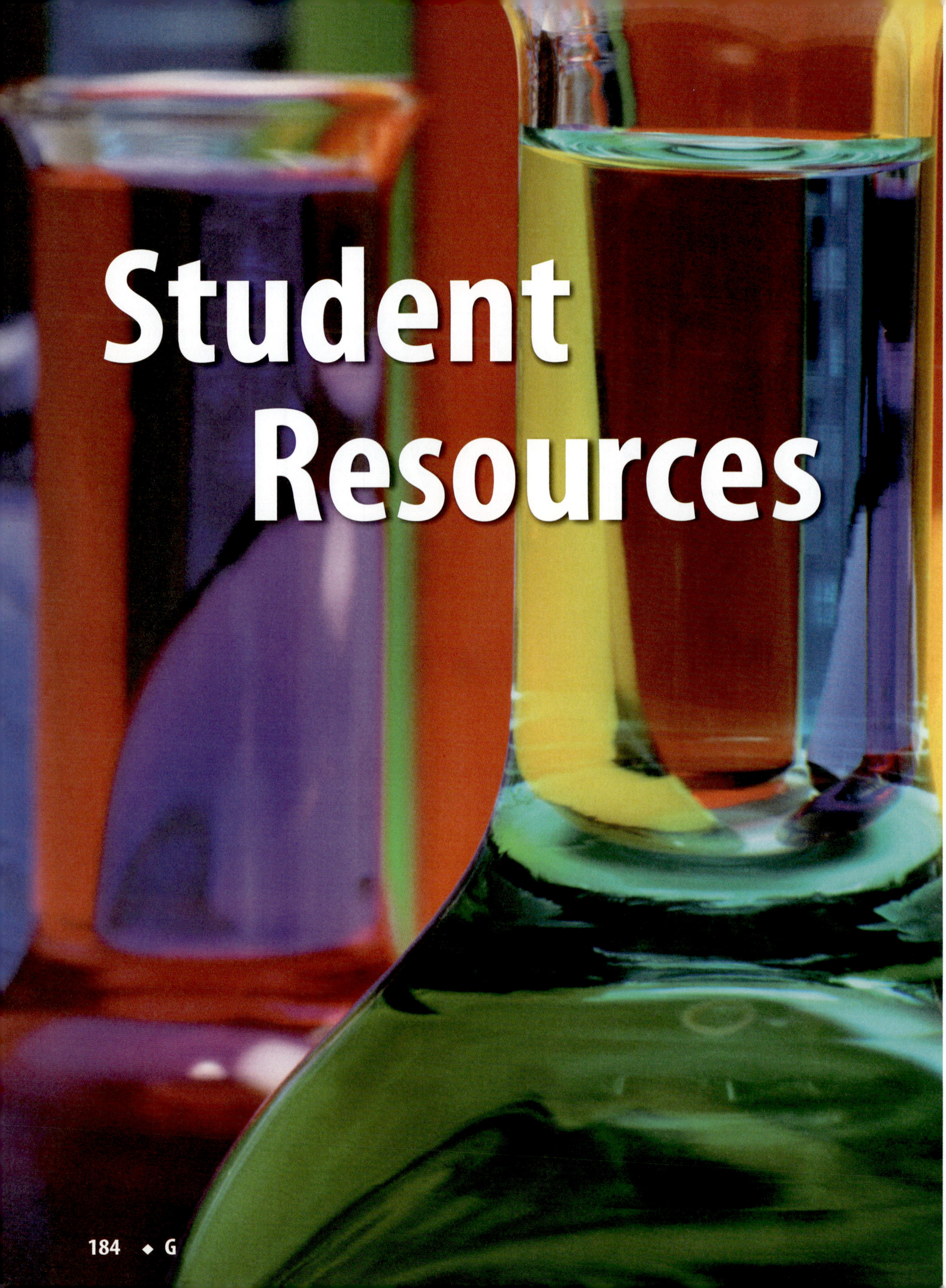

CONTENTS

Scientific Methods

Scientists use an orderly approach called scientific methods to solve problems. These include organizing and recording data so others can understand them. Scientists use many variations in these methods when they solve problems.

Identify a Question

The first step in a scientific investigation or experiment is to identify a question to be answered or a problem to be solved. For example, you might ask which gasoline is the most efficient.

Gather and Organize Information

After you have identified your question, begin gathering and organizing information. There are many ways to gather information, such as researching in a library, interviewing those knowledgeable about the subject, testing and working in the laboratory and field. Fieldwork is investigations and observations done outside of a laboratory.

Researching Information Before moving in a new direction, it is important to gather the information that already is known about the subject. Start by asking yourself questions to determine exactly what you need to know. Then you will look for the information in various reference sources, like the student is doing in **Figure 1.** Some sources may include textbooks, encyclopedias, government documents, professional journals, science magazines, and the Internet. Always list the sources of your information.

Figure 1 The Internet can be a valuable research tool.

Evaluate Sources of Information Not all sources of information are reliable. You should evaluate all of your sources of information, and use only those you know to be dependable. For example, if you are researching ways to make homes more energy efficient, a site written by the U.S. Department of Energy would be more reliable than a site written by a company that is trying to sell a new type of weatherproofing material. Also, remember that research always is changing. Consult the most current resources available to you. For example, a 1985 resource about saving energy would not reflect the most recent findings.

Sometimes scientists use data that they did not collect themselves, or conclusions drawn by other researchers. This data must be evaluated carefully. Ask questions about how the data were obtained, if the investigation was carried out properly, and if it has been duplicated exactly with the same results. Would you reach the same conclusion from the data? Only when you have confidence in the data can you believe it is true and feel comfortable using it.

Interpret Scientific Illustrations As you research a topic in science, you will see drawings, diagrams, and photographs to help you understand what you read. Some illustrations are included to help you understand an idea that you can't see easily by yourself, like the tiny particles in an atom in **Figure 2.** A drawing helps many people to remember details more easily and provides examples that clarify difficult concepts or give additional information about the topic you are studying. Most illustrations have labels or a caption to identify or to provide more information.

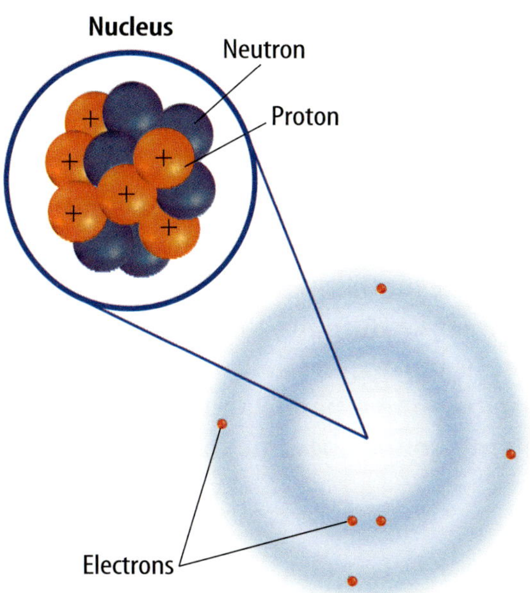

Figure 2 This drawing shows an atom of carbon with its six protons, six neutrons, and six electrons.

Concept Maps One way to organize data is to draw a diagram that shows relationships among ideas (or concepts). A concept map can help make the meanings of ideas and terms more clear, and help you understand and remember what you are studying. Concept maps are useful for breaking large concepts down into smaller parts, making learning easier.

Network Tree A type of concept map that not only shows a relationship, but how the concepts are related is a network tree, shown in **Figure 3.** In a network tree, the words are written in the ovals, while the description of the type of relationship is written across the connecting lines.

When constructing a network tree, write down the topic and all major topics on separate pieces of paper or notecards. Then arrange them in order from general to specific. Branch the related concepts from the major concept and describe the relationship on the connecting line. Continue to more specific concepts until finished.

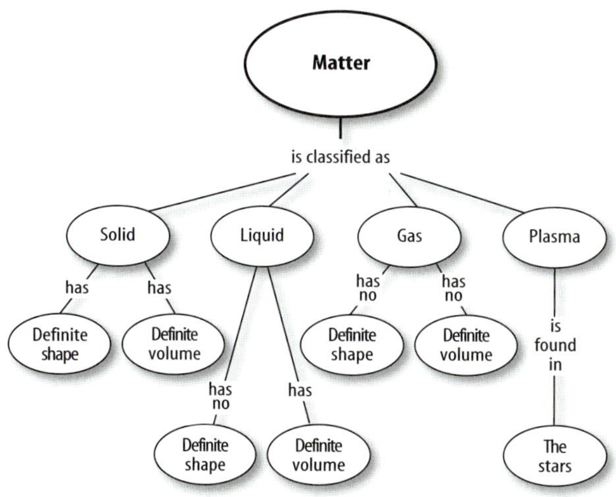

Figure 3 A network tree shows how concepts or objects are related.

Events Chain Another type of concept map is an events chain. Sometimes called a flow chart, it models the order or sequence of items. An events chain can be used to describe a sequence of events, the steps in a procedure, or the stages of a process.

When making an events chain, first find the one event that starts the chain. This event is called the initiating event. Then, find the next event and continue until the outcome is reached, as shown in **Figure 4.**

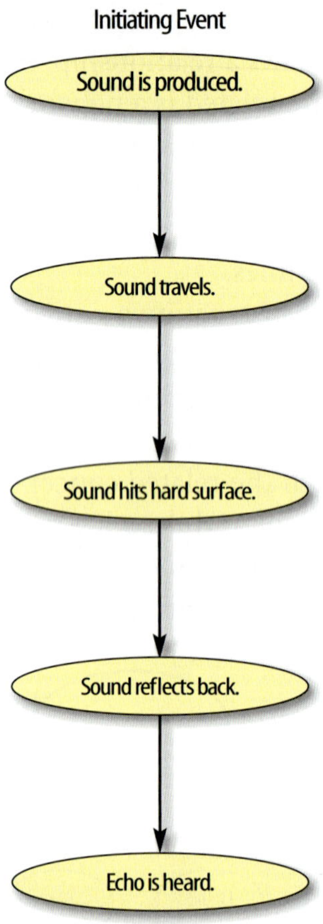

Figure 4 Events-chain concept maps show the order of steps in a process or event. This concept map shows how a sound makes an echo.

Cycle Map A specific type of events chain is a cycle map. It is used when the series of events do not produce a final outcome, but instead relate back to the beginning event, such as in **Figure 5.** Therefore, the cycle repeats itself.

To make a cycle map, first decide what event is the beginning event. This is also called the initiating event. Then list the next events in the order that they occur, with the last event relating back to the initiating event. Words can be written between the events that describe what happens from one event to the next. The number of events in a cycle map can vary, but usually contain three or more events.

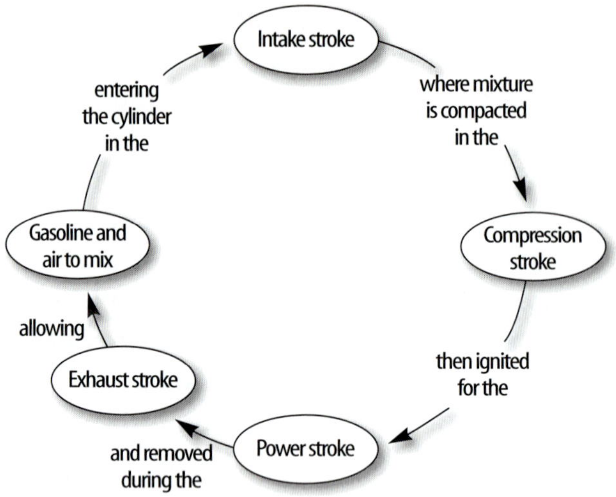

Figure 5 A cycle map shows events that occur in a cycle.

Spider Map A type of concept map that you can use for brainstorming is the spider map. When you have a central idea, you might find that you have a jumble of ideas that relate to it but are not necessarily clearly related to each other. The spider map on sound in **Figure 6** shows that if you write these ideas outside the main concept, then you can begin to separate and group unrelated terms so they become more useful.

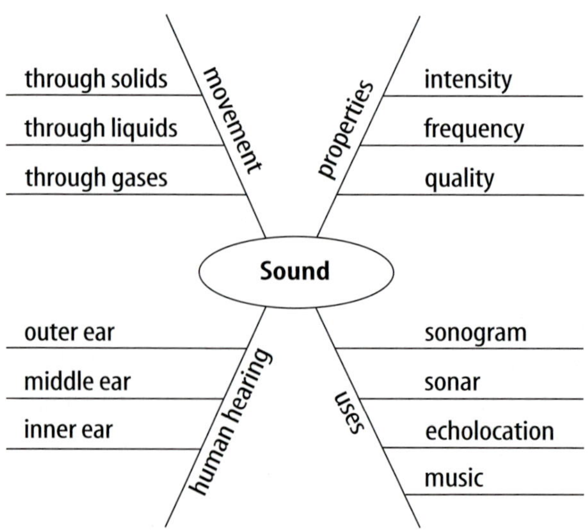

Figure 6 A spider map allows you to list ideas that relate to a central topic but not necessarily to one another.

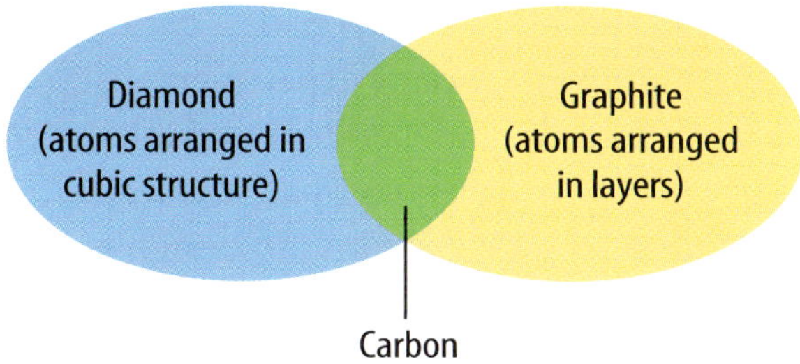

Figure 7 This Venn diagram compares and contrasts two substances made from carbon.

Venn Diagram To illustrate how two subjects compare and contrast you can use a Venn diagram. You can see the characteristics that the subjects have in common and those that they do not, shown in **Figure 7.**

To create a Venn diagram, draw two overlapping ovals that that are big enough to write in. List the characteristics unique to one subject in one oval, and the characteristics of the other subject in the other oval. The characteristics in common are listed in the overlapping section.

Make and Use Tables One way to organize information so it is easier to understand is to use a table. Tables can contain numbers, words, or both.

To make a table, list the items to be compared in the first column and the characteristics to be compared in the first row. The title should clearly indicate the content of the table, and the column or row heads should be clear. Notice that in **Table 1** the units are included.

Table 1 Recyclables Collected During Week

Day of Week	Paper (kg)	Aluminum (kg)	Glass (kg)
Monday	5.0	4.0	12.0
Wednesday	4.0	1.0	10.0
Friday	2.5	2.0	10.0

Make a Model One way to help you better understand the parts of a structure, the way a process works, or to show things too large or small for viewing is to make a model. For example, an atomic model made of a plastic-ball nucleus and pipe-cleaner electron shells can help you visualize how the parts of an atom relate to each other. Other types of models can by devised on a computer or represented by equations.

Form a Hypothesis

A possible explanation based on previous knowledge and observations is called a hypothesis. After researching gasoline types and recalling previous experiences in your family's car you form a hypothesis—our car runs more efficiently because we use premium gasoline. To be valid, a hypothesis has to be something you can test by using an investigation.

Predict When you apply a hypothesis to a specific situation, you predict something about that situation. A prediction makes a statement in advance, based on prior observation, experience, or scientific reasoning. People use predictions to make everyday decisions. Scientists test predictions by performing investigations. Based on previous observations and experiences, you might form a prediction that cars are more efficient with premium gasoline. The prediction can be tested in an investigation.

Design an Experiment A scientist needs to make many decisions before beginning an investigation. Some of these include: how to carry out the investigation, what steps to follow, how to record the data, and how the investigation will answer the question. It also is important to address any safety concerns.

Science Skill Handbook

Test the Hypothesis

Now that you have formed your hypothesis, you need to test it. Using an investigation, you will make observations and collect data, or information. This data might either support or not support your hypothesis. Scientists collect and organize data as numbers and descriptions.

Follow a Procedure In order to know what materials to use, as well as how and in what order to use them, you must follow a procedure. **Figure 8** shows a procedure you might follow to test your hypothesis.

Procedure

1. Use regular gasoline for two weeks.
2. Record the number of kilometers between fill-ups and the amount of gasoline used.
3. Switch to premium gasoline for two weeks.
4. Record the number of kilometers between fill-ups and the amount of gasoline used.

Figure 8 A procedure tells you what to do step by step.

Identify and Manipulate Variables and Controls In any experiment, it is important to keep everything the same except for the item you are testing. The one factor you change is called the independent variable. The change that results is the dependent variable. Make sure you have only one independent variable, to assure yourself of the cause of the changes you observe in the dependent variable. For example, in your gasoline experiment the type of fuel is the independent variable. The dependent variable is the efficiency.

Many experiments also have a control—an individual instance or experimental subject for which the independent variable is not changed. You can then compare the test results to the control results. To design a control you can have two cars of the same type. The control car uses regular gasoline for four weeks. After you are done with the test, you can compare the experimental results to the control results.

Collect Data

Whether you are carrying out an investigation or a short observational experiment, you will collect data, as shown in **Figure 9.** Scientists collect data as numbers and descriptions and organize it in specific ways.

Observe Scientists observe items and events, then record what they see. When they use only words to describe an observation, it is called qualitative data. Scientists' observations also can describe how much there is of something. These observations use numbers, as well as words, in the description and are called quantitative data. For example, if a sample of the element gold is described as being "shiny and very dense" the data are qualitative. Quantitative data on this sample of gold might include "a mass of 30 g and a density of 19.3 g/cm^3."

Figure 9 Collecting data is one way to gather information directly.

Figure 10 Record data neatly and clearly so it is easy to understand.

When you make observations you should examine the entire object or situation first, and then look carefully for details. It is important to record observations accurately and completely. Always record your notes immediately as you make them, so you do not miss details or make a mistake when recording results from memory. Never put unidentified observations on scraps of paper. Instead they should be recorded in a notebook, like the one in **Figure 10.** Write your data neatly so you can easily read it later. At each point in the experiment, record your observations and label them. That way, you will not have to determine what the figures mean when you look at your notes later. Set up any tables that you will need to use ahead of time, so you can record any observations right away. Remember to avoid bias when collecting data by not including personal thoughts when you record observations. Record only what you observe.

Estimate Scientific work also involves estimating. To estimate is to make a judgment about the size or the number of something without measuring or counting. This is important when the number or size of an object or population is too large or too difficult to accurately count or measure.

Sample Scientists may use a sample or a portion of the total number as a type of estimation. To sample is to take a small, representative portion of the objects or organisms of a population for research. By making careful observations or manipulating variables within that portion of the group, information is discovered and conclusions are drawn that might apply to the whole population. A poorly chosen sample can be unrepresentative of the whole. If you were trying to determine the rainfall in an area, it would not be best to take a rainfall sample from under a tree.

Measure You use measurements everyday. Scientists also take measurements when collecting data. When taking measurements, it is important to know how to use measuring tools properly. Accuracy also is important.

Length To measure length, the distance between two points, scientists use meters. Smaller measurements might be measured in centimeters or millimeters.

Length is measured using a metric ruler or meter stick. When using a metric ruler, line up the 0-cm mark with the end of the object being measured and read the number of the unit where the object ends. Look at the metric ruler shown in **Figure 11.** The centimeter lines are the long, numbered lines, and the shorter lines are millimeter lines. In this instance, the length would be 4.50 cm.

Figure 11 This metric ruler has centimeter and millimeter divisions.

Mass The SI unit for mass is the kilogram (kg). Scientists can measure mass using units formed by adding metric prefixes to the unit gram (g), such as milligram (mg). To measure mass, you might use a triple-beam balance similar to the one shown in **Figure 12.** The balance has a pan on one side and a set of beams on the other side. Each beam has a rider that slides on the beam.

When using a triple-beam balance, place an object on the pan. Slide the largest rider along its beam until the pointer drops below zero. Then move it back one notch. Repeat the process for each rider proceeding from the larger to smaller until the pointer swings an equal distance above and below the zero point. Sum the masses on each beam to find the mass of the object. Move all riders back to zero when finished.

Instead of putting materials directly on the balance, scientists often take a tare of a container. A tare is the mass of a container into which objects or substances are placed for measuring their masses. To mass objects or substances, find the mass of a clean container. Remove the container from the pan, and place the object or substances in the container. Find the mass of the container with the materials in it. Subtract the mass of the empty container from the mass of the filled container to find the mass of the materials you are using.

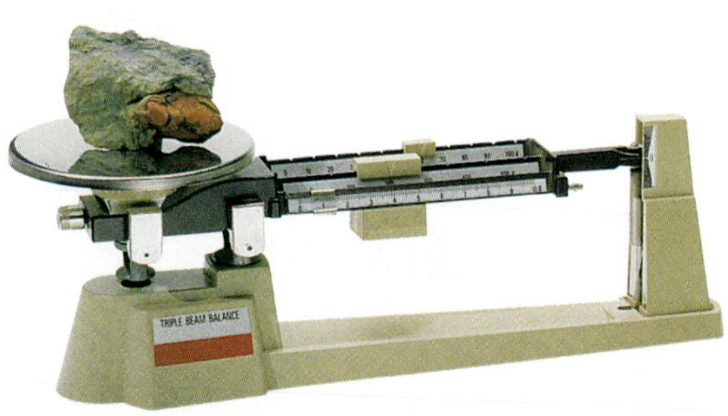

Figure 12 A triple-beam balance is used to determine the mass of an object.

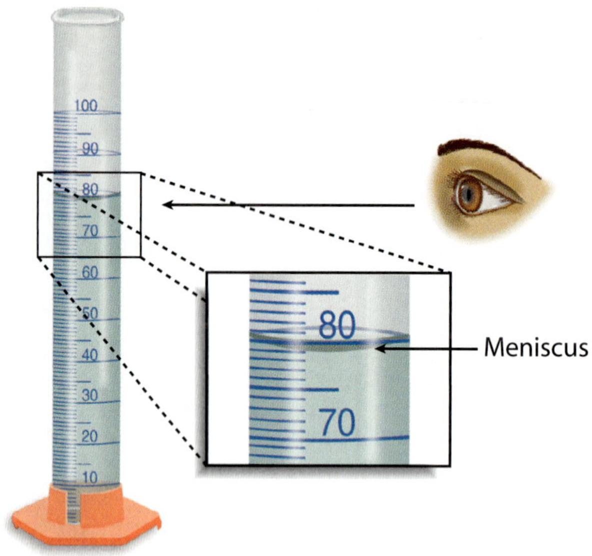

Figure 13 Graduated cylinders measure liquid volume.

Liquid Volume To measure liquids, the unit used is the liter. When a smaller unit is needed, scientists might use a milliliter. Because a milliliter takes up the volume of a cube measuring 1 cm on each side it also can be called a cubic centimeter ($cm^3 = cm \times cm \times cm$).

You can use beakers and graduated cylinders to measure liquid volume. A graduated cylinder, shown in **Figure 13,** is marked from bottom to top in milliliters. In lab, you might use a 10-mL graduated cylinder or a 100-mL graduated cylinder. When measuring liquids, notice that the liquid has a curved surface. Look at the surface at eye level, and measure the bottom of the curve. This is called the meniscus. The graduated cylinder in **Figure 13** contains 79.0 mL, or 79.0 cm^3, of a liquid.

Temperature Scientists often measure temperature using the Celsius scale. Pure water has a freezing point of 0°C and boiling point of 100°C. The unit of measurement is degrees Celsius. Two other scales often used are the Fahrenheit and Kelvin scales.

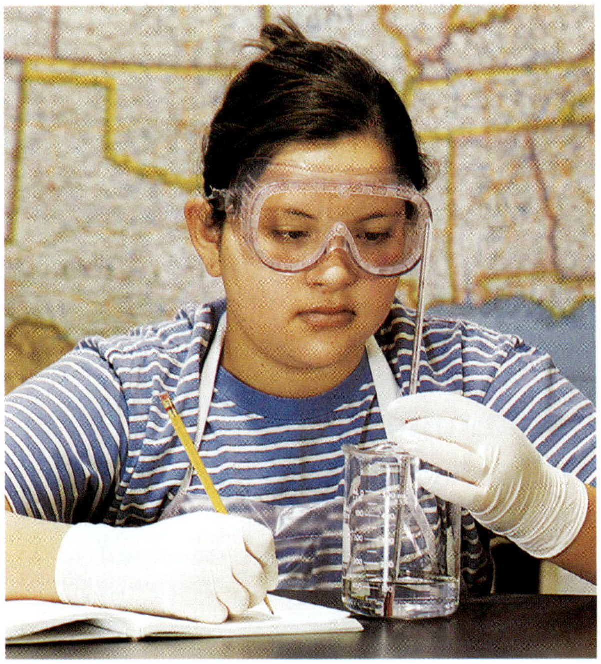

Figure 14 A thermometer measures the temperature of an object.

Scientists use a thermometer to measure temperature. Most thermometers in a laboratory are glass tubes with a bulb at the bottom end containing a liquid such as colored alcohol. The liquid rises or falls with a change in temperature. To read a glass thermometer like the thermometer in **Figure 14,** rotate it slowly until a red line appears. Read the temperature where the red line ends.

Form Operational Definitions An operational definition defines an object by how it functions, works, or behaves. For example, when you are playing hide and seek and a tree is home base, you have created an operational definition for a tree.

Objects can have more than one operational definition. For example, a ruler can be defined as a tool that measures the length of an object (how it is used). It can also be a tool with a series of marks used as a standard when measuring (how it works).

Analyze the Data

To determine the meaning of your observations and investigation results, you will need to look for patterns in the data. Then you must think critically to determine what the data mean. Scientists use several approaches when they analyze the data they have collected and recorded. Each approach is useful for identifying specific patterns.

Interpret Data The word *interpret* means "to explain the meaning of something." When analyzing data from an experiement, try to find out what the data show. Identify the control group and the test group to see whether or not changes in the independent variable have had an effect. Look for differences in the dependent variable between the control and test groups.

Classify Sorting objects or events into groups based on common features is called classifying. When classifying, first observe the objects or events to be classified. Then select one feature that is shared by some members in the group, but not by all. Place those members that share that feature in a subgroup. You can classify members into smaller and smaller subgroups based on characteristics. Remember that when you classify, you are grouping objects or events for a purpose. Keep your purpose in mind as you select the features to form groups and subgroups.

Compare and Contrast Observations can be analyzed by noting the similarities and differences between two more objects or events that you observe. When you look at objects or events to see how they are similar, you are comparing them. Contrasting is looking for differences in objects or events.

Recognize Cause and Effect A cause is a reason for an action or condition. The effect is that action or condition. When two events happen together, it is not necessarily true that one event caused the other. Scientists must design a controlled investigation to recognize the exact cause and effect.

Draw Conclusions

When scientists have analyzed the data they collected, they proceed to draw conclusions about the data. These conclusions are sometimes stated in words similar to the hypothesis that you formed earlier. They may confirm a hypothesis, or lead you to a new hypothesis.

Infer Scientists often make inferences based on their observations. An inference is an attempt to explain observations or to indicate a cause. An inference is not a fact, but a logical conclusion that needs further investigation. For example, you may infer that a fire has caused smoke. Until you investigate, however, you do not know for sure.

Apply When you draw a conclusion, you must apply those conclusions to determine whether the data supports the hypothesis. If your data do not support your hypothesis, it does not mean that the hypothesis is wrong. It means only that the result of the investigation did not support the hypothesis. Maybe the experiment needs to be redesigned, or some of the initial observations on which the hypothesis was based were incomplete or biased. Perhaps more observation or research is needed to refine your hypothesis. A successful investigation does not always come out the way you originally predicted.

Avoid Bias Sometimes a scientific investigation involves making judgments. When you make a judgment, you form an opinion. It is important to be honest and not to allow any expectations of results to bias your judgments. This is important throughout the entire investigation, from researching to collecting data to drawing conclusions.

Communicate

The communication of ideas is an important part of the work of scientists. A discovery that is not reported will not advance the scientific community's understanding or knowledge. Communication among scientists also is important as a way of improving their investigations.

Scientists communicate in many ways, from writing articles in journals and magazines that explain their investigations and experiments, to announcing important discoveries on television and radio. Scientists also share ideas with colleagues on the Internet or present them as lectures, like the student is doing in **Figure 15.**

Figure 15 A student communicates to his peers about his investigation.

SAFETY SYMBOLS	HAZARD	EXAMPLES	PRECAUTION	REMEDY
DISPOSAL	Special disposal procedures need to be followed.	certain chemicals, living organisms	Do not dispose of these materials in the sink or trash can.	Dispose of wastes as directed by your teacher.
BIOLOGICAL	Organisms or other biological materials that might be harmful to humans	bacteria, fungi, blood, unpreserved tissues, plant materials	Avoid skin contact with these materials. Wear mask or gloves.	Notify your teacher if you suspect contact with material. Wash hands thoroughly.
EXTREME TEMPERATURE	Objects that can burn skin by being too cold or too hot	boiling liquids, hot plates, dry ice, liquid nitrogen	Use proper protection when handling.	Go to your teacher for first aid.
SHARP OBJECT	Use of tools or glassware that can easily puncture or slice skin	razor blades, pins, scalpels, pointed tools, dissecting probes, broken glass	Practice common-sense behavior and follow guidelines for use of the tool.	Go to your teacher for first aid.
FUME	Possible danger to respiratory tract from fumes	ammonia, acetone, nail polish remover, heated sulfur, moth balls	Make sure there is good ventilation. Never smell fumes directly. Wear a mask.	Leave foul area and notify your teacher immediately.
ELECTRICAL	Possible danger from electrical shock or burn	improper grounding, liquid spills, short circuits, exposed wires	Double-check setup with teacher. Check condition of wires and apparatus.	Do not attempt to fix electrical problems. Notify your teacher immediately.
IRRITANT	Substances that can irritate the skin or mucous membranes of the respiratory tract	pollen, moth balls, steel wool, fiberglass, potassium permanganate	Wear dust mask and gloves. Practice extra care when handling these materials.	Go to your teacher for first aid.
CHEMICAL	Chemicals can react with and destroy tissue and other materials	bleaches such as hydrogen peroxide; acids such as sulfuric acid, hydrochloric acid; bases such as ammonia, sodium hydroxide	Wear goggles, gloves, and an apron.	Immediately flush the affected area with water and notify your teacher.
TOXIC	Substance may be poisonous if touched, inhaled, or swallowed.	mercury, many metal compounds, iodine, poinsettia plant parts	Follow your teacher's instructions.	Always wash hands thoroughly after use. Go to your teacher for first aid.
FLAMMABLE	Flammable chemicals may be ignited by open flame, spark, or exposed heat.	alcohol, kerosene, potassium permanganate	Avoid open flames and heat when using flammable chemicals.	Notify your teacher immediately. Use fire safety equipment if applicable.
OPEN FLAME	Open flame in use, may cause fire.	hair, clothing, paper, synthetic materials	Tie back hair and loose clothing. Follow teacher's instruction on lighting and extinguishing flames.	Notify your teacher immediately. Use fire safety equipment if applicable.

Eye Safety
Proper eye protection should be worn at all times by anyone performing or observing science activities.

Clothing Protection
This symbol appears when substances could stain or burn clothing.

Animal Safety
This symbol appears when safety of animals and students must be ensured.

Handwashing
After the lab, wash hands with soap and water before removing goggles.

Safety in the Science Laboratory

The science laboratory is a safe place to work if you follow standard safety procedures. Being responsible for your own safety helps to make the entire laboratory a safer place for everyone. When performing any lab, read and apply the caution statements and safety symbol listed at the beginning of the lab.

General Safety Rules

1. Obtain your teacher's permission to begin all investigations and use laboratory equipment.
2. Study the procedure. Ask your teacher any questions. Be sure you understand safety symbols shown on the page.
3. Notify your teacher about allergies or other health conditions which can affect your participation in a lab.
4. Learn and follow use and safety procedures for your equipment. If unsure, ask your teacher.
5. Never eat, drink, chew gum, apply cosmetics, or do any personal grooming in the lab. Never use lab glassware as food or drink containers. Keep your hands away from your face and mouth.
6. Know the location and proper use of the safety shower, eye wash, fire blanket, and fire alarm.

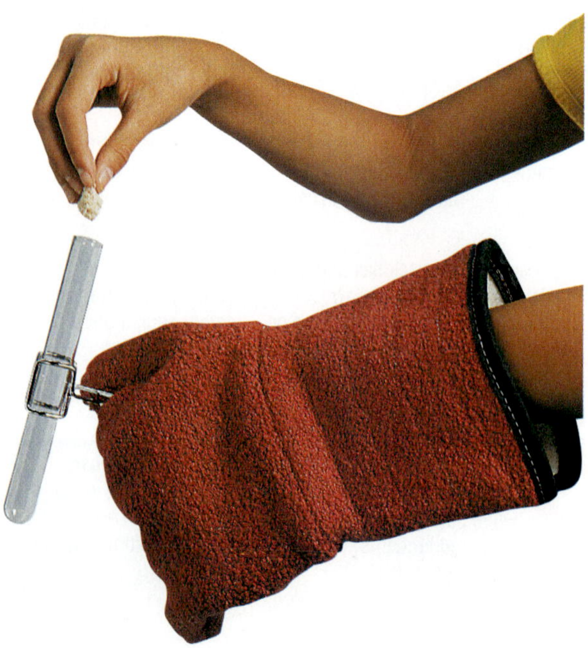

Prevent Accidents

1. Use the safety equipment provided to you. Goggles and a safety apron should be worn during investigations.
2. Do NOT use hair spray, mousse, or other flammable hair products. Tie back long hair and tie down loose clothing.
3. Do NOT wear sandals or other open-toed shoes in the lab.
4. Remove jewelry on hands and wrists. Loose jewelry, such as chains and long necklaces, should be removed to prevent them from getting caught in equipment.
5. Do not taste any substances or draw any material into a tube with your mouth.
6. Proper behavior is expected in the lab. Practical jokes and fooling around can lead to accidents and injury.
7. Keep your work area uncluttered.

Laboratory Work

1. Collect and carry all equipment and materials to your work area before beginning a lab.
2. Remain in your own work area unless given permission by your teacher to leave it.

3. Always slant test tubes away from yourself and others when heating them, adding substances to them, or rinsing them.
4. If instructed to smell a substance in a container, hold the container a short distance away and fan vapors towards your nose.
5. Do NOT substitute other chemicals/substances for those in the materials list unless instructed to do so by your teacher.
6. Do NOT take any materials or chemicals outside of the laboratory.
7. Stay out of storage areas unless instructed to be there and supervised by your teacher.

Laboratory Cleanup

1. Turn off all burners, water, and gas, and disconnect all electrical devices.
2. Clean all pieces of equipment and return all materials to their proper places.
3. Dispose of chemicals and other materials as directed by your teacher. Place broken glass and solid substances in the proper containers. Never discard materials in the sink.
4. Clean your work area.
5. Wash your hands with soap and water thoroughly BEFORE removing your goggles.

Emergencies

1. Report any fire, electrical shock, glassware breakage, spill, or injury, no matter how small, to your teacher immediately. Follow his or her instructions.
2. If your clothing should catch fire, STOP, DROP, and ROLL. If possible, smother it with the fire blanket or get under a safety shower. NEVER RUN.
3. If a fire should occur, turn off all gas and leave the room according to established procedures.
4. In most instances, your teacher will clean up spills. Do NOT attempt to clean up spills unless you are given permission and instructions to do so.
5. If chemicals come into contact with your eyes or skin, notify your teacher immediately. Use the eyewash or flush your skin or eyes with large quantities of water.
6. The fire extinguisher and first-aid kit should only be used by your teacher unless it is an extreme emergency and you have been given permission.
7. If someone is injured or becomes ill, only a professional medical provider or someone certified in first aid should perform first-aid procedures.

1 3-D Maps

Time Required 45 minutes

Materials
- Students can use 11-in×17-in sheets of paper instead of poster board.
- Metric tape measures are easier to use during this lab than metersticks.

Safety Precaution Caution students not to climb on top of furniture.

Teaching Strategies
- Instruct students to include the scale they used in the corner of their maps. They should also provide a title for their maps.
- Encourage students to choose box-shaped furniture pieces to make the model construction easier.

Conclude and Apply
1. Answers will vary but 1 cm: 1 m is a common scale.
2. Answers may include studying the topography of an ecosystem in preparation for field work.

Extra Try at Home Labs

Extra Try at Home Labs

EXTRA Try at Home Labs

From Your Kitchen, Junk Drawer, or Yard

1 3-D Maps

Real-World Question
How can you make a topographic map of a room in your house?

Possible Materials
- meterstick
- metric ruler
- metric tape measure
- poster board
- black marker
- construction paper
- transparent tape

Procedure
1. Measure the length and width of your room in meters. Include the measurements of any odd shapes or angles in the room, as well as elevations.
2. Decide upon a scale for your map.
3. Using your scale, draw the outline of your room on the poster board.
4. Measure the length, width, and height of a piece of furniture.
5. Using your scale, measure and cut out the sides for a model of the furniture piece from construction paper. Tape the pieces of the model together.
6. Place your furniture model on your map to match the actual piece's location in your room.
7. Construct two or three other models of furniture for your map.

Conclude and Apply
1. What scale did you use for your map?
2. Infer how a biologist might use a topographic map.

2 Rock and Roll

Real-World Question
How can we model the weathering of rock?

Possible Materials
- white glue
- sand
- plastic bowl
- plastic spoon
- cookie tray
- barbecue brush
- aluminum foil
- cooking oil
- empty coffee can with lid
- water
- measuring cup
- transparent packing tape

Procedure
1. Make your own sedimentary rocks by adding equal amounts of white glue and sand to a bowl. Stir the sand and glue together until you make several small lumps.
2. Lay aluminum foil on the bottom of a cookie tray and coat the foil with cooking oil.
3. Lay your rocks on the tray in direct sunlight for three days until they dry.
4. Place your rocks in a coffee can and pour 50 mL of water into the can.
5. Place the lid on the can and secure the lid with thick transparent tape.
6. Shake the contents of the can for 4 min, open the lid, and observe your rocks.

Conclude and Apply
1. Describe what happened to your sedimentary rocks.
2. Infer how this activity modeled the weathering of rocks.

Adult supervision required for all labs.

These labs are available at bookg.msscience.com.

2 Rock and Roll

Time Required 20 minutes, plus five-minute session after three days

Materials Students should use a plastic bowl and spoon to mix the glue and sand.

Safety Precaution Caution students to hold the lids of the coffee cans while shaking them in case the tape does not hold.

Teaching Strategy To save time, consider having students make sedimentary rocks at home and then complete the weathering part of the lab in class.

Conclude and Apply
1. Students will observe some of the rocks worn away into sediments.
2. Weathering agents such as water wear away particles from rocks to form sediments.

198 ◆ G STUDENT RESOURCES

3 Modeling Mudslides

Real-World Question
How can mudslides be prevented?

Possible Materials
- deep, rectangular basin
- bowl
- measuring cup
- hose
- water
- soil (potting or garden)
- blocks of wood or bricks
- protractor
- sod
- several houseplants

Procedure
1. Pour soil into one-half of a deep, rectangular basin. Lightly pack the soil down.
2. Prop the container up on wooden blocks until the end with the soil in it is raised to a 60° angle.
3. Fill a bowl with water and slowly pour the water over the soil near the edge of the basin.
4. Continue pouring water onto the soil until a mudslide is created.
5. Clean the basin and repeat the lab, but plant sod or houseplants in the soil before watering it.

Conclude and Apply
1. Describe what happened to the soil without plants and with plants.
2. Explain how this lab modeled a mudslide.
3. Describe the relationship between vegetation and mudslides.

4 In Deep Water

Real-World Question
How can the water table and a well be modeled?

Possible Materials
- clear-plastic drink bottle (500-mL)
- aquarium gravel
- water
- blue food dye
- measuring cup
- long dropper

Procedure
1. Fill a clear-plastic bottle with aquarium gravel.
2. Pour 450 mL of water into the the measuring cup and add several drops of blue food dye.
3. Pour 300 mL of the blue water into the bottle with the gravel. Insert the dropper down into the gravel and try to suck out some of the water.
4. Pour another 150 mL of water into the bottle and try to suck out some of the water with the dropper.

Conclude and Apply
1. Describe how this lab models the water table and a well.
2. Infer how deep a well must be dug for it to yield water.
3. Infer why some wells only yield water at certain times of the year.

Adult supervision required for all labs.

3 Modeling Mudslides

Time Required 20 minutes

Materials
- Students may have to empty the water from their basins before a mudslide is created if they pack their soil too tight.
- Grass sod works better than houseplants.

Safety Precaution Dispose of the soil where it will not pollute water.

Teaching Strategy Have students leave the soil near the edges of the container loose.

Conclude and Apply
1. Without the plants, the entire body of soil slides down the slope. Plants help prevent mudslides.
2. Mudflows are triggered when exposed soil becomes saturated with water from heavy rains.
3. The roots of vegetation help stabilize soil. When vegetation is removed, it is easier for mudslides to occur.

4 In Deep Water

Time Required 15 minutes

Materials
- White gravel provides an excellent contrast to the blue water.
- Larger plastic bottles can be used but the amounts of water added to the bottle will need to be adjusted.

Safety Precaution Students should never drink anything in the science lab.

Teaching Strategy Once the lab is completed, draw a cross-sectional diagram on the board showing the soil beneath the surface of the ground, the water table, a well, and the well's shaft.

Conclude and Apply
1. The level of the blue water represents the water table, and the eyedropper represents a well.
2. A well shaft must be dug below the water table.
3. The water table rises during wet seasons and falls during dry seasons. Some wells can only be dug deep enough to be below the water table during the wet season.

5 Making Burrows

Time Required 30 minutes

Materials A bright color gelatin works best for the top layer.

Safety Precaution Caution students not to mix the powder too vigorously.

Teaching Strategy Remind students that many marine organisms burrow into the sediment and that these burrows are a type of trace fossil.

Conclude and Apply

1. The two layers became more mixed as more burrows were made.
2. At first, distinct burrows are visible. After a lot of burrowing, individual burrows no longer can be discerned. The layers have a mottled appearance.
3. Burrowing animals mix thin layers of sediment on the ocean floor. This burrowing can be recognized in rock, such as limestone. The amount of mottling indicates how intensely the sediment was burrowed.

Extra Try at Home Labs

Extra Try at Home Labs

5 Making Burrows

Real-World Question

How does burrowing affect sediment layers?

Possible Materials

- clear-glass bowl
- white flour
- colored gelatin powder (3 packages)
- paintbrush
- pencil

Procedure

1. Add 3 cm of white flour to the bowl. Flatten the top of the flour layer.
2. Carefully sprinkle gelatin powder over the flour to form a colored layer about 0.25 cm thick.
3. The two layers represent two different layers of sediment.
4. Use a paintbrush or pencil to make "burrows" in the "sediment."
5. Make sure to make some of the burrows at the edge of the bowl so that you can see how it affects the sediment.
6. Continue to make more burrows and observe the effect on the two layers.

Conclude and Apply

1. How did the two layers of powder change as you continued to make burrows?
2. Were the "trace fossils" easy to recognize at first? How about after a lot of burrowing?
3. How do you think burrowing animals affect layers of sediment on the ocean floor? How could this burrowing be recognized in rock?

6 History in a Bottle

Real-World Question

What does the geologic column look like?

Possible Materials

- clear-plastic 2-liter bottle
- scissors
- 3-in × 5-in index cards
- colored markers
- permanent marker
- transparent tape
- metric ruler
- sand (3 different colors)
- aquarium gravel (3 different colors)

Procedure

1. Cut the top 5 cm off a clear-plastic 2-L soda bottle. Remove the label.
2. Cut 12 square cards measuring 2 cm × 2 cm.
3. Draw a picture of a trilobite, coral, fish, amphibian, insect, reptile, mouse, coniferous tree, dinosaur, bird, flower, large mammal, and human on the 12 cards.
4. Starting at the bottom, inside of the bottle and working up, tape the trilobite, coral, fish, amphibian, insect, and reptile pictures face out in that order. The reptile picture should be about half way up the bottle.
5. Pour red sand into the bottle until it covers your reptile picture.
6. Tape the mouse, conifer tree, dinosaur, bird, and flower pictures above the red sand in that order. Pour in blue sand until it covers the flower picture.
7. Tape the large mammal and human pictures above the blue sand. Pour green sand into the bottle until it covers the person.

Conclude and Apply

1. Research what era each color of sand represents.
2. Infer why few fossils of organisms living before the Paleozoic Era are found.

Adult supervision required for all labs.

These labs are available at bookg.msscience.com.

6 History in a Bottle

Time Required 45 minutes

Materials

- Colored gravel from a pet store can also be used for the sediment layers.
- Explain to students that the organism pictures are taped one above the other in a column.

Safety Precaution Caution students to be careful with the scissors.

Teaching Strategies

- The Paleozoic Era (red sand) will fill about 12 cm of the bottle, the Mesozoic Era (blue sand) about 8 cm, and the Cenozoic Era (green sand) about 3 cm.
- Encourage students to find pictures of organisms online.

Conclude and Apply

1. red: Paleozoic, blue: Mesozoic, and green: Cenozoic
2. Much of the fossil record has been destroyed by changes in Earth's crust.

Computer Skills

People who study science rely on computers, like the one in **Figure 16,** to record and store data and to analyze results from investigations. Whether you work in a laboratory or just need to write a lab report with tables, good computer skills are a necessity.

Using the computer comes with responsibility. Issues of ownership, security, and privacy can arise. Remember, if you did not author the information you are using, you must provide a source for your information. Also, anything on a computer can be accessed by others. Do not put anything on the computer that you would not want everyone to know. To add more security to your work, use a password.

Use a Word Processing Program

A computer program that allows you to type your information, change it as many times as you need to, and then print it out is called a word processing program. Word processing programs also can be used to make tables.

Figure 16 A computer will make reports neater and more professional looking.

Learn the Skill To start your word processing program, a blank document, sometimes called "Document 1," appears on the screen. To begin, start typing. To create a new document, click the ***New*** button on the standard tool bar. These tips will help you format the document.

- The program will automatically move to the next line; press *Enter* if you wish to start a new paragraph.
- Symbols, called non-printing characters, can be hidden by clicking the *Show/Hide* button on your toolbar.
- To insert text, move the cursor to the point where you want the insertion to go, click on the mouse once, and type the text.
- To move several lines of text, select the text and click the *Cut* button on your toolbar. Then position your cursor in the location that you want to move the cut text and click *Paste.* If you move to the wrong place, click *Undo.*
- The spell check feature does not catch words that are misspelled to look like other words, like "cold" instead of "gold." Always reread your document to catch all spelling mistakes.
- To learn about other word processing methods, read the user's manual or click on the *Help* button.
- You can integrate databases, graphics, and spreadsheets into documents by copying from another program and pasting it into your document, or by using desktop publishing (DTP). DTP software allows you to put text and graphics together to finish your document with a professional look. This software varies in how it is used and its capabilities.

Use a Database

A collection of facts stored in a computer and sorted into different fields is called a database. A database can be reorganized in any way that suits your needs.

Learn the Skill A computer program that allows you to create your own database is a database management system (DBMS). It allows you to add, delete, or change information. Take time to get to know the features of your database software.

- Determine what facts you would like to include and research to collect your information.
- Determine how you want to organize the information.
- Follow the instructions for your particular DBMS to set up fields. Then enter each item of data in the appropriate field.
- Follow the instructions to sort the information in order of importance.
- Evaluate the information in your database, and add, delete, or change as necessary.

Use the Internet

The Internet is a global network of computers where information is stored and shared. To use the Internet, like the students in **Figure 17,** you need a modem to connect your computer to a phone line and an Internet Service Provider account.

Learn the Skill To access internet sites and information, use a "Web browser," which lets you view and explore pages on the World Wide Web. Each page is its own site, and each site has its own address, called a URL. Once you have found a Web browser, follow these steps for a search (this also is how you search a database).

Figure 17 The Internet allows you to search a global network for a variety of information.

- Be as specific as possible. If you know you want to research "gold," don't type in "elements." Keep narrowing your search until you find what you want.
- Web sites that end in *.com* are commercial Web sites; *.org, .edu,* and *.gov* are non-profit, educational, or government Web sites.
- Electronic encyclopedias, almanacs, indexes, and catalogs will help locate and select relevant information.
- Develop a "home page" with relative ease. When developing a Web site, NEVER post pictures or disclose personal information such as location, names, or phone numbers. Your school or community usually can host your Web site. A basic understanding of HTML (hypertext mark-up language), the language of Web sites, is necessary. Software that creates HTML code is called authoring software, and can be downloaded free from many Web sites. This software allows text and pictures to be arranged as the software is writing the HTML code.

Use a Spreadsheet

A spreadsheet, shown in **Figure 18,** can perform mathematical functions with any data arranged in columns and rows. By entering a simple equation into a cell, the program can perform operations in specific cells, rows, or columns.

Learn the Skill Each column (vertical) is assigned a letter, and each row (horizontal) is assigned a number. Each point where a row and column intersect is called a cell, and is labeled according to where it is located—Column A, Row 1 (A1).

- Decide how to organize the data, and enter it in the correct row or column.
- Spreadsheets can use standard formulas or formulas can be customized to calculate cells.
- To make a change, click on a cell to make it activate, and enter the edited data or formula.
- Spreadsheets also can display your results in graphs. Choose the style of graph that best represents the data.

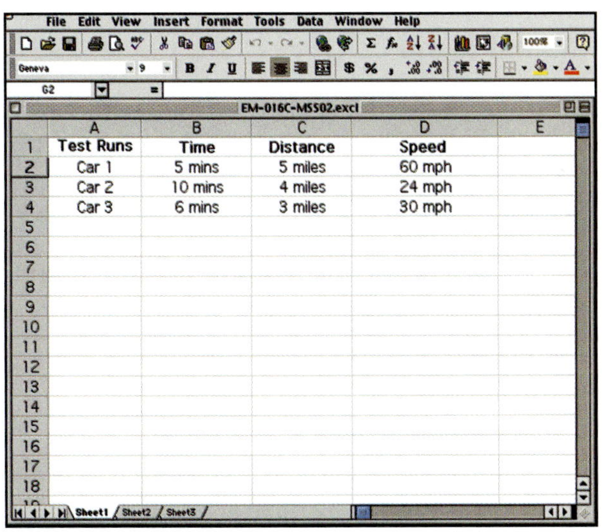

	A	B	C	D	E
1	Test Runs	Time	Distance	Speed	
2	Car 1	5 mins	5 miles	60 mph	
3	Car 2	10 mins	4 miles	24 mph	
4	Car 3	6 mins	3 miles	30 mph	

Figure 18 A spreadsheet allows you to perform mathematical operations on your data.

Use Graphics Software

Adding pictures, called graphics, to your documents is one way to make your documents more meaningful and exciting. This software adds, edits, and even constructs graphics. There is a variety of graphics software programs. The tools used for drawing can be a mouse, keyboard, or other specialized devices. Some graphics programs are simple. Others are complicated, called computer-aided design (CAD) software.

Learn the Skill It is important to have an understanding of the graphics software being used before starting. The better the software is understood, the better the results. The graphics can be placed in word-processing document.

- Clip art can be found on a variety of internet sites, and on CDs. These images can be copied and pasted into your document.
- When beginning, try editing existing drawings, then work up to creating drawings.
- The images are made of tiny rectangles of color called pixels. Each pixel can be altered.
- Digital photography is another way to add images. The photographs in the memory of a digital camera can be downloaded into a computer, then edited and added to the document.
- Graphics software also can allow animation. The software allows drawings to have the appearance of movement by connecting basic drawings automatically. This is called in-betweening, or tweening.
- Remember to save often.

Presentation Skills

Develop Multimedia Presentations

Most presentations are more dynamic if they include diagrams, photographs, videos, or sound recordings, like the one shown in **Figure 19.** A multimedia presentation involves using stereos, overhead projectors, televisions, computers, and more.

Learn the Skill Decide the main points of your presentation, and what types of media would best illustrate those points.

- Make sure you know how to use the equipment you are working with.
- Practice the presentation using the equipment several times.
- Enlist the help of a classmate to push play or turn lights out for you. Be sure to practice your presentation with him or her.
- If possible, set up all of the equipment ahead of time, and make sure everything is working properly.

Figure 19 These students are engaging the audience using a variety of tools.

Computer Presentations

There are many different interactive computer programs that you can use to enhance your presentation. Most computers have a compact disc (CD) drive that can play both CDs and digital video discs (DVDs). Also, there is hardware to connect a regular CD, DVD, or VCR. These tools will enhance your presentation.

Another method of using the computer to aid in your presentation is to develop a slide show using a computer program. This can allow movement of visuals at the presenter's pace, and can allow for visuals to build on one another.

Learn the Skill In order to create multimedia presentations on a computer, you need to have certain tools. These may include traditional graphic tools and drawing programs, animation programs, and authoring systems that tie everything together. Your computer will tell you which tools it supports. The most important step is to learn about the tools that you will be using.

- Often, color and strong images will convey a point better than words alone. Use the best methods available to convey your point.
- As with other presentations, practice many times.
- Practice your presentation with the tools you and any assistants will be using.
- Maintain eye contact with the audience. The purpose of using the computer is not to prompt the presenter, but to help the audience understand the points of the presentation.

Math Review

Use Fractions

A fraction compares a part to a whole. In the fraction $\frac{2}{3}$, the 2 represents the part and is the numerator. The 3 represents the whole and is the denominator.

Reduce Fractions To reduce a fraction, you must find the largest factor that is common to both the numerator and the denominator, the greatest common factor (GCF). Divide both numbers by the GCF. The fraction has then been reduced, or it is in its simplest form.

Example Twelve of the 20 chemicals in the science lab are in powder form. What fraction of the chemicals used in the lab are in powder form?

Step 1 Write the fraction.

$$\frac{\text{part}}{\text{whole}} = \frac{12}{20}$$

Step 2 To find the GCF of the numerator and denominator, list all of the factors of each number.

Factors of 12: 1, 2, 3, 4, 6, 12 (the numbers that divide evenly into 12)

Factors of 20: 1, 2, 4, 5, 10, 20 (the numbers that divide evenly into 20)

Step 3 List the common factors.

1, 2, 4.

Step 4 Choose the greatest factor in the list.

The GCF of 12 and 20 is 4.

Step 5 Divide the numerator and denominator by the GCF.

$$\frac{12 \div 4}{20 \div 4} = \frac{3}{5}$$

In the lab, $\frac{3}{5}$ of the chemicals are in powder form.

Practice Problem At an amusement park, 66 of 90 rides have a height restriction. What fraction of the rides, in its simplest form, has a height restriction?

Add and Subtract Fractions To add or subtract fractions with the same denominator, add or subtract the numerators and write the sum or difference over the denominator. After finding the sum or difference, find the simplest form for your fraction.

Example 1 In the forest outside your house, $\frac{1}{8}$ of the animals are rabbits, $\frac{3}{8}$ are squirrels, and the remainder are birds and insects. How many are mammals?

Step 1 Add the numerators.

$$\frac{1}{8} + \frac{3}{8} = \frac{(1 + 3)}{8} = \frac{4}{8}$$

Step 2 Find the GCF.

$$\frac{4}{8} \quad \text{(GCF, 4)}$$

Step 3 Divide the numerator and denominator by the GCF.

$$\frac{4}{4} = 1, \quad \frac{8}{4} = 2$$

$\frac{1}{2}$ of the animals are mammals.

Example 2 If $\frac{7}{16}$ of the Earth is covered by freshwater, and $\frac{1}{16}$ of that is in glaciers, how much freshwater is not frozen?

Step 1 Subtract the numerators.

$$\frac{7}{16} - \frac{1}{16} = \frac{(7 - 1)}{16} = \frac{6}{16}$$

Step 2 Find the GCF.

$$\frac{6}{16} \quad \text{(GCF, 2)}$$

Step 3 Divide the numerator and denominator by the GCF.

$$\frac{6}{2} = 3, \quad \frac{16}{2} = 8$$

$\frac{3}{8}$ of the freshwater is not frozen.

Practice Problem A bicycle rider is going 15 km/h for $\frac{4}{9}$ of his ride, 10 km/h for $\frac{2}{9}$ of his ride, and 8 km/h for the remainder of the ride. How much of his ride is he going over 8 km/h?

Math Skill Handbook

Reduce Fractions

$$\frac{66 \div 6}{90 \div 6} = \frac{11}{15}$$

Add and Subtract Fractions

$$\frac{4}{9} + \frac{2}{9} = \frac{6}{9}$$

$$\frac{6 \div 3}{9 \div 3} = \frac{2}{3}$$

Math Skill Handbook

MATH SKILL HANDBOOK G ◆ 205

Unlike Denominators

Problem 1

$1 \times 5 = 5, 8 \times 5 = 40$

$1 \times 4 = 4, 10 \times 4 = 40$

$\frac{5}{40} + \frac{4}{40} = \frac{9}{40}$

Problem 2

If $\frac{7}{10}$ are involuntary, the remainder are voluntary.

$\frac{10}{10} - \frac{7}{10} = \frac{3}{10}$

Unlike Denominators To add or subtract fractions with unlike denominators, first find the least common denominator (LCD). This is the smallest number that is a common multiple of both denominators. Rename each fraction with the LCD, and then add or subtract. Find the simplest form if necessary.

Example 1 A chemist makes a paste that is $\frac{1}{2}$ table salt (NaCl), $\frac{1}{3}$ sugar ($C_6H_{12}O_6$), and the rest water (H_2O). How much of the paste is a solid?

Step 1 Find the LCD of the fractions.

$\frac{1}{2} + \frac{1}{3}$ (LCD, 6)

Step 2 Rename each numerator and each denominator with the LCD.

$1 \times 3 = 3, \quad 2 \times 3 = 6$

$1 \times 2 = 2, \quad 3 \times 2 = 6$

Step 3 Add the numerators.

$\frac{3}{6} + \frac{2}{6} = \frac{(3 + 2)}{6} = \frac{5}{6}$

$\frac{5}{6}$ of the paste is a solid.

Example 2 The average precipitation in Grand Junction, CO, is $\frac{7}{10}$ inch in November, and $\frac{3}{5}$ inch in December. What is the total average precipitation?

Step 1 Find the LCD of the fractions.

$\frac{7}{10} + \frac{3}{5}$ (LCD, 10)

Step 2 Rename each numerator and each denominator with the LCD.

$7 \times 1 = 7, \quad 10 \times 1 = 10$

$3 \times 2 = 6, \quad 5 \times 2 = 10$

Step 3 Add the numerators.

$\frac{7}{10} + \frac{6}{10} = \frac{(7 + 6)}{10} = \frac{13}{10}$

$\frac{13}{10}$ inches total precipitation, or $1\frac{3}{10}$ inches.

Practice Problem On an electric bill, about $\frac{1}{8}$ of the energy is from solar energy and about $\frac{1}{10}$ is from wind power. How much of the total bill is from solar energy and wind power combined?

Example 3 In your body, $\frac{7}{10}$ of your muscle contractions are involuntary (cardiac and smooth muscle tissue). Smooth muscle makes $\frac{3}{15}$ of your muscle contractions. How many of your muscle contractions are made by cardiac muscle?

Step 1 Find the LCD of the fractions.

$\frac{7}{10} - \frac{3}{15}$ (LCD, 30)

Step 2 Rename each numerator and each denominator with the LCD.

$7 \times 3 = 21, \quad 10 \times 3 = 30$

$3 \times 2 = 6, \quad 15 \times 2 = 30$

Step 3 Subtract the numerators.

$\frac{21}{30} - \frac{6}{30} = \frac{(21 - 6)}{30} = \frac{15}{30}$

Step 4 Find the GCF.

$\frac{15}{30}$ (GCF, 15)

$\frac{1}{2}$

$\frac{1}{2}$ of all muscle contractions are cardiac muscle.

Example 4 Tony wants to make cookies that call for $\frac{3}{4}$ of a cup of flour, but he only has $\frac{1}{3}$ of a cup. How much more flour does he need?

Step 1 Find the LCD of the fractions.

$\frac{3}{4} - \frac{1}{3}$ (LCD, 12)

Step 2 Rename each numerator and each denominator with the LCD.

$3 \times 3 = 9, \quad 4 \times 3 = 12$

$1 \times 4 = 4, \quad 3 \times 4 = 12$

Step 3 Subtract the numerators.

$\frac{9}{12} - \frac{4}{12} = \frac{(9 - 4)}{12} = \frac{5}{12}$

$\frac{5}{12}$ of a cup of flour.

Practice Problem Using the information provided to you in Example 3 above, determine how many muscle contractions are voluntary (skeletal muscle).

Math Skill Handbook

206 ◆ G STUDENT RESOURCES

Multiply Fractions To multiply with fractions, multiply the numerators and multiply the denominators. Find the simplest form if necessary.

Example Multiply $\frac{3}{5}$ by $\frac{1}{3}$.

Step 1 Multiply the numerators and denominators.

$$\frac{3}{5} \times \frac{1}{3} = \frac{(3 \times 1)}{(5 \times 3)} = \frac{3}{15}$$

Step 2 Find the GCF.

$\frac{3}{15}$ (GCF, 3)

Step 3 Divide the numerator and denominator by the GCF.

$\frac{3}{3} = 1,\ \frac{15}{3} = 5$

$\frac{1}{5}$

$\frac{3}{5}$ multiplied by $\frac{1}{3}$ is $\frac{1}{5}$.

Practice Problem Multiply $\frac{3}{14}$ by $\frac{5}{16}$.

Find a Reciprocal Two numbers whose product is 1 are called multiplicative inverses, or reciprocals.

Example Find the reciprocal of $\frac{3}{8}$.

Step 1 Inverse the fraction by putting the denominator on top and the numerator on the bottom.

$\frac{8}{3}$

The reciprocal of $\frac{3}{8}$ is $\frac{8}{3}$.

Practice Problem Find the reciprocal of $\frac{4}{9}$.

Divide Fractions To divide one fraction by another fraction, multiply the dividend by the reciprocal of the divisor. Find the simplest form if necessary.

Example 1 Divide $\frac{1}{9}$ by $\frac{1}{3}$.

Step 1 Find the reciprocal of the divisor.

The reciprocal of $\frac{1}{3}$ is $\frac{3}{1}$.

Step 2 Multiply the dividend by the reciprocal of the divisor.

$$\frac{\frac{1}{9}}{\frac{1}{3}} = \frac{1}{9} \times \frac{3}{1} = \frac{(1 \times 3)}{(9 \times 1)} = \frac{3}{9}$$

Step 3 Find the GCF.

$\frac{3}{9}$ (GCF, 3)

Step 4 Divide the numerator and denominator by the GCF.

$\frac{3}{3} = 1,\ \frac{9}{3} = 3$

$\frac{1}{3}$

$\frac{1}{9}$ divided by $\frac{1}{3}$ is $\frac{1}{3}$.

Example 2 Divide $\frac{3}{5}$ by $\frac{1}{4}$.

Step 1 Find the reciprocal of the divisor.

The reciprocal of $\frac{1}{4}$ is $\frac{4}{1}$.

Step 2 Multiply the dividend by the reciprocal of the divisor.

$$\frac{\frac{3}{5}}{\frac{1}{4}} = \frac{3}{5} \times \frac{4}{1} = \frac{(3 \times 4)}{(5 \times 1)} = \frac{12}{5}$$

$\frac{3}{5}$ divided by $\frac{1}{4}$ is $\frac{12}{5}$ or $2\frac{2}{5}$.

Practice Problem Divide $\frac{3}{11}$ by $\frac{7}{10}$.

Math Skill Handbook

Multiply Fractions

$$\frac{3}{14} \times \frac{5}{16} = \frac{(3 \times 5)}{(14 \times 16)} = \frac{15}{224}$$

Find a Reciprocal

$\frac{9}{4}$

Divide Fractions

The reciprocal of $\frac{7}{10}$ is $\frac{10}{7}$.

$$\frac{3}{11} \times \frac{10}{7} = \frac{(3 \times 10)}{(11 \times 7)} = \frac{30}{77}$$

Math Skill Handbook

MATH SKILL HANDBOOK G ◆ 207

Use Ratios

When you compare two numbers by division, you are using a ratio. Ratios can be written 3 to 5, 3:5, or $\frac{3}{5}$. Ratios, like fractions, also can be written in simplest form.

Ratios can represent probabilities, also called odds. This is a ratio that compares the number of ways a certain outcome occurs to the number of outcomes. For example, if you flip a coin 100 times, what are the odds that it will come up heads? There are two possible outcomes, heads or tails, so the odds of coming up heads are 50:100. Another way to say this is that 50 out of 100 times the coin will come up heads. In its simplest form, the ratio is 1:2.

Example 1 A chemical solution contains 40 g of salt and 64 g of baking soda. What is the ratio of salt to baking soda as a fraction in simplest form?

Step 1 Write the ratio as a fraction.

$$\frac{\text{salt}}{\text{baking soda}} = \frac{40}{64}$$

Step 2 Express the fraction in simplest form.

The GCF of 40 and 64 is 8.

$$\frac{40}{64} = \frac{40 \div 8}{64 \div 8} = \frac{5}{8}$$

The ratio of salt to baking soda in the sample is 5:8.

Example 2 Sean rolls a 6-sided die 6 times. What are the odds that the side with a 3 will show?

Step 1 Write the ratio as a fraction.

$$\frac{\text{number of sides with a 3}}{\text{number of sides}} = \frac{1}{6}$$

Step 2 Multiply by the number of attempts.

$$\frac{1}{6} \times 6 \text{ attempts} = \frac{6}{6} \text{ attempts} = 1 \text{ attempt}$$

1 attempt out of 6 will show a 3.

Practice Problem Two metal rods measure 100 cm and 144 cm in length. What is the ratio of their lengths in simplest form?

Use Decimals

A fraction with a denominator that is a power of ten can be written as a decimal. For example, 0.27 means $\frac{27}{100}$. The decimal point separates the ones place from the tenths place.

Any fraction can be written as a decimal using division. For example, the fraction $\frac{5}{8}$ can be written as a decimal by dividing 5 by 8. Written as a decimal, it is 0.625.

Add or Subtract Decimals When adding and subtracting decimals, line up the decimal points before carrying out the operation.

Example 1 Find the sum of 47.68 and 7.80.

Step 1 Line up the decimal places when you write the numbers.

$$\begin{array}{r} 47.68 \\ +\ 7.80 \\ \hline \end{array}$$

Step 2 Add the decimals.

$$\begin{array}{r} 47.68 \\ +\ 7.80 \\ \hline 55.48 \end{array}$$

The sum of 47.68 and 7.80 is 55.48.

Example 2 Find the difference of 42.17 and 15.85.

Step 1 Line up the decimal places when you write the number.

$$\begin{array}{r} 42.17 \\ -15.85 \\ \hline \end{array}$$

Step 2 Subtract the decimals.

$$\begin{array}{r} 42.17 \\ -15.85 \\ \hline 26.32 \end{array}$$

The difference of 42.17 and 15.85 is 26.32.

Practice Problem Find the sum of 1.245 and 3.842.

Use Ratios

$$\frac{100 \text{ cm}}{144 \text{ cm}} = \frac{100 \div 4}{144 \div 4} = \frac{25}{36}$$

25:36

Add or Subtract Decimals

$$\begin{array}{r} \scriptstyle 1 \\ 1.245 \\ +\ 3.842 \\ \hline 5.087 \end{array}$$

Math Skill Handbook

208 ◆ G STUDENT RESOURCES

Multiply Decimals To multiply decimals, multiply the numbers like any other number, ignoring the decimal point. Count the decimal places in each factor. The product will have the same number of decimal places as the sum of the decimal places in the factors.

Example Multiply 2.4 by 5.9.

Step 1 Multiply the factors like two whole numbers.
$24 \times 59 = 1416$

Step 2 Find the sum of the number of decimal places in the factors. Each factor has one decimal place, for a sum of two decimal places.

Step 3 The product will have two decimal places.
14.16

The product of 2.4 and 5.9 is 14.16.

Practice Problem Multiply 4.6 by 2.2.

Divide Decimals When dividing decimals, change the divisor to a whole number. To do this, multiply both the divisor and the dividend by the same power of ten. Then place the decimal point in the quotient directly above the decimal point in the dividend. Then divide as you do with whole numbers.

Example Divide 8.84 by 3.4.

Step 1 Multiply both factors by 10.
$3.4 \times 10 = 34,\ 8.84 \times 10 = 88.4$

Step 2 Divide 88.4 by 34.

$$\begin{array}{r} 2.6 \\ 34\overline{)88.4} \\ -68 \\ \hline 20\,4 \\ -20\,4 \\ \hline 0 \end{array}$$

8.84 divided by 3.4 is 2.6.

Practice Problem Divide 75.6 by 3.6.

Use Proportions

An equation that shows that two ratios are equivalent is a proportion. The ratios $\frac{2}{4}$ and $\frac{5}{10}$ are equivalent, so they can be written as $\frac{2}{4} = \frac{5}{10}$. This equation is a proportion.

When two ratios form a proportion, the cross products are equal. To find the cross products in the proportion $\frac{2}{4} = \frac{5}{10}$, multiply the 2 and the 10, and the 4 and the 5. Therefore $2 \times 10 = 4 \times 5$, or $20 = 20$.

Because you know that both proportions are equal, you can use cross products to find a missing term in a proportion. This is known as solving the proportion.

Example The heights of a tree and a pole are proportional to the lengths of their shadows. The tree casts a shadow of 24 m when a 6-m pole casts a shadow of 4 m. What is the height of the tree?

Step 1 Write a proportion.
$$\frac{\text{height of tree}}{\text{height of pole}} = \frac{\text{length of tree's shadow}}{\text{length of pole's shadow}}$$

Step 2 Substitute the known values into the proportion. Let h represent the unknown value, the height of the tree.
$$\frac{h}{6} = \frac{24}{4}$$

Step 3 Find the cross products.
$h \times 4 = 6 \times 24$

Step 4 Simplify the equation.
$4h = 144$

Step 5 Divide each side by 4.
$$\frac{4h}{4} = \frac{144}{4}$$
$h = 36$

The height of the tree is 36 m.

Practice Problem The ratios of the weights of two objects on the Moon and on Earth are in proportion. A rock weighing 3 N on the Moon weighs 18 N on Earth. How much would a rock that weighs 5 N on the Moon weigh on Earth?

Multiply Decimals

Multiply 4.6 and 2.2 by 10.

$46 \times 22 = 1012$

Each factor had one decimal place.

10.12

Divide Decimals

Multiply both factors by 10.

Divide 756 by 36.

$$\begin{array}{r} 21 \\ 36\overline{)756} \\ 72 \\ \hline 36 \\ 36 \\ \hline 0 \end{array}$$

Use Proportions

$$\frac{3}{18} = \frac{5}{w}$$

$w \times 3 = 5 \times 18$

$$\frac{3w}{3} = \frac{90}{3}$$

$w = 30$

Math Skill Handbook

MATH SKILL HANDBOOK G ◆ 209

Use Percentages

$\frac{73}{365} = \frac{x}{100}$

$\frac{7300}{365} = \frac{365x}{365}$

$20\% = x$

Solve One-Step Equations

$h = gd$

$\frac{17.4}{12.3} = \frac{12.3d}{12.3}$

$1.41 = d$

Use Percentages

The word *percent* means "out of one hundred." It is a ratio that compares a number to 100. Suppose you read that 77 percent of the Earth's surface is covered by water. That is the same as reading that the fraction of the Earth's surface covered by water is $\frac{77}{100}$. To express a fraction as a percent, first find the equivalent decimal for the fraction. Then, multiply the decimal by 100 and add the percent symbol.

Example Express $\frac{13}{20}$ as a percent.

Step 1 Find the equivalent decimal for the fraction.

$$\begin{array}{r} 0.65 \\ 20\overline{)13.00} \\ \underline{12\,0} \\ 1\,00 \\ \underline{1\,00} \\ 0 \end{array}$$

Step 2 Rewrite the fraction $\frac{13}{20}$ as 0.65.

Step 3 Multiply 0.65 by 100 and add the % sign.

$0.65 \times 100 = 65 = 65\%$

So, $\frac{13}{20} = 65\%$.

This also can be solved as a proportion.

Example Express $\frac{13}{20}$ as a percent.

Step 1 Write a proportion.

$\frac{13}{20} = \frac{x}{100}$

Step 2 Find the cross products.

$1300 = 20x$

Step 3 Divide each side by 20.

$\frac{1300}{20} = \frac{20x}{20}$

$65\% = x$

Practice Problem In one year, 73 of 365 days were rainy in one city. What percent of the days in that city were rainy?

Solve One-Step Equations

A statement that two things are equal is an equation. For example, $A = B$ is an equation that states that A is equal to B.

An equation is solved when a variable is replaced with a value that makes both sides of the equation equal. To make both sides equal the inverse operation is used. Addition and subtraction are inverses, and multiplication and division are inverses.

Example 1 Solve the equation $x - 10 = 35$.

Step 1 Find the solution by adding 10 to each side of the equation.

$x - 10 = 35$

$x - 10 + 10 = 35 + 10$

$x = 45$

Step 2 Check the solution.

$x - 10 = 35$

$45 - 10 = 35$

$35 = 35$

Both sides of the equation are equal, so $x = 45$.

Example 2 In the formula $a = bc$, find the value of c if $a = 20$ and $b = 2$.

Step 1 Rearrange the formula so the unknown value is by itself on one side of the equation by dividing both sides by b.

$a = bc$

$\frac{a}{b} = \frac{bc}{b}$

$\frac{a}{b} = c$

Step 2 Replace the variables a and b with the values that are given.

$\frac{a}{b} = c$

$\frac{20}{2} = c$

$10 = c$

Step 3 Check the solution.

$a = bc$

$20 = 2 \times 10$

$20 = 20$

Both sides of the equation are equal, so $c = 10$ is the solution when $a = 20$ and $b = 2$.

Practice Problem In the formula $h = gd$, find the value of d if $g = 12.3$ and $h = 17.4$.

Math Skill Handbook

210 ◆ G STUDENT RESOURCES

Use Statistics

The branch of mathematics that deals with collecting, analyzing, and presenting data is statistics. In statistics, there are three common ways to summarize data with a single number—the mean, the median, and the mode.

The **mean** of a set of data is the arithmetic average. It is found by adding the numbers in the data set and dividing by the number of items in the set.

The **median** is the middle number in a set of data when the data are arranged in numerical order. If there were an even number of data points, the median would be the mean of the two middle numbers.

The **mode** of a set of data is the number or item that appears most often.

Another number that often is used to describe a set of data is the range. The **range** is the difference between the largest number and the smallest number in a set of data.

A **frequency table** shows how many times each piece of data occurs, usually in a survey. **Table 2** below shows the results of a student survey on favorite color.

Table 2 Student Color Choice

Color	Tally	Frequency
red	||||	4
blue	卌	5
black	||	2
green	|||	3
purple	卌 ||	7
yellow	卌 |	6

Based on the frequency table data, which color is the favorite?

Example The speeds (in m/s) for a race car during five different time trials are 39, 37, 44, 36, and 44.

To find the mean:

Step 1 Find the sum of the numbers.
$39 + 37 + 44 + 36 + 44 = 200$

Step 2 Divide the sum by the number of items, which is 5.
$200 \div 5 = 40$

The mean is 40 m/s.

To find the median:

Step 1 Arrange the measures from least to greatest.
36, 37, 39, 44, 44

Step 2 Determine the middle measure.
36, 37, 39, 44, 44

The median is 39 m/s.

To find the mode:

Step 1 Group the numbers that are the same together.
44, 44, 36, 37, 39

Step 2 Determine the number that occurs most in the set.
44, 44, 36, 37, 39

The mode is 44 m/s.

To find the range:

Step 1 Arrange the measures from largest to smallest.
44, 44, 39, 37, 36

Step 2 Determine the largest and smallest measures in the set.
44, 44, 39, 37, 36

Step 3 Find the difference between the largest and smallest measures.
$44 - 36 = 8$

The range is 8 m/s.

Practice Problem Find the mean, median, mode, and range for the data set 8, 4, 12, 8, 11, 14, 16.

Use Statistics

mean
$8 + 4 + 12 + 8 + 11 + 14 + 16 = 73$
$73 \div 7 = 10.4$

median
4, 8, 8, 11, 12, 14, 16

mode
4, 8, 8, 11, 12, 14, 16

range
4, 8, 8, 11, 12, 14, 16
$16 - 4 = 12$

Math Skill Handbook

MATH SKILL HANDBOOK G ◆ 211

Perimeter

Problem 1

$P = 2(18 \text{ m} + 7 \text{ m})$

$P = 2(25 \text{ m})$

$P = 50 \text{ m}$

Problem 2

$P = 1.6 \text{ cm} + 2.4 \text{ cm} + 2.4 \text{ cm}$

$P = 6.4 \text{ cm}$

Area of a Rectangle

$A = (4 \text{ m} \times 4 \text{ m})$

$A = 16 \text{ m}^2$

Area of a Triangle

$A = \frac{1}{2}(27 \text{ cm} \times 17 \text{ cm})$

$A = \frac{1}{2}(459 \text{ cm}^2)$

$A = 229.5 \text{ cm}^2$

Math Skill Handbook

Use Geometry

The branch of mathematics that deals with the measurement, properties, and relationships of points, lines, angles, surfaces, and solids is called geometry.

Perimeter The **perimeter** (P) is the distance around a geometric figure. To find the perimeter of a rectangle, add the length and width and multiply that sum by two, or $2(l + w)$. To find perimeters of irregular figures, add the length of the sides.

Example 1 Find the perimeter of a rectangle that is 3 m long and 5 m wide.

Step 1 You know that the perimeter is 2 times the sum of the width and length.
$P = 2(3 \text{ m} + 5 \text{ m})$

Step 2 Find the sum of the width and length.
$P = 2(8 \text{ m})$

Step 3 Multiply by 2.
$P = 16 \text{ m}$

The perimeter is 16 m.

Example 2 Find the perimeter of a shape with sides measuring 2 cm, 5 cm, 6 cm, 3 cm.

Step 1 You know that the perimeter is the sum of all the sides.
$P = 2 + 5 + 6 + 3$

Step 2 Find the sum of the sides.
$P = 2 + 5 + 6 + 3$
$P = 16$

The perimeter is 16 cm.

Practice Problem Find the perimeter of a rectangle with a length of 18 m and a width of 7 m.

Practice Problem Find the perimeter of a triangle measuring 1.6 cm by 2.4 cm by 2.4 cm.

Area of a Rectangle The **area** (A) is the number of square units needed to cover a surface. To find the area of a rectangle, multiply the length times the width, or $l \times w$. When finding area, the units also are multiplied. Area is given in square units.

Example Find the area of a rectangle with a length of 1 cm and a width of 10 cm.

Step 1 You know that the area is the length multiplied by the width.
$A = (1 \text{ cm} \times 10 \text{ cm})$

Step 2 Multiply the length by the width. Also multiply the units.
$A = 10 \text{ cm}^2$

The area is 10 cm².

Practice Problem Find the area of a square whose sides measure 4 m.

Area of a Triangle To find the area of a triangle, use the formula:

$$A = \frac{1}{2}(\text{base} \times \text{height})$$

The base of a triangle can be any of its sides. The height is the perpendicular distance from a base to the opposite endpoint, or vertex.

Example Find the area of a triangle with a base of 18 m and a height of 7 m.

Step 1 You know that the area is $\frac{1}{2}$ the base times the height.
$A = \frac{1}{2}(18 \text{ m} \times 7 \text{ m})$

Step 2 Multiply $\frac{1}{2}$ by the product of 18×7. Multiply the units.
$A = \frac{1}{2}(126 \text{ m}^2)$
$A = 63 \text{ m}^2$

The area is 63 m².

Practice Problem Find the area of a triangle with a base of 27 cm and a height of 17 cm.

Math Skill Handbook

Circumference of a Circle The **diameter** (d) of a circle is the distance across the circle through its center, and the **radius** (r) is the distance from the center to any point on the circle. The radius is half of the diameter. The distance around the circle is called the **circumference** (C). The formula for finding the circumference is:

$$C = 2\pi r \text{ or } C = \pi d$$

The circumference divided by the diameter is always equal to 3.1415926... This nonterminating and nonrepeating number is represented by the Greek letter π (pi). An approximation often used for π is 3.14.

Example 1 Find the circumference of a circle with a radius of 3 m.

Step 1 You know the formula for the circumference is 2 times the radius times π.
$C = 2\pi(3)$

Step 2 Multiply 2 times the radius.
$C = 6\pi$

Step 3 Multiply by π.
$C = 19$ m

The circumference is 19 m.

Example 2 Find the circumference of a circle with a diameter of 24.0 cm.

Step 1 You know the formula for the circumference is the diameter times π.
$C = \pi(24.0)$

Step 2 Multiply the diameter by π.
$C = 75.4$ cm

The circumference is 75.4 cm.

Practice Problem Find the circumference of a circle with a radius of 19 cm.

Area of a Circle The formula for the area of a circle is:
$A = \pi r^2$

Example 1 Find the area of a circle with a radius of 4.0 cm.

Step 1 $A = \pi(4.0)^2$

Step 2 Find the square of the radius.
$A = 16\pi$

Step 3 Multiply the square of the radius by π.
$A = 50 \text{ cm}^2$

The area of the circle is 50 cm^2.

Example 2 Find the area of a circle with a radius of 225 m.

Step 1 $A = \pi(225)^2$

Step 2 Find the square of the radius.
$A = 50625\pi$

Step 3 Multiply the square of the radius by π.
$A = 158962.5$

The area of the circle is 158,962 m^2.

Example 3 Find the area of a circle whose diameter is 20.0 mm.

Step 1 You know the formula for the area of a circle is the square of the radius times π, and that the radius is half of the diameter.
$A = \pi\left(\frac{20.0}{2}\right)^2$

Step 2 Find the radius.
$A = \pi(10.0)^2$

Step 3 Find the square of the radius.
$A = 100\pi$

Step 4 Multiply the square of the radius by π.
$A = 314 \text{ mm}^2$

The area is 314 mm^2.

Practice Problem Find the area of a circle with a radius of 16 m.

Circumference of a Circle

$C = 2\pi r$
$C = 2\pi(19)$
$C = 38\pi$
$C = 119.3$

Area of a Circle

$A = \pi r^2$
$A = \pi(16 \text{ m})^2$
$A = \pi\ 256 \text{ m}^2$
$A = 803.8 \text{ m}^2$

Math Skill Handbook

MATH SKILL HANDBOOK G ◆ 213

Volume

Problem 1

$V = 8\text{ m} \times 4\text{ m} \times 4\text{ m}$

$V = 128\text{ m}^3$

Problem 2

$V = \pi r^2 \times \text{height}$

$V = \left[\pi\left(\frac{1}{2} \times 7\right)^2\right] \times 16$

$V = [\pi(3.5)^2] \times 16$

$V = [\pi(12.25)] \times 16$

$V = 38.46 \times 16$

$V = 615.36$

Volume The measure of space occupied by a solid is the **volume** (V). To find the volume of a rectangular solid multiply the length times width times height, or $V = l \times w \times h$. It is measured in cubic units, such as cubic centimeters (cm^3).

Example Find the volume of a rectangular solid with a length of 2.0 m, a width of 4.0 m, and a height of 3.0 m.

Step 1 You know the formula for volume is the length times the width times the height.

$V = 2.0\text{ m} \times 4.0\text{ m} \times 3.0\text{ m}$

Step 2 Multiply the length times the width times the height.

$V = 24\text{ m}^3$

The volume is 24 m^3.

Practice Problem Find the volume of a rectangular solid that 8 m long, 4 m wide, and 4 m high.

To find the volume of other solids, multiply the area of the base times the height.

Example 1 Find the volume of a solid that has a triangular base with a length of 8.0 m and a height of 7.0 m. The height of the entire solid is 15.0 m.

Step 1 You know that the base is a triangle, and the area of a triangle is $\frac{1}{2}$ the base times the height, and the volume is the area of the base times the height.

$V = \left[\frac{1}{2}(b \times h)\right] \times 15$

Step 2 Find the area of the base.

$V = \left[\frac{1}{2}(8 \times 7)\right] \times 15$

$V = \left(\frac{1}{2} \times 56\right) \times 15$

Step 3 Multiply the area of the base by the height of the solid.

$V = 28 \times 15$

$V = 420\text{ m}^3$

The volume is 420 m^3.

Example 2 Find the volume of a cylinder that has a base with a radius of 12.0 cm, and a height of 21.0 cm.

Step 1 You know that the base is a circle, and the area of a circle is the square of the radius times π, and the volume is the area of the base times the height.

$V = (\pi r^2) \times 21$

$V = (\pi 12^2) \times 21$

Step 2 Find the area of the base.

$V = 144\pi \times 21$

$V = 452 \times 21$

Step 3 Multiply the area of the base by the height of the solid.

$V = 9490\text{ cm}^3$

The volume is 9490 cm^3.

Example 3 Find the volume of a cylinder that has a diameter of 15 mm and a height of 4.8 mm.

Step 1 You know that the base is a circle with an area equal to the square of the radius times π. The radius is one-half the diameter. The volume is the area of the base times the height.

$V = (\pi r^2) \times 4.8$

$V = \left[\pi\left(\frac{1}{2} \times 15\right)^2\right] \times 4.8$

$V = (\pi 7.5^2) \times 4.8$

Step 2 Find the area of the base.

$V = 56.25\pi \times 4.8$

$V = 176.63 \times 4.8$

Step 3 Multiply the area of the base by the height of the solid.

$V = 847.8$

The volume is 847.8 mm^3.

Practice Problem Find the volume of a cylinder with a diameter of 7 cm in the base and a height of 16 cm.

Math Skill Handbook

214 ◆ G STUDENT RESOURCES

Science Applications

Measure in SI

The metric system of measurement was developed in 1795. A modern form of the metric system, called the International System (SI), was adopted in 1960 and provides the standard measurements that all scientists around the world can understand.

The SI system is convenient because unit sizes vary by powers of 10. Prefixes are used to name units. Look at **Table 3** for some common SI prefixes and their meanings.

Table 3 Some SI Prefixes

Prefix	Symbol	Meaning	
kilo-	k	1,000	thousand
hecto-	h	100	hundred
deka-	da	10	ten
deci-	d	0.1	tenth
centi-	c	0.01	hundredth
milli-	m	0.001	thousandth

Example How many grams equal one kilogram?

Step 1 Find the prefix *kilo* in **Table 3.**

Step 2 Using **Table 3,** determine the meaning of *kilo.* According to the table, it means 1,000. When the prefix *kilo* is added to a unit, it means that there are 1,000 of the units in a "*kilo*unit."

Step 3 Apply the prefix to the units in the question. The units in the question are grams. There are 1,000 grams in a kilogram.

Practice Problem Is a milligram larger or smaller than a gram? How many of the smaller units equal one larger unit? What fraction of the larger unit does one smaller unit represent?

Dimensional Analysis

Convert SI Units In science, quantities such as length, mass, and time sometimes are measured using different units. A process called dimensional analysis can be used to change one unit of measure to another. This process involves multiplying your starting quantity and units by one or more conversion factors. A conversion factor is a ratio equal to one and can be made from any two equal quantities with different units. If 1,000 mL equal 1 L then two ratios can be made.

$$\frac{1{,}000 \text{ mL}}{1 \text{ L}} = \frac{1 \text{ L}}{1{,}000 \text{ mL}} = 1$$

One can covert between units in the SI system by using the equivalents in **Table 3** to make conversion factors.

Example 1 How many cm are in 4 m?

Step 1 Write conversion factors for the units given. From **Table 3,** you know that 100 cm = 1 m. The conversion factors are

$$\frac{100 \text{ cm}}{1 \text{ m}} \text{ and } \frac{1 \text{ m}}{100 \text{ cm}}$$

Step 2 Decide which conversion factor to use. Select the factor that has the units you are converting from (m) in the denominator and the units you are converting to (cm) in the numerator.

$$\frac{100 \text{ cm}}{1 \text{ m}}$$

Step 3 Multiply the starting quantity and units by the conversion factor. Cancel the starting units with the units in the denominator. There are 400 cm in 4 m.

$$4 \cancel{\text{m}} \times \frac{100 \text{ cm}}{1 \cancel{\text{m}}} = 400 \text{ cm}$$

Practice Problem How many milligrams are in one kilogram? (Hint: You will need to use two conversion factors from **Table 3.**)

Measure in SI

smaller; 1000; one thousandth

Dimensional Analysis

$$x \text{ mg} = 1 \cancel{\text{kg}} \times \frac{1000 \cancel{\text{g}}}{1 \cancel{\text{kg}}} \times \frac{1000 \text{ mg}}{1 \cancel{\text{g}}} =$$

1,000,000 mg

1,000,000 mg = 1 kg

MATH SKILL HANDBOOK G ◆ 215

Convert Between Unit Systems

$$\frac{(1\ \text{in})^3}{(2.54\ \text{cm})^3}$$

$$= \frac{1\ \text{in} \times 1\ \text{in} \times 1\ \text{in}}{2.54\ \text{cm} \times 2.54\ \text{cm} \times 2.54\ \text{cm}}$$

$$= \frac{1\ \text{in}^3}{16.39\ \text{cm}^3}$$

Math Skill Handbook

Table 4 Unit System Equivalents

Type of Measurement	Equivalent
Length	1 in = 2.54 cm 1 yd = 0.91 m 1 mi = 1.61 km
Mass and Weight*	1 oz = 28.35 g 1 lb = 0.45 kg 1 ton (short) = 0.91 tonnes (metric tons) 1 lb = 4.45 N
Volume	1 in^3 = 16.39 cm^3 1 qt = 0.95 L 1 gal = 3.78 L
Area	1 in^2 = 6.45 cm^2 1 yd^2 = 0.83 m^2 1 mi^2 = 2.59 km^2 1 acre = 0.40 hectares
Temperature	$°C = \frac{(°F - 32)}{1.8}$ K = °C + 273

*Weight is measured in standard Earth gravity.

Convert Between Unit Systems Table 4 gives a list of equivalents that can be used to convert between English and SI units.

Example If a meterstick has a length of 100 cm, how long is the meterstick in inches?

Step 1 Write the conversion factors for the units given. From **Table 4,** 1 in = 2.54 cm.

$$\frac{1\ \text{in}}{2.54\ \text{cm}} \quad \textit{and} \quad \frac{2.54\ \text{cm}}{1\ \text{in}}$$

Step 2 Determine which conversion factor to use. You are converting from cm to in. Use the conversion factor with cm on the bottom.

$$\frac{1\ \text{in}}{2.54\ \text{cm}}$$

Step 3 Multiply the starting quantity and units by the conversion factor. Cancel the starting units with the units in the denominator. Round your answer based on the number of significant figures in the conversion factor.

$$100\ \cancel{\text{cm}} \times \frac{1\ \text{in}}{2.54\ \cancel{\text{cm}}} = 39.37\ \text{in}$$

The meterstick is 39.4 in long.

Practice Problem A book has a mass of 5 lbs. What is the mass of the book in kg?

Practice Problem Use the equivalent for in and cm (1 in = 2.54 cm) to show how 1 in^3 = 16.39 cm^3.

Math Skill Handbook

Math Skill Handbook

Precision and Significant Digits

When you make a measurement, the value you record depends on the precision of the measuring instrument. This precision is represented by the number of significant digits recorded in the measurement. When counting the number of significant digits, all digits are counted except zeros at the end of a number with no decimal point such as 2,050, and zeros at the beginning of a decimal such as 0.03020. When adding or subtracting numbers with different precision, round the answer to the smallest number of decimal places of any number in the sum or difference. When multiplying or dividing, the answer is rounded to the smallest number of significant digits of any number being multiplied or divided.

Example The lengths 5.28 and 5.2 are measured in meters. Find the sum of these lengths and record your answer using the correct number of significant digits.

Step 1 Find the sum.

5.28 m	2 digits after the decimal
+ 5.2 m	1 digit after the decimal
10.48 m	

Step 2 Round to one digit after the decimal because the least number of digits after the decimal of the numbers being added is 1.

The sum is 10.5 m.

Practice Problem How many significant digits are in the measurement 7,071,301 m? How many significant digits are in the measurement 0.003010 g?

Practice Problem Multiply 5.28 and 5.2 using the rule for multiplying and dividing. Record the answer using the correct number of significant digits.

Scientific Notation

Many times numbers used in science are very small or very large. Because these numbers are difficult to work with scientists use scientific notation. To write numbers in scientific notation, move the decimal point until only one non-zero digit remains on the left. Then count the number of places you moved the decimal point and use that number as a power of ten. For example, the average distance from the Sun to Mars is 227,800,000,000 m. In scientific notation, this distance is 2.278×10^{11} m. Because you moved the decimal point to the left, the number is a positive power of ten.

The mass of an electron is about 0.000 000 000 000 000 000 000 000 000 000 911 kg. Expressed in scientific notation, this mass is 9.11×10^{-31} kg. Because the decimal point was moved to the right, the number is a negative power of ten.

Example Earth is 149,600,000 km from the Sun. Express this in scientific notation.

Step 1 Move the decimal point until one non-zero digit remains on the left.

1.496 000 00

Step 2 Count the number of decimal places you have moved. In this case, eight.

Step 3 Show that number as a power of ten, 10^8.

The Earth is 1.496×10^8 km from the Sun.

Practice Problem How many significant digits are in 149,600,000 km? How many significant digits are in 1.496×10^8 km?

Practice Problem Parts used in a high performance car must be measured to 7×10^{-6} m. Express this number as a decimal.

Practice Problem A CD is spinning at 539 revolutions per minute. Express this number in scientific notation.

Precision and Significant Digits

Problem 1
7; 4

Problem 2
$5.28 \times 5.2 = 27.456$

5.28 has 3 significant digits.

5.2 has 2 significant digits.

When multiplying and dividing, the answer is rounded to the smallest number of significant digits of the numbers being multiplied or divided—in this case, 2.

27.456 is rounded to 27.

Scientific Notation

Problem 1
4; 4

Problem 2
0.000007

Problem 3
5.39×10^2

Math Skill Handbook

MATH SKILL HANDBOOK G ◆ 217

Line Graph

x	y
3	52
6	72
9	83
12	86

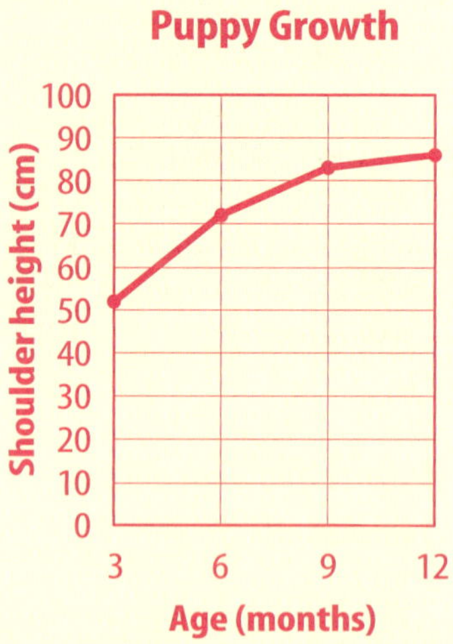

Make and Use Graphs

Data in tables can be displayed in a graph—a visual representation of data. Common graph types include line graphs, bar graphs, and circle graphs.

Line Graph A line graph shows a relationship between two variables that change continuously. The independent variable is changed and is plotted on the *x*-axis. The dependent variable is observed, and is plotted on the *y*-axis.

Example Draw a line graph of the data below from a cyclist in a long-distance race.

Table 5 Bicycle Race Data

Time (h)	Distance (km)
0	0
1	8
2	16
3	24
4	32
5	40

Step 1 Determine the *x*-axis and *y*-axis variables. Time varies independently of distance and is plotted on the *x*-axis. Distance is dependent on time and is plotted on the *y*-axis.

Step 2 Determine the scale of each axis. The *x*-axis data ranges from 0 to 5. The *y*-axis data ranges from 0 to 40.

Step 3 Using graph paper, draw and label the axes. Include units in the labels.

Step 4 Draw a point at the intersection of the time value on the *x*-axis and corresponding distance value on the *y*-axis. Connect the points and label the graph with a title, as shown in **Figure 20.**

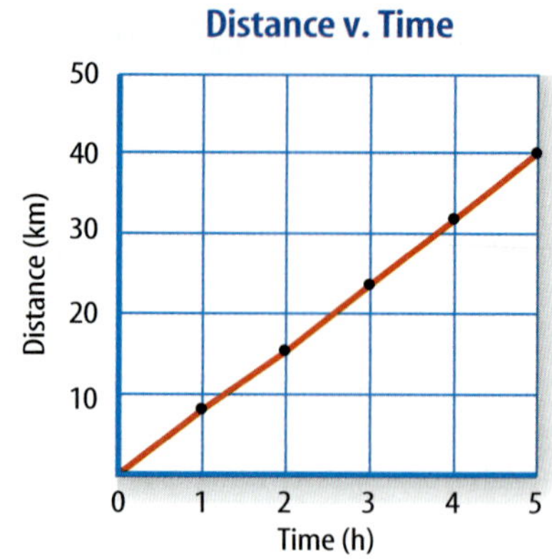

Figure 20 This line graph shows the relationship between distance and time during a bicycle ride.

Practice Problem A puppy's shoulder height is measured during the first year of her life. The following measurements were collected: (3 mo, 52 cm), (6 mo, 72 cm), (9 mo, 83 cm), (12 mo, 86). Graph this data.

Find a Slope The slope of a straight line is the ratio of the vertical change, rise, to the horizontal change, run.

$$\text{Slope} = \frac{\text{vertical change (rise)}}{\text{horizontal change (run)}} = \frac{\text{change in } y}{\text{change in } x}$$

Example Find the slope of the graph in **Figure 20.**

Step 1 You know that the slope is the change in *y* divided by the change in *x*.

$$\text{Slope} = \frac{\text{change in } y}{\text{change in } x}$$

Step 2 Determine the data points you will be using. For a straight line, choose the two sets of points that are the farthest apart.

$$\text{Slope} = \frac{(40-0)\text{ km}}{(5-0)\text{ hr}}$$

Step 3 Find the change in *y* and *x*.

$$\text{Slope} = \frac{40\text{ km}}{5\text{h}}$$

Step 4 Divide the change in *y* by the change in *x*.

$$\text{Slope} = \frac{8\text{ km}}{\text{h}}$$

The slope of the graph is 8 km/h.

Bar Graph To compare data that does not change continuously you might choose a bar graph. A bar graph uses bars to show the relationships between variables. The x-axis variable is divided into parts. The parts can be numbers such as years, or a category such as a type of animal. The y-axis is a number and increases continuously along the axis.

Example A recycling center collects 4.0 kg of aluminum on Monday, 1.0 kg on Wednesday, and 2.0 kg on Friday. Create a bar graph of this data.

Step 1 Select the x-axis and y-axis variables. The measured numbers (the masses of aluminum) should be placed on the y-axis. The variable divided into parts (collection days) is placed on the x-axis.

Step 2 Create a graph grid like you would for a line graph. Include labels and units.

Step 3 For each measured number, draw a vertical bar above the x-axis value up to the y-axis value. For the first data point, draw a vertical bar above Monday up to 4.0 kg.

Aluminum Collected During Week

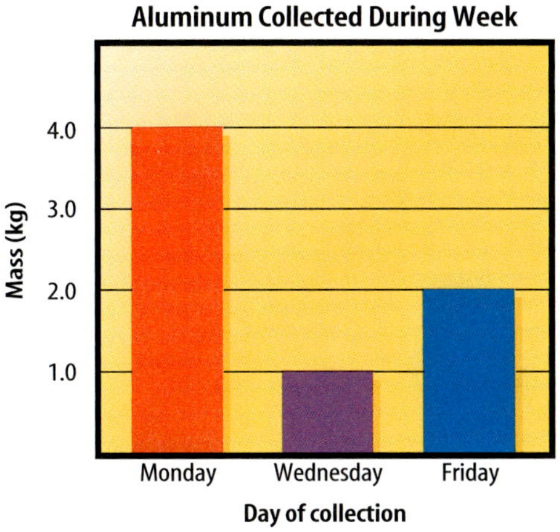

Practice Problem Draw a bar graph of the gases in air: 78% nitrogen, 21% oxygen, 1% other gases.

Circle Graph To display data as parts of a whole, you might use a circle graph. A circle graph is a circle divided into sections that represent the relative size of each piece of data. The entire circle represents 100%, half represents 50%, and so on.

Example Air is made up of 78% nitrogen, 21% oxygen, and 1% other gases. Display the composition of air in a circle graph.

Step 1 Multiply each percent by 360° and divide by 100 to find the angle of each section in the circle.

$$78\% \times \frac{360°}{100} = 280.8°$$

$$21\% \times \frac{360°}{100} = 75.6°$$

$$1\% \times \frac{360°}{100} = 3.6°$$

Step 2 Use a compass to draw a circle and to mark the center of the circle. Draw a straight line from the center to the edge of the circle.

Step 3 Use a protractor and the angles you calculated to divide the circle into parts. Place the center of the protractor over the center of the circle and line the base of the protractor over the straight line.

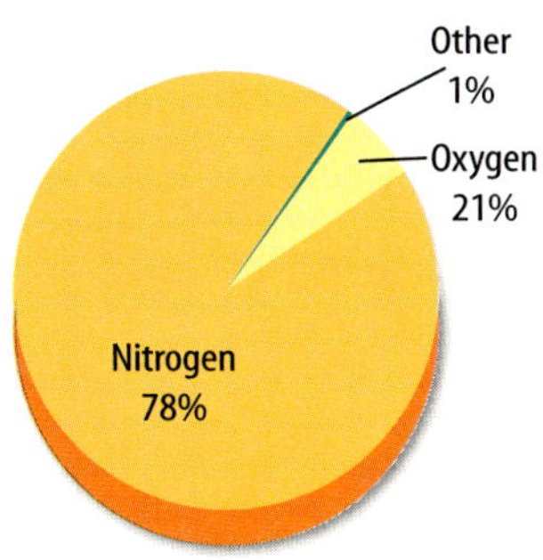

Practice Problem Draw a circle graph to represent the amount of aluminum collected during the week shown in the bar graph to the left.

Bar Graph

Composition of Air

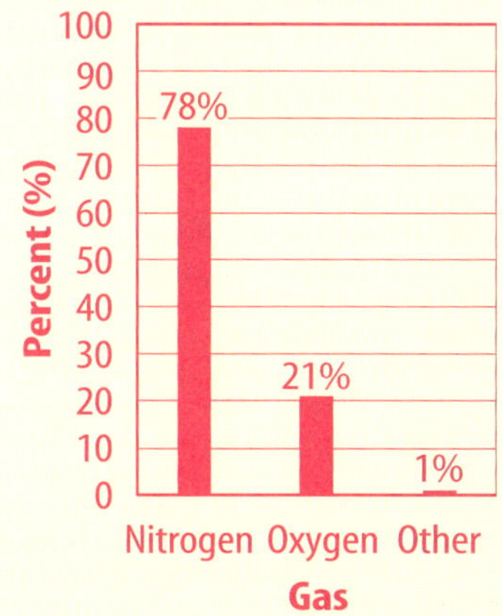

Circle Graph

The total amount of aluminum collected is:

$$4.0 \text{ kg} + 1.0 \text{ kg} + 2.0 \text{ kg} = 7.0 \text{ kg}$$

$$\frac{4.0 \text{ kg}}{7.0 \text{ kg}} = \frac{x}{360°};\ x = 206°$$

$$\frac{1.0 \text{ kg}}{7.0 \text{ kg}} = \frac{x}{360°};\ x = 51°$$

$$\frac{2.0 \text{ kg}}{7.0 \text{ kg}} = \frac{x}{360°};\ x = 103°$$

MATH SKILL HANDBOOK G ◆ 219

Weather Map Symbols

Sample Station Model

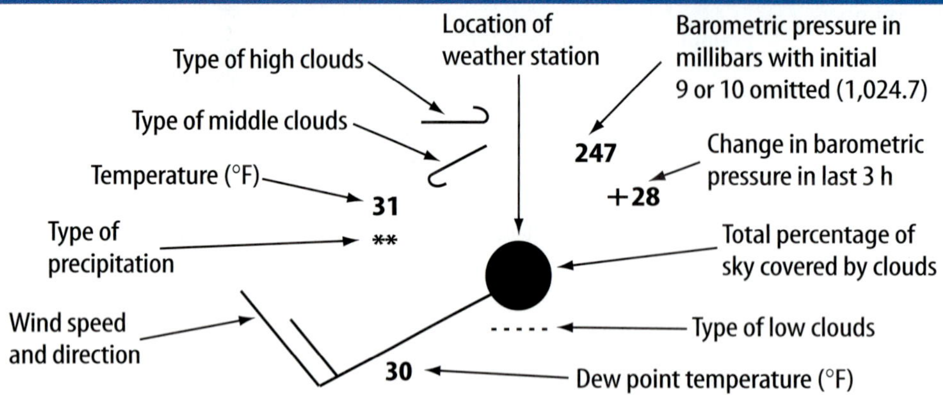

Sample Plotted Report at Each Station

Precipitation		Wind Speed and Direction		Sky Coverage		Some Types of High Clouds	
≡	Fog	○	0 calm	○	No cover		Scattered cirrus
★	Snow		1–2 knots		1/10 or less		Dense cirrus in patches
●	Rain		3–7 knots		2/10 to 3/10		Veil of cirrus covering entire sky
	Thunderstorm		8–12 knots		4/10		Cirrus not covering entire sky
,	Drizzle		13–17 knots		–		
▽	Showers		18–22 knots		6/10		
			23–27 knots		7/10		
			48–52 knots		Overcast with openings		
		1 knot = 1.852 km/h		●	Completely overcast		

Some Types of Middle Clouds		Some Types of Low Clouds		Fronts and Pressure Systems	
	Thin altostratus layer		Cumulus of fair weather	(H) or High (L) or Low	Center of high- or low-pressure system
	Thick altostratus layer		Stratocumulus	▲▲▲▲	Cold front
	Thin altostratus in patches	-----	Fractocumulus of bad weather		Warm front
	Thin altostratus in bands	—	Stratus of fair weather		Occluded front
					Stationary front

Rocks

Rocks		
Rock Type	**Rock Name**	**Characteristics**
Igneous (intrusive)	Granite	Large mineral grains of quartz, feldspar, hornblende, and mica. Usually light in color.
	Diorite	Large mineral grains of feldspar, hornblende, and mica. Less quartz than granite. Intermediate in color.
	Gabbro	Large mineral grains of feldspar, augite, and olivine. No quartz. Dark in color.
Igneous (extrusive)	Rhyolite	Small mineral grains of quartz, feldspar, hornblende, and mica, or no visible grains. Light in color.
	Andesite	Small mineral grains of feldspar, hornblende, and mica or no visible grains. Intermediate in color.
	Basalt	Small mineral grains of feldspar, augite, and possibly olivine or no visible grains. No quartz. Dark in color.
	Obsidian	Glassy texture. No visible grains. Volcanic glass. Fracture looks like broken glass.
	Pumice	Frothy texture. Floats in water. Usually light in color.
Sedimentary (detrital)	Conglomerate	Coarse grained. Gravel or pebble-size grains.
	Sandstone	Sand-sized grains 1/16 to 2 mm.
	Siltstone	Grains are smaller than sand but larger than clay.
	Shale	Smallest grains. Often dark in color. Usually platy.
Sedimentary (chemical or organic)	Limestone	Major mineral is calcite. Usually forms in oceans and lakes. Often contains fossils.
	Coal	Forms in swampy areas. Compacted layers of organic material, mainly plant remains.
Sedimentary (chemical)	Rock Salt	Commonly forms by the evaporation of seawater.
Metamorphic (foliated)	Gneiss	Banding due to alternate layers of different minerals, of different colors. Parent rock often is granite.
	Schist	Parallel arrangement of sheetlike minerals, mainly micas. Forms from different parent rocks.
	Phyllite	Shiny or silky appearance. May look wrinkled. Common parent rocks are shale and slate.
	Slate	Harder, denser, and shinier than shale. Common parent rock is shale.
Metamorphic (nonfoliated)	Marble	Calcite or dolomite. Common parent rock is limestone.
	Soapstone	Mainly of talc. Soft with greasy feel.
	Quartzite	Hard with interlocking quartz crystals. Common parent rock is sandstone.

Minerals

Minerals					
Mineral (formula)	**Color**	**Streak**	**Hardness**	**Breakage Pattern**	**Uses and Other Properties**
Graphite (C)	black to gray	black to gray	1–1.5	basal cleavage (scales)	pencil lead, lubricants for locks, rods to control some small nuclear reactions, battery poles
Galena (PbS)	gray	gray to black	2.5	cubic cleavage perfect	source of lead, used for pipes, shields for X rays, fishing equipment sinkers
Hematite (Fe_2O_3)	black or reddish-brown	reddish-brown	5.5–6.5	irregular fracture	source of iron; converted to pig iron, made into steel
Magnetite (Fe_3O_4)	black	black	6	conchoidal fracture	source of iron, attracts a magnet
Pyrite (FeS_2)	light, brassy, yellow	greenish-black	6–6.5	uneven fracture	fool's gold
Talc ($Mg_3Si_4O_{10}(OH)_2$)	white, greenish	white	1	cleavage in one direction	used for talcum powder, sculptures, paper, and tabletops
Gypsum ($CaSO_4 \cdot 2H_2O$)	colorless, gray, white, brown	white	2	basal cleavage	used in plaster of paris and dry wall for building construction
Sphalerite (ZnS)	brown, reddish-brown, greenish	light to dark brown	3.5–4	cleavage in six directions	main ore of zinc; used in paints, dyes, and medicine
Muscovite ($KAl_3Si_3O_{10}(OH)_2$)	white, light gray, yellow, rose, green	colorless	2–2.5	basal cleavage	occurs in large, flexible plates; used as an insulator in electrical equipment, lubricant
Biotite ($K(Mg,Fe)_3(AlSi_3O_{10})(OH)_2$)	black to dark brown	colorless	2.5–3	basal cleavage	occurs in large, flexible plates
Halite (NaCl)	colorless, red, white, blue	colorless	2.5	cubic cleavage	salt; soluble in water; a preservative

Minerals

Minerals					
Mineral (formula)	**Color**	**Streak**	**Hardness**	**Breakage Pattern**	**Uses and Other Properties**
Calcite ($CaCO_3$)	colorless, white, pale blue	colorless, white	3	cleavage in three directions	fizzes when HCl is added; used in cements and other building materials
Dolomite ($CaMg(CO_3)_2$)	colorless, white, pink, green, gray, black	white	3.5–4	cleavage in three directions	concrete and cement; used as an ornamental building stone
Fluorite (CaF_2)	colorless, white, blue, green, red, yellow, purple	colorless	4	cleavage in four directions	used in the manufacture of optical equipment; glows under ultraviolet light
Hornblende ($(CaNa)_{2-3}(Mg,Al,Fe)_5$-$(Al,Si)_2Si_6O_{22}(OH)_2$)	green to black	gray to white	5–6	cleavage in two directions	will transmit light on thin edges; 6-sided cross section
Feldspar ($(KAlSi_3O_8)$ $(NaAlSi_3O_8)$, $(CaAl_2Si_2O_8)$)	colorless, white to gray, green	colorless	6	two cleavage planes meet at 90° angle	used in the manufacture of ceramics
Augite ($((Ca,Na)(Mg,Fe,Al)(Al,Si)_2O_6)$)	black	colorless	6	cleavage in two directions	square or 8-sided cross section
Olivine ($((Mg,Fe)_2SiO_4)$)	olive, green	none	6.5–7	conchoidal fracture	gemstones, refractory sand
Quartz (SiO_2)	colorless, various colors	none	7	conchoidal fracture	used in glass manufacture, electronic equipment, radios, computers, watches, gemstones

PERIODIC TABLE OF THE ELEMENTS

Columns of elements are called groups. Elements in the same group have similar chemical properties.

Element — Hydrogen
Atomic number — 1
Symbol — H
Atomic mass — 1.008
State of matter

Gas
Liquid
Solid
Synthetic

The first three symbols tell you the state of matter of the element at room temperature. The fourth symbol identifies elements that are not present in significant amounts on Earth. Useful amounts are made synthetically.

Period	1	2	3	4	5	6	7	8	9
1	Hydrogen 1 H 1.008 [Gas]								
2	Lithium 3 Li 6.941 [Solid]	Beryllium 4 Be 9.012 [Solid]							
3	Sodium 11 Na 22.990 [Solid]	Magnesium 12 Mg 24.305 [Solid]							
4	Potassium 19 K 39.098 [Solid]	Calcium 20 Ca 40.078 [Solid]	Scandium 21 Sc 44.956 [Solid]	Titanium 22 Ti 47.867 [Solid]	Vanadium 23 V 50.942 [Solid]	Chromium 24 Cr 51.996 [Solid]	Manganese 25 Mn 54.938 [Solid]	Iron 26 Fe 55.845 [Solid]	Cobalt 27 Co 58.933 [Solid]
5	Rubidium 37 Rb 85.468 [Solid]	Strontium 38 Sr 87.62 [Solid]	Yttrium 39 Y 88.906 [Solid]	Zirconium 40 Zr 91.224 [Solid]	Niobium 41 Nb 92.906 [Solid]	Molybdenum 42 Mo 95.94 [Solid]	Technetium 43 Tc (98) [Synthetic]	Ruthenium 44 Ru 101.07 [Solid]	Rhodium 45 Rh 102.906 [Solid]
6	Cesium 55 Cs 132.905 [Solid]	Barium 56 Ba 137.327 [Solid]	Lanthanum 57 La 138.906 [Solid]	Hafnium 72 Hf 178.49 [Solid]	Tantalum 73 Ta 180.948 [Solid]	Tungsten 74 W 183.84 [Solid]	Rhenium 75 Re 186.207 [Solid]	Osmium 76 Os 190.23 [Solid]	Iridium 77 Ir 192.217 [Solid]
7	Francium 87 Fr (223) [Solid]	Radium 88 Ra (226) [Solid]	Actinium 89 Ac (227) [Solid]	Rutherfordium 104 Rf (261) [Synthetic]	Dubnium 105 Db (262) [Synthetic]	Seaborgium 106 Sg (266) [Synthetic]	Bohrium 107 Bh (264) [Synthetic]	Hassium 108 Hs (277) [Synthetic]	Meitnerium 109 Mt (268) [Synthetic]

The number in parentheses is the mass number of the longest-lived isotope for that element.

Rows of elements are called periods. Atomic number increases across a period.

The arrow shows where these elements would fit into the periodic table. They are moved to the bottom of the table to save space.

Lanthanide series	Cerium 58 Ce 140.116 [Solid]	Praseodymium 59 Pr 140.908 [Solid]	Neodymium 60 Nd 144.24 [Solid]	Promethium 61 Pm (145) [Synthetic]	Samarium 62 Sm 150.36 [Solid]
Actinide series	Thorium 90 Th 232.038 [Solid]	Protactinium 91 Pa 231.036 [Solid]	Uranium 92 U 238.029 [Solid]	Neptunium 93 Np (237) [Synthetic]	Plutonium 94 Pu (244) [Synthetic]

Reference Handbooks

The color of an element's block tells you if the element is a metal, nonmetal, or metalloid.

Science Online
Visit bookg.msscience.com for updates to the periodic table.

10	11	12	13	14	15	16	17	18
								Helium 2 He 4.003
			Boron 5 B 10.811	Carbon 6 C 12.011	Nitrogen 7 N 14.007	Oxygen 8 O 15.999	Fluorine 9 F 18.998	Neon 10 Ne 20.180
			Aluminum 13 Al 26.982	Silicon 14 Si 28.086	Phosphorus 15 P 30.974	Sulfur 16 S 32.065	Chlorine 17 Cl 35.453	Argon 18 Ar 39.948
Nickel 28 Ni 58.693	Copper 29 Cu 63.546	Zinc 30 Zn 65.409	Gallium 31 Ga 69.723	Germanium 32 Ge 72.64	Arsenic 33 As 74.922	Selenium 34 Se 78.96	Bromine 35 Br 79.904	Krypton 36 Kr 83.798
Palladium 46 Pd 106.42	Silver 47 Ag 107.868	Cadmium 48 Cd 112.411	Indium 49 In 114.818	Tin 50 Sn 118.710	Antimony 51 Sb 121.760	Tellurium 52 Te 127.60	Iodine 53 I 126.904	Xenon 54 Xe 131.293
Platinum 78 Pt 195.078	Gold 79 Au 196.967	Mercury 80 Hg 200.59	Thallium 81 Tl 204.383	Lead 82 Pb 207.2	Bismuth 83 Bi 208.980	Polonium 84 Po (209)	Astatine 85 At (210)	Radon 86 Rn (222)
Darmstadtium 110 Ds (281)	Unununium * 111 Uuu (272)	Ununbium * 112 Uub (285)		Ununquadium * 114 Uuq (289)		** 116		** 118

* The names and symbols for elements 111–114 are temporary. Final names will be selected when the elements' discoveries are verified.
** Elements 116 and 118 were thought to have been created. The claim was retracted because the experimental results could not be repeated.

Europium 63 Eu 151.964	Gadolinium 64 Gd 157.25	Terbium 65 Tb 158.925	Dysprosium 66 Dy 162.500	Holmium 67 Ho 164.930	Erbium 68 Er 167.259	Thulium 69 Tm 168.934	Ytterbium 70 Yb 173.04	Lutetium 71 Lu 174.967
Americium 95 Am (243)	Curium 96 Cm (247)	Berkelium 97 Bk (247)	Californium 98 Cf (251)	Einsteinium 99 Es (252)	Fermium 100 Fm (257)	Mendelevium 101 Md (258)	Nobelium 102 No (259)	Lawrencium 103 Lr (262)

Topographic Map Symbols

Topographic Map Symbols	
Primary highway, hard surface	Index contour
Secondary highway, hard surface	Supplementary contour
Light-duty road, hard or improved surface	Intermediate contour
Unimproved road	Depression contours
Railroad: single track	
Railroad: multiple track	Boundaries: national
Railroads in juxtaposition	State
	County, parish, municipal
Buildings	Civil township, precinct, town, barrio
Schools, church, and cemetery	Incorporated city, village, town, hamlet
Buildings (barn, warehouse, etc.)	Reservation, national or state
Wells other than water (labeled as to type)	Small park, cemetery, airport, etc.
Tanks: oil, water, etc. (labeled only if water)	Land grant
Located or landmark object; windmill	Township or range line, U.S. land survey
Open pit, mine, or quarry; prospect	
	Township or range line, approximate location
Marsh (swamp)	
Wooded marsh	Perennial streams
Woods or brushwood	Elevated aqueduct
Vineyard	Water well and spring
Land subject to controlled inundation	Small rapids
Submerged marsh	Large rapids
Mangrove	Intermittent lake
Orchard	Intermittent stream
Scrub	Aqueduct tunnel
Urban area	Glacier
	Small falls
x7369 Spot elevation	Large falls
670 Water elevation	Dry lake bed

Cómo usar el glosario en español:
1. Busca el término en inglés que desees encontrar.
2. El término en español, junto con la definición, se encuentran en la columna de la derecha.

Pronunciation Key

Use the following key to help you sound out words in the glossary.

a	**back** (BAK)	**ew**	**food** (FEWD)
ay	**day** (DAY)	**yoo**	**pure** (PYOOR)
ah	**father** (FAH thur)	**yew**	**few** (FYEW)
ow	**flower** (FLOW ur)	**uh**	**comma** (CAH muh)
ar	**car** (CAR)	**u** (+ con)	**rub** (RUB)
e	**less** (LES)	**sh**	**shelf** (SHELF)
ee	**leaf** (LEEF)	**ch**	**nature** (NAY chur)
ih	**trip** (TRIHP)	**g**	**gift** (GIHFT)
i (i + con + e)	**idea** (i DEE uh)	**j**	**gem** (JEM)
oh	**go** (GOH)	**ing**	**sing** (SING)
aw	**soft** (SAWFT)	**zh**	**vision** (VIH zhun)
or	**orbit** (OR buht)	**k**	**cake** (KAYK)
oy	**coin** (COYN)	**s**	**seed, cent** (SEED, SENT)
oo	**foot** (FOOT)	**z**	**zone, raise** (ZOHN, RAYZ)

English — A — Español

abrasion: a type of erosion that occurs when wind-blown sediments strike rocks and sediments, polishing and pitting their surface. (p. 76)

abrasión: tipo de erosión que ocurre cuando los sedimentos arrastrados por el viento golpean las rocas y los sedimentos, puliendo y llenando de hoyos su superficie. (p. 76)

absolute age: age, in years, of a rock or other object; can be determined by using properties of the atoms that make up materials. (p. 139)

edad absoluta: edad, en años, de una roca u otro objeto; puede determinarse utilizando las propiedades de los átomos de los materiales. (p. 139)

aquifer (AK wuh fur): layer of permeable rock that allows water to flow through. (p. 104)

acuífero: capa de roca permeable que permite que el agua fluya a través de ella. (p. 104)

B

beach: deposit of sediment whose materials vary in size, color, and composition and is most commonly found on a smooth, gently sloped shoreline. (p. 111)

playa: depósito de sedimentos cuyos materiales varían en tamaño, color y composición y que comúnmente se encuentran en las líneas costeras planas y poco inclinadas. (p. 111)

C

carbon film: thin film of carbon residue preserved as a fossil. (p. 126)

película de carbono: capa delgada de residuos de carbono preservada como un fósil. (p. 126)

cast: a type of body fossil that forms when crystals fill a mold or sediments wash into a mold and harden into rock. (p. 127)

vaciado: tipo de cuerpo fósil que se forma cuando los cristales llenan un molde o los sedimentos son lavados hacia un molde y se endurecen convirtiéndose en roca. (p. 127)

cave: underground opening that can form when acidic groundwater dissolves limestone. (p. 107)

cueva: apertura subterránea que puede formarse cuando el agua subterránea acidificada disuelve la piedra caliza. (p. 107)

Cenozoic (seh nuh ZOH ihk) Era: era of recent life that began about 66 million years ago and continues today; includes the first appearance of *Homo sapiens* about 400,000 years ago. (p. 174)

Era Cenozoica: era de vida reciente que comenzó hace aproximadamente 66 millones de años y continúa hasta hoy; incluye la aparición del *Homo sapiens* cerca de 400,000 años atrás. (p. 174)

channel: groove created by water moving down the same path. (p. 94)

canal: surco creado por el agua cuando se mueve cuesta abajo por el mismo curso. (p. 94)

chemical weathering: occurs when chemical reactions dissolve the minerals in rocks or change them into different minerals. (p. 39)

erosión química: ocurre cuando las reacciones químicas disuelven los minerales en las rocas o los convierten en diferentes minerales. (p. 39)

climate: average weather pattern in an area over a long period of time; can be classified by temperature, humidity, precipitation, and vegetation. (p. 40)

clima: modelo meteorológico promedio en un área durante un periodo de tiempo largo; puede clasificarse por temperatura, humedad, precipitación y vegetación. (p. 40)

conic projection: map made by projecting points and lines from a globe onto a cone. (p. 19)

proyección cónica: mapa hecho por la proyección de puntos y líneas desde un globo a un cono. (p. 19)

contour farming: planting along the natural contours of the land to reduce soil ersosion. (p. 53)

cultivo de contorno: plantación a lo largo de los contornos naturales de la tierra para reducir la erosión de los suelos. (p. 53)

contour line: line on a map that connects points of equal elevation. (p. 20)

curva de nivel: línea en un mapa que conecta puntos de la misma elevación. (p. 20)

creep: a type of mass movement in which sediments move down-slope very slowly; is common in areas of freezing and thawing, and can cause walls, trees, and fences to lean downhill. (p. 66)

reptación: tipo de movimiento en masa en el que los sedimentos se mueven hacia abajo muy lentamente; es común en áreas sujetas a congelación y descongelación y puede causar que los muros, los árboles y los cercos se inclinen hacia abajo. (p. 66)

cyanobacteria: chlorophyll-containing, photosynthetic bacteria thought to be one of Earth's earliest life-forms. (p. 163)

cianobacteria: bacteria fotosintética que contiene clorofila; se cree que es una de las primeras formas de vida que surgió en la tierra. (p. 163)

D

deflation: a type of erosion that occurs when wind blows over loose sediments, removes small particles, and leaves coarser sediments behind. (p. 76)

deflación: tipo de erosión que ocurre cuando el viento sopla sobre los sedimentos sueltos, retira partículas pequeñas y deja los sedimentos grandes. (p. 76)

deposition: dropping of sediments that occurs when an agent of erosion, such as gravity, a glacier, wind, or water, loses its energy and can no longer carry its load. (p. 65)

deposición: caída de sedimentos que ocurre cuando un agente erosivo como la gravedad, un glaciar, el viento o el agua, pierde su energía y ya no puede continuar con su carga. (p. 65)

drainage basin: land area from which a river or stream collects runoff. (p. 96)

cuenca de drenaje: terreno del que un río o arroyo recolecta sus aguas. (p. 96)

dune (DOON): mound formed when windblown sediments pile up behind an obstacle; common landform in desert areas. (p. 79)

duna: amontonamiento de tierra formado cuando los sedimentos arrastrados por el aire se apilan detrás de un obstáculo; forma de terreno común en las áreas desérticas. (p. 79)

E

eon: longest subdivision in the geologic time scale that is based on the abundance of certain types of fossils and is subdivided into eras, periods, and epochs. (p. 155)

epoch: next-smaller division of geologic time after the period; is characterized by differences in life-forms that may vary regionally. (p. 155)

equator: imaginary line that wraps around Earth at 0° latitude, halfway between the north and south poles. (p. 14)

era: second-longest division of geologic time; is subdivided into periods and is based on major worldwide changes in types of fossils. (p. 155)

erosion: process in which surface materials are worn away and transported from one place to another by agents such as gravity, water, wind, and glaciers. (p. 64)

eón: la más grande subdivisión en la escala del tiempo geológico; se basa en la abundancia de cierto tipo de fósiles y está dividida en eras, periodos y épocas. (p. 155)

época: la siguiente división más pequeña del tiempo geológico después del periodo; está caracterizada por diferencias en las formas de vida que pueden variar regionalmente. (p. 155)

ecuador: línea imaginaria que rodea a la Tierra en el punto de latitud 0°, a la mitad de la distancia entre el polo norte y el polo sur. (p. 14)

era: la segunda división más grande del tiempo geológico; está subdividida en periodos y se basa en cambios mayores en todo el mundo con respecto a los tipos de fósiles. (p. 155)

erosión: proceso mediante el cual los materiales de la superficie son desgastados y transportados de un lugar a otro por agentes como la gravedad, el agua, el viento o los glaciares. (p. 64)

F

fault-block mountains: mountains formed from huge, tilted blocks of rock that are separated from surrounding rocks by faults. (p. 12)

folded mountains: mountains formed when horizontal rock layers are squeezed from opposite sides, causing them to buckle and fold. (p. 11)

fossils: remains, imprints, or traces of prehistoric organisms that can tell when and where organisms once lived and how they lived. (p. 125)

montañas de fallas: montañas formadas por bloques rocosos grandes e inclinados separados de las rocas circundantes por fracturas. (p. 12)

montañas de plegamiento: montañas formadas cuando las capas rocosas horizontales son comprimidas desde lados opuestos, causando que se colapsen y plieguen. (p. 11)

fósiles: restos, huellas o trazas de organismos prehistóricos que pueden informar cuándo, dónde y cómo vivieron tales organismos. (p. 125)

G

geologic time scale: division of Earth's history into time units based largely on the types of life-forms that lived only during certain periods. (p. 154)

geyser: hot spring that erupts periodically and shoots water and steam into the air—for example, Old Faithful in Yellowstone National Park. (p. 107)

glaciers: large, moving masses of ice and snow that change large areas of Earth's surface through erosion and deposition. (p. 69)

escala del tiempo geológico: división de la historia de la Tierra en unidades de tiempo; se basa en los tipos de formas de vida que vivieron sólo durante ciertos periodos. (p. 154)

géiser: aguas termales que erupcionan periódicamente arrojando agua y vapor al aire –por ejemplo, Old Faithful en el Parque Nacional Yellowstone. (p. 107)

glaciares: grandes masas de hielo y nieve en movimiento que cambian extensas áreas de la superficie terrestre a través de la erosión y la deposición. (p. 69)

groundwater: water that soaks into the ground and collects in pores and empty spaces and is an important source of drinking water. (p. 103)

agua subterránea: agua que se difunde en el suelo y se acumula en poros y espacios vacíos siendo una fuente importante de agua potable. (p. 103)

H

half-life: time it takes for half the atoms of an isotope to decay. (p. 140)

horizon: each layer in a soil profile—horizon A (top layer of soil), horizon B (middle layer), and horizon C (bottom layer). (p. 44)

humus (HYEW mus): dark-colored, decayed organic matter that supplies nutrients to plants and is found mainly in topsoil. (p. 44)

vida media: tiempo que le toma a la mitad de los átomos de un isótopo para desintegrarse. (p. 140)

horizonte: cada capa en un perfil de suelos: horizonte A (la capa superior del suelo), horizonte B (la capa media) y horizonte C (la capa inferior). (p. 44)

humus: materia orgánica en descomposición, de color oscuro, que suministra nutrientes a las plantas y se encuentra principalmente en la parte superior del suelo. (p. 44)

I

ice wedging: mechanical weathering process that occurs when water freezes in the cracks of rocks and expands, causing the rock to break apart. (p. 38)

impermeable: describes materials that water cannot pass through. (p. 104)

index fossils: remains of species that existed on Earth for a relatively short period of time, were abundant and widespread geographically, and can be used by geologists to assign the ages of rock layers. (p. 129)

gelifracción: proceso de erosión mecánica que ocurre cuando el agua se congela en las grietas de las rocas y luego se expande, causando que la roca de fraccione. (p. 38)

impermeable: describe materiales que impiden el paso del agua a través de ellos. (p. 104)

fósiles índice: restos de especies que existieron sobre la Tierra durante un periodo de tiempo relativamente corto y que fueron abundantes y ampliamente diseminadas geográficamente; los geólogos pueden usarlos para inferir las edades de las capas rocosas. (p. 129)

L

latitude: distance in degrees north or south of the equator. (p. 14)

leaching: removal of minerals that have been dissolved in water. (p. 45)

litter: twigs, leaves, and other organic matter that help prevent erosion and hold water and may eventually be changed into humus by decomposing organisms. (p. 45)

loess (LES): windblown deposit of tightly packed, fine-grained sediments. (p. 79)

longitude: distance in degrees east or west of the prime meridian. (p. 15)

latitud: distancia en grados al norte o sur del ecuador. (p. 14)

lixiviación: remoción de minerales que han sido disueltos en el agua. (p. 45)

hojarasca: ramitas, hojas y demás material orgánico que ayuda a prevenir la erosión y a mantener el agua, y que eventualmente puede ser transformado en humus por los organismos descomponedores. (p. 45)

loes: depósito arrastrado por el viento que se compone de sedimentos de partículas finas y se encuentran muy compactados. (p. 79)

longitud: distancia en grados al este u oeste del meridiano inicial. (p. 15)

longshore current: current that runs parallel to the shoreline, is caused by waves colliding with the shore at slight angles, and moves tons of loose sediment. (p. 110)

corriente costera: corriente que corre paralela a la línea costera, es causada por olas que chocan contra la orilla en ángulos tenues y mueve toneladas de sedimentos sueltos. (p. 110)

M

map legend: explains the meaning of symbols used on a map. (p. 22)

map scale: relationship between distances on a map and distances on Earth's surface that can be represented as a ratio or as a small bar divided into sections. (p. 22)

mass movement: any type of erosion that occurs as gravity moves materials down-slope. (p. 65)

meander (mee AN dur): broad, c-shaped curve in a river or stream, formed by erosion of its outer bank. (p. 97)

mechanical weathering: physical processes that break rock apart without changing its chemical makeup; can be caused by ice wedging, animals, and plant roots. (p. 37)

Mesozoic (mez uh ZOH ihk) Era: middle era of Earth's history, during which Pangaea broke apart, dinosaurs appeared, and reptiles and gymnosperms were the dominant land life-forms. (p. 170)

mold: a type of body fossil that forms in rock when an organism with hard parts is buried, decays or dissolves, and leaves a cavity in the rock. (p. 127)

moraine: large ridge of rocks and soil deposited by a glacier when it stops moving forward. (p. 71)

leyenda del mapa: explica el significado de los símbolos utilizados en un mapa. (p. 22)

escala del mapa: relación entre las distancias en un mapa y las distancias sobre la superficie terrestre, que puede representarse como una relación o como una barra pequeña dividida en secciones. (p. 22)

movimiento en masa: cualquier tipo de erosión que ocurre cuando la gravedad mueve materiales cuesta abajo. (p. 65)

meandro: curva amplia en forma de C en un río o arroyo, formada por la erosión de su rivera externa. (p. 97)

erosión mecánica: procesos físicos que fraccionan la roca sin cambiar su composición química; puede ser causada por gelifracción, animales y raíces de las plantas. (p. 37)

Era Mesozoica: era media de la historia de la Tierra durante la cual se escindió la Pangea y aparecieron los dinosaurios; los reptiles y gimnospermas fueron las formas de vida que dominaron la tierra. (p. 170)

moldura: tipo de cuerpo fósil que se formó en la roca cuando un organismo con partes duras fue enterrado, descompuesto o disuelto, dejando una cavidad en la roca. (p. 127)

morrena: grandes cúmulos de rocas y suelo depositados por un glaciar cuando deja de moverse hacia adelante. (p. 71)

N

natural selection: process by which organisms that are suited to a particular environment are better able to survive and reproduce than organisms that are not. (p. 157)

no-till farming: method for reducing soil erosion; plant stalks are left in the field after harvesting and the next year's crop is planted within the stalks without plowing. (p. 52)

selección natural: proceso mediante el cual los organismos que están adaptados a un ambiente particular están mejor capacitados para sobrevivir y reproducirse que los organismos que no están adaptados. (p. 157)

cultivo sin labranza: método para reducir la erosión del suelo; los tallos de las plantas se dejan en el terreno después de la cosecha y el cultivo del siguiente año se siembra entre los tallos sin hacer labranza alguna. (p. 52)

O

organic evolution: change of organisms over geologic time. (p. 156)

outwash: material deposited by meltwater from a glacier. (p. 71)

oxidation (ahk sih DAY shun): chemical weathering process that occurs when some minerals are exposed to oxygen and water over time. (p. 40)

evolución orgánica: cambio de los organismos a través del tiempo geológico. (p. 156)

derrubio: material depositado por la corriente de agua del hielo derretido de un glaciar. (p. 71)

oxidación: proceso de erosión química que ocurre cuando algunos minerales son expuestos al oxígeno y al agua. (p. 40)

P

Paleozoic Era: era of ancient life, which began about 544 million years ago, when organisms developed hard parts, and ended with mass extinctions about 245 million years ago. (p. 164)

Pangaea (pan JEE uh): large, ancient landmass that was composed of all the continents joined together. (p. 161)

period: third-longest division of geologic time; is subdivided into epochs and is characterized by the types of life that existed worldwide. (p. 155)

permeable (PUR mee uh bul): describes soil and rock with connecting pores through which water can flow. (p. 104)

permineralized remains: fossils in which the spaces inside are filled with minerals from groundwater. (p. 126)

plain: large, flat landform that often has thick, fertile soil and is usually found in the interior region of a continent. (p. 8)

plateau (pla TOH): flat, raised landform made up of nearly horizontal rocks that have been uplifted. (p. 10)

plucking: process that adds gravel, sand, and boulders to a glacier's bottom and sides as water freezes and thaws, breaking off pieces of surrounding rock. (p. 70)

Precambrian (pree KAM bree un) time: longest part of Earth's history, lasting from 4.0 billion to about 544 million years ago. (p. 162)

prime meridian: imaginary line that represents 0° longitude and runs from the north pole through Greenwich, England, to the south pole. (p. 15)

Era Paleozoica: era de la vida antigua que comenzó hace 544 millones de años, cuando los organismos desarrollaron partes duras; terminó con extinciones en masa hace unos 245 millones de años. (p. 164)

Pangea: masa terrestre antigua que estaba compuesta por todos los continentes unidos. (p. 161)

periodo: la tercera división más grande del tiempo geológico; está subdividido en épocas y se caracteriza por los tipos de vida que existieron en todo el mundo. (p. 155)

permeable: describe el suelo y la roca con poros conectados a través de los cuales el agua puede fluir. (p. 104)

restos permineralizados: fósiles en los que los espacios interiores son llenados con minerales de aguas subterráneas. (p. 126)

planicie: formación de terreno extenso y plano que a menudo tiene suelos gruesos y fértiles; generalmente se encuentra en la región interior de un continente. (p. 8)

meseta: formación de terreno plano y elevado constituida por rocas casi horizontales que han sido levantadas. (p. 10)

gelivación: proceso que agrega grava, arena y cantos a la parte inferior y lateral de un glaciar conforme el agua se congela y descongela, fraccionando las piezas de las rocas circundantes. (p. 70)

tiempo precámbrico: la parte más duradera de la historia de la Tierra; duró desde hace 4.0 billones de años hasta hace aproximadamente 544 millones de años. (p. 162)

meridiano inicial: línea imaginaria que representa los cero grados de longitud y va desde el polo norte pasando por Greenwich, Inglaterra, hasta el polo sur. (p. 15)

principle of superposition: states that in undisturbed rock layers, the oldest rocks are on the bottom and the rocks become progressively younger toward the top. (p. 132)

principio de superposición: establece que en las capas rocosas no perturbadas, las rocas más antiguas están en la parte inferior y las rocas son más jóvenes conforme están más cerca de la superficie. (p. 132)

R

radioactive decay: process in which some isotopes break down into other isotopes and particles. (p. 139)

desintegración radiactiva: proceso en el que algunos isótopos se desintegran en otros isótopos y partículas. (p. 139)

radiometric dating: process used to calculate the absolute age of rock by measuring the ratio of parent isotope to daughter product in a mineral and knowing the half-life of the parent. (p. 141)

fechado radiométrico: proceso utilizado para calcular la edad absoluta de las rocas midiendo la relación isótopo parental a producto derivado en un mineral y conociendo la vida media del parental. (p. 141)

relative age: the age of something compared with other things. (p. 133)

edad relativa: la edad de algo comparado con otras cosas. (p. 133)

runoff: any rainwater that does not soak into the ground or evaporate but flows over Earth's surface; generally flows into streams and has the ability to erode and carry sediments. (p. 92)

escorrentía: agua de lluvia que no se difunde en el suelo ni se evapora pero que fluye sobre la superficie terrestre; generalmente fluye hacia los arroyos y tiene la capacidad de causar erosión y transportar sedimentos. (p. 92)

S

sheet erosion: a type of surface water erosion caused by runoff that occurs when water flowing as sheets picks up sediments and carries them away. (p. 95)

erosión laminar: tipo de erosión causada por las corrientes de agua de lluvia; ocurre cuando el agua fluye laminarmente recogiendo sedimentos y llevándolos a otro lugar. (p. 95)

slump: a type of mass movement that occurs when a mass of material moves down a curved slope. (p. 65)

desprendimiento: tipo de movimiento en masa que ocurre cuando un volumen de material se mueve hacia abajo de una cuesta curvada. (p. 65)

soil: mixture of weathered rock and mineral fragments, decayed organic matter, mineral fragments, water, and air that can take thousands of years to develop. (p. 42)

suelo: mezcla de roca erosionada y fragmentos minerales, materia orgánica en descomposición, fragmentos minerales, agua y aire, y que puede tardar miles de años para formarse. (p. 42)

soil profile: vertical section of soil layers, each of which is a horizon. (p. 44)

perfil de suelos: sección vertical de las capas del suelo, cada una de las cuales es un horizonte. (p. 44)

species: group of organisms that reproduces only with other members of their own group. (p. 156)

especie: grupo de organismos que se reproduce sólo entre los miembros de su mismo grupo. (p. 156)

spring: forms when the water table meets Earth's surface; often found on hillsides and used as a freshwater source. (p. 107)

manantial: se forma cuando el nivel freático alcanza la superficie terrestre; a menudo se encuentran en las laderas y se usan como fuente de agua potable. (p. 107)

T

terracing: farming method used to reduce erosion on steep slopes. (p. 53)

terraceo: método de siembra usado para reducir la erosión en cuestas inclinadas. (p. 53)

till: mixture of different-sized sediments that is dropped from the base of a retreating glacier and can cover huge areas of land. (p. 70)

tillita: mezcla de sedimentos de diferentes tamaños que ha caído de la base de un glaciar en retroceso y puede cubrir grandes extensiones de terreno. (p. 70)

topographic map: map that shows the changes in elevation of Earth's surface and indicates such features as roads and cities. (p. 20)

mapa topográfico: mapa que muestra los cambios en la elevación de la superficie terrestre que puede ser representado como una relación e indica características como carreteras y ciudades. (p. 20)

trilobite (TRI luh bite): organism with a three-lobed exoskeleton that was abundant in Paleozoic oceans and is considered to be an index fossil. (p. 155)

trilobite: organismo con un exoesqueleto trilobulado que fue abundante en los océanos del Paleozoico y es considerado como un fósil índice. (p. 155)

U

unconformity (un kun FOR mih tee): gap in the rock layer that is due to erosion or periods without any deposition. (p. 134)

discordancia: brecha en la capa rocosa que es debida a la erosión o a periodos sin deposición. (p. 134)

uniformitarianism: principle stating that Earth processes occurring today are similar to those that occurred in the past. (p. 143)

uniformitarianismo: principio que establece que los procesos de la Tierra que ocurren actualmente son similares a los que ocurrieron en el pasado. (p. 143)

upwarped mountains: mountains formed when blocks of Earth's crust are pushed up by forces inside Earth. (p. 12)

montañas de levantamiento: montañas que se forman cuando los bloques de la corteza terrestre son empujados hacia arriba por fuerzas del interior de la Tierra. (p. 12)

V

volcanic mountains: mountains formed when molten material reaches Earth's surface through a weak crustal area and piles up into a cone-shaped structure. (p. 13)

montañas volcánicas: montañas formadas cuando material derretido alcanza la superficie a través de un área débil de la corteza terrestre y se acumula formando una estructura en forma de cono. (p. 13)

W

water table: upper surface of the zone of saturation; drops during a drought. (p. 104)

nivel freático: parte superior de la zona de saturación; desciende durante las sequías. (p. 104)

weathering: mechanical or chemical surface processes that break rock into smaller and smaller pieces. (p. 36)

erosión: proceso superficial químico o mecánico que fracciona la roca en trozos cada vez más pequeños. (p. 36)

Italic numbers = illustration/photo　**Bold numbers = vocabulary term**
lab = a page on which the entry is used in a lab
act = a page on which the entry is used in an activity

A

B

C

Index

T

U

V

W

Magnification Key: Magnifications listed are the magnifications at which images were originally photographed.
LM–Light Microscope
SEM–Scanning Electron Microscope
TEM–Transmission Electron Microscope

Acknowledgments: Glencoe would like to acknowledge the artists and agencies who participated in illustrating this program: Absolute Science Illustration; Andrew Evansen; Argosy; Articulate Graphics; Craig Attebery, represented by Frank & Jeff Lavaty; CHK America; John Edwards and Associates; Gagliano Graphics; Pedro Julio Gonzalez, represented by Melissa Turk & The Artist Network; Robert Hynes, represented by Mendola Ltd.; Morgan Cain & Associates; JTH Illustration; Laurie O'Keefe; Matthew Pippin, represented by Beranbaum Artist's Representative; Precision Graphics; Publisher's Art; Rolin Graphics, Inc.; Wendy Smith, represented by Melissa Turk & The Artist Network; Kevin Torline, represented by Berendsen and Associates, Inc.; WILDlife ART; Phil Wilson, represented by Cliff Knecht Artist Representative; Zoo Botanica.

Photo Credits

Cover Jack Dykinga/Getty Images; **i ii** Jack Dykinga/Getty Images; **iv** (bkgd)John Evans, (inset)Jack Dykinga/Getty Images; **v** (t)PhotoDisc, (b)John Evans; **vi** (l)John Evans, (r)Geoff Butler; **vii** (l)John Evans, (r)PhotoDisc; **viii** PhotoDisc; **ix** Aaron Haupt Photography; **x** Sylvester Allred/Visuals Unlimited; **xi** (t)PhotoTake NYC/PictureQuest, (b)Curt Schieber; **xii** John Giustina/Getty Images; **2** Charles Palek/Earth Scenes; **3** (t)Tom Bean/Stone, (b)JC Marchak/AP/Wide World Photos; **4** C. C. Lockwood/Earth Scenes; **5** (t)David Ulmer/Stock Boston, (b)Mark Burnett; **6–7** GSFC/NASA; **9** (tl)Alan Maichrowicz/Peter Arnold, Inc., (tr)Carr Clifton/Minden Pictures, (b)Stephen G. Maka/DRK Photo; **10** Ron Mellot; **11** John Lemker/Earth Scenes; **12** (t)John Kieffer/Peter Arnold, Inc., (b)Carr Clifton/Minden Pictures; **13** David Muench/CORBIS; **16** Dominic Oldershaw; **21** (t)Rob & Ann Simpson, (b)courtesy Maps a la Carte, Inc. and TopoZone.com; **24** CORBIS; **26** (t)Layne Kennedy/CORBIS, (b)John Evans; **27** John Evans; **28** (tl)Culver Pictures, (tcl)PhotoDisc, (tcr)William Manning/The Stock Market/CORBIS, (tr)Kunio Owaki/The Stock Market/CORBIS, (c)Pictor, (b)PhotoDisc; **29** (l)Tom Bean/DRK Photo, (r)Marc Muench; **30** William Weber; **32** Aaron Haupt; **34–35** Andrew Brown, Ecoscene/CORBIS; **37** (l)StudiOhio, (r)Tom Bean/DRK Photo; **38** W. Perry Conway/CORBIS; **39** Hans Strand/Stone; **40** (tl)Craig Kramer, (tr)A.J. Copley/Visuals Unlimited, (bl br)John Evans; **41** (l)William Johnson/Stock Boston, (r)Runk/Schoenberger from Grant Heilman; **43** (bkgd)Stephen R. Wagner, (t)James D. Balog, (c)Martin Miller, (b)Steven C. Wilson/Entheos; **44** (l)Bonnie Heidel/Visuals Unlimited, (r)John Bova/Photo Researchers; **50** (l)Gary Braasch/CORBIS, (r)Donna Ikenberry/Earth Scenes; **51** Chip & Jill Isenhart/Tom Stack & Associates; **52** (t)Dr. Russ Utgard, (b)Denny Eilers from Grant Heilman; **53** Georg Custer/Photo Researchers; **54** (t)George H. Harrison from Grant Heilman, (b)Bob Daemmrich; **55** KS Studios; **56** Larry Hamill; **57** (l)Tom Bean/DRK Photo, (r)David M. Dennis/Earth Scenes; **59** Matt Meadows; **60** Georg Custer/Photo Researchers; **62–63** Paul A. Souders/CORBIS; **64** Robert L. Schuster/USGS; **65** Martin G. Miller/Visuals Unlimited; **66** (t)John D. Cunningham/Visuals Unlimited, (bl)Sylvester Allred/Visuals Unlimited, (br)Tom Uhlman/Visuals Unlimited; **67** AP/Wide World Photos; **68** Martin G. Miller/Visuals Unlimited; **70** James N. Westwater; **71** (t)Tom Bean/Stone/Getty Images, (b)Tom Bean/CORBIS; **72** John Gerlach/Visuals Unlimited; **73** Gregory G. Dimijian/Photo Researchers; **74** Mark E. Gibson/Visuals Unlimited; **75** Timothy Fuller; **76** Galen Rowell/CORBIS; **78** Fletcher & Baylis/Photo Researchers; **79** (t)John D. Cunningham/Visuals Unlimited, (b)file photo; **80** (bkgd)Breck P. Kent/Earth Scenes, (tl)Stephen J. Krasemann/Photo Researchers, (tr)Steve McCurry, (b)Wyman P. Meinzer; **81** John Giustina/FPG/Getty Images; **82** (t)Greg Vaughn/Tom Stack & Assoc., (b)Matt Meadows; **84** (t)World Class Images, (c)Yann Arthus-Bertrand/CORBIS, (b)AP/Wide World Photos; **86** John D. Cunningham/Visuals Unlimited; **90–91** William Manning/The Stock Market/CORBIS; **91** Aaron Haupt; **92** (l)Michael Busselle/Stone/Getty Images, (r)David Woodfall/DRK Photo; **93** Tim Davis/Stone/Getty Images; **94** (t)Grant Heilman Photography, (b)KS Studios; **96** Mel Allen/ICL/Panoramic Images; **98** CORBIS/PictureQuest; **99** (l)Harald Sund/The Image Bank/Getty Images, (r)Loren McIntyre; **100** C. Davidson/Comstock; **101** James L. Amos/CORBIS; **102** (l)Wolfgang Kaehler, (r)Nigel Press/Stone/Getty Images; **103** First Image; **105** CORBIS; **106** file photo; **107** Barbara Filet; **108** Chad Ehlers/Stone/Getty Images; **110** Macduff Everton/The Image Bank/Getty Images; **111** (tl)Steve Bentsen, (tr)SuperStock, (bl)Runk/Schoenberger from Grant Heilman, (br)Breck P. Kent/Earth Scenes; **112** Bruce Roberts/Photo Researchers; **114 115** KS Studios; **116** Gary Bogdon/CORBIS Sygma; **117** (l)Todd Powell/Index Stock, (r)J. Wengle/DRK Photo; **120** Barbara Filet; **121** Grant Heilman Photography; **122–123** Hugh Sitton/Getty Images; **124** (t)Mark E. Gibson/Visuals Unlimited, (b)D.E. Hurlbert & James DiLoreto/Smithsonian Institution; **125** Jeffrey Rotman/CORBIS; **126** (t)Dr. John A. Long, (b)A.J. Copley/Visuals Unlimited; **128** (t)PhotoTake, NYC/PictureQuest, (bl br)Louis Psihoyos/Matrix; **130** David M. Dennis; **131** (l)Gary Retherford/Photo Researchers, (r)Lawson Wood/CORBIS; **132** Aaron Haupt; **135** (bkgd)Lyle Rosbotham, (l)IPR/12-18 T. Bain, British Geological Survey/NERC. All rights reserved, (r)Tom Bean/CORBIS; **136** Jim Hughes/PhotoVenture/Visuals Unlimited; **137** (l)Michael T. Sedam/CORBIS, (r)Pat O'Hara/CORBIS; **139** Aaron Haupt; **141** James King-Holmes/Science Photo Library/Photo Researchers; **142** Kenneth Garrett; **143** WildCountry/CORBIS; **144** (t)A.J. Copley/Visuals Unlimited, (b)Lawson Wood/CORBIS; **145** Matt Meadows; **146** Jacques Bredy; **147** (tl)Francois Gohier/Photo Researchers, (tr)Sinclair Stammers/Photo Researchers, (b)Mark E. Gibson/DRK Photo; **150** Tom Bean/CORBIS; **152–153** Roger Garwood & Trish Ainslie/CORBIS; **153** KS Studios; **154** Tom & Therisa Stack/Tom Stack & Assoc.; **156** (l)Gerald & Buff Corsi/Visuals Unlimited, (r)John Gerlach/Animals Animals; **158** (tl)Mark Boulron/Photo Researchers, (others)Walter

Chandoha; **159** Jeff Lepore/Photo Researchers; **163** (l)Mitsuaki Iwago/Minden Pictures, (r)R. Calentine/ Visuals Unlimited; **165** J.G. Gehling/Discover Magazine; **166** Gerry Ellis/ENP Images; **169** Matt Meadows; **172** (l)David Burnham/Fossilworks, Inc., (r)François Gohier/Photo Researchers; **174** Michael Andrews/Earth Scenes; **175** Tom J. Ulrich/Visuals Unlimited; **176** David M. Dennis; **177** Mark Burnett; **179** (l)E. Webber/Visuals Unlimited, (r)Len Rue, Jr./Animals Animals; **181** John Cancalosi/Stock Boston; **184** PhotoDisc; **186** Tom Pantages; **190** Michell D. Bridwell/PhotoEdit, Inc.; **191** (t)Mark Burnett, (b)Dominic Oldershaw; **192** StudiOhio; **193** Timothy Fuller; **194** Aaron Haupt; **196** KS Studios; **197** Matt Meadows; **198** PhotoDisc; **199** Giboux/Getty Images; **201** Amanita Pictures; **202** Bob Daemmrich; **204** Davis Barber/PhotoEdit, Inc.

PERIODIC TABLE OF THE ELEMENTS

Columns of elements are called groups. Elements in the same group have similar chemical properties.

Element — Hydrogen
Atomic number — 1
Symbol — H
Atomic mass — 1.008
State of matter

Gas
Liquid
Solid
Synthetic

The first three symbols tell you the state of matter of the element at room temperature. The fourth symbol identifies elements that are not present in significant amounts on Earth. Useful amounts are made synthetically.

Period	1	2	3	4	5	6	7	8	9
1	Hydrogen 1 **H** 1.008								
2	Lithium 3 **Li** 6.941	Beryllium 4 **Be** 9.012							
3	Sodium 11 **Na** 22.990	Magnesium 12 **Mg** 24.305							
4	Potassium 19 **K** 39.098	Calcium 20 **Ca** 40.078	Scandium 21 **Sc** 44.956	Titanium 22 **Ti** 47.867	Vanadium 23 **V** 50.942	Chromium 24 **Cr** 51.996	Manganese 25 **Mn** 54.938	Iron 26 **Fe** 55.845	Cobalt 27 **Co** 58.933
5	Rubidium 37 **Rb** 85.468	Strontium 38 **Sr** 87.62	Yttrium 39 **Y** 88.906	Zirconium 40 **Zr** 91.224	Niobium 41 **Nb** 92.906	Molybdenum 42 **Mo** 95.94	Technetium 43 **Tc** (98)	Ruthenium 44 **Ru** 101.07	Rhodium 45 **Rh** 102.906
6	Cesium 55 **Cs** 132.905	Barium 56 **Ba** 137.327	Lanthanum 57 **La** 138.906	Hafnium 72 **Hf** 178.49	Tantalum 73 **Ta** 180.948	Tungsten 74 **W** 183.84	Rhenium 75 **Re** 186.207	Osmium 76 **Os** 190.23	Iridium 77 **Ir** 192.217
7	Francium 87 **Fr** (223)	Radium 88 **Ra** (226)	Actinium 89 **Ac** (227)	Rutherfordium 104 **Rf** (261)	Dubnium 105 **Db** (262)	Seaborgium 106 **Sg** (266)	Bohrium 107 **Bh** (264)	Hassium 108 **Hs** (277)	Meitnerium 109 **Mt** (268)

The number in parentheses is the mass number of the longest-lived isotope for that element.

Rows of elements are called periods. Atomic number increases across a period.

The arrow shows where these elements would fit into the periodic table. They are moved to the bottom of the table to save space.

Lanthanide series	Cerium 58 **Ce** 140.116	Praseodymium 59 **Pr** 140.908	Neodymium 60 **Nd** 144.24	Promethium 61 **Pm** (145)	Samarium 62 **Sm** 150.36
Actinide series	Thorium 90 **Th** 232.038	Protactinium 91 **Pa** 231.036	Uranium 92 **U** 238.029	Neptunium 93 **Np** (237)	Plutonium 94 **Pu** (244)